Microbial Processing of Metal Sulfides

Microbial Processing of Metal Sulfides

edited by

Edgardo R. Donati
University of La Plata, Argentina

and

Wolfgang Sand
University of Duisburg-Essen, Germany

Springer

A C.I.P. Catalogue record for this book is available from the Library of Congress.

ISBN 978-1-4020-5588-1 (HB)
ISBN 978-1-4020-5589-8 (e-book)

Published by Springer,
P.O. Box 17, 3300 AA Dordrecht, The Netherlands.

www.springer.com

Printed on acid-free paper

TABLE OF CONTENTS

LIST OF CONTRIBUTORS

Heloisa A. Acciari Universidade Estadual Paulista, Instituto de Química, Departamento de Bioquímica e Tecnologia Química, Araraquara-SP, Brazil.

Fernando Acevedo School of Biochemical Engineering, Pontifical Catholic University of Valparaiso, Valparaiso, Chile.

Antonio Ballester Departamento de Ciencia de Materiales e Ingeniería Metalúrgica, Facultad de Ciencias Químicas, Universidad Complutense de Madrid, Madrid, Spain.

Assis V. Benedetti Universidade Estadual Paulista, Instituto de Química, Departamento de Fisico-Química, Araraquara-SP, Brazil.

María Luisa Blázquez Departamento de Ciencia de Materiales e Ingeniería Metalúrgica, Facultad de Ciencias Químicas, Universidad Complutense de Madrid, Madrid, Spain.

Alvaro Banderas Unit of Bacterial Cell Communication, Laboratory of Molecular Microbiology and Biotechnology, Department of Biology, Faculty of Sciences, University of Chile, Santiago, Chile.

Denise Bevilaqua Universidade Estadual Paulista, Instituto de Química, Departamento de Bioquímica e Tecnologia Química, Araraquara-SP, Brazil

Domingo Cantero Universidad de Cádiz, Cádiz, Spain.

Patricia Chiacchiarini Facultad de Ingeniería, Universidad Nacional del Comahue, Neuquén, Argentina.

David G. Dixon Department of Materials Engineering, University of British Columbia, Vancouver, Canada.

Edgardo Donati Cindefi (CONICET-UNLP), Universidad Nacional de La Plata, La Plata, Argentina.

Oswaldo Garcia Jr. Universidade Estadual Paulista, Instituto de Química, Departamento de Bioquímica e Tecnologia Química, Araraquara-SP, Brazil.

Juan Carlos Gentina School of Biochemical Engineering, Pontifical Catholic University of Valparaiso, Valparaiso, Chile.

Alejandra Giaveno Facultad de Ingeniería, Universidad Nacional del Comahue, Neuquén, Argentina.

Nicolas Guiliani Unit of Bacterial Cell Communication, Laboratory of Molecular Microbiology and Biotechnology, Department of Biology, Faculty of Sciences, University of Chile, Santiago, Chile.

Jose Manuel Gómez Montes de Oca Universidad de Cádiz, Cádiz, Spain.

Felisa González Departamento de Ciencia de Materiales e Ingeniería Metalúrgica, Facultad de Ciencias Químicas, Universidad Complutense de Madrid, Madrid, Spain.

K. Hanumantha Rao Division of Mineral Processing, Luleå University of Technology, Luleå, Sweden.

David S. Holmes Center of Bioinformatics and Genome Biology, Andrés Bello University (UNAB), Life Science Foundation and Millennium Institute of Fundamental and Applied Biology, Santiago, Chile.

Eugenia Jedlicki Program of Cellular and Molecular Biology, I.C.B.M., Faculty of Medicine, University of Chile, Santiago, Chile.

Carlos A. Jerez Laboratory of Molecular Microbiology and Biotechnology, Department of Biology, Faculty of Sciences, University of Chile, Santiago, Chile.

Laura Lavalle Facultad de Ingeniería, Universidad Nacional del Comahue, Neuquén, Argentina.

Renata Matlakowska Warsaw University, Faculty of Biology, Warsaw, Poland.

Dominique Henri Roger Morin BRGM, Orléans, France.

Jesús A. Muñoz Departamento de Ciencia de Materiales e Ingeniería Metalúrgica, Facultad de Ciencias Químicas, Universidad Complutense de Madrid, Madrid, Spain.

Jochen Petersen Department of Chemical Engineering, University of Cape Town, South Africa.

Cristina Pogliani Cindefi (CONICET-UNLP), Universidad Nacional de La Plata, La Plata, Argentina.

Raquel Quatrini Center of Bioinformatics and Genome Biology, Andrés Bello University (UNAB), Life Science Foundation and Millennium Institute of Fundamental and Applied Biology, Santiago, Chile.

Thore Rohwerder University of Duisburg-Essen, Biofilm Centre, Aquatic Biotechnology, Duisburg, Germany.

Wolfgang Sand University of Duisburg-Essen, Biofilm Centre, Aquatic Biotechnology, Duisburg, Germany.

Axel Schippers Geomicrobiology, Federal Institute for Geosciences and Natural Resources, Hannover, Germany.

Aleksandra Sklodowska Warsaw University, Faculty of Biology, Warsaw, Poland.

S. Subramanian Department of Metallurgy, Indian Institute of Science, Bangalore, India.

Jorge Valdés Center of Bioinformatics and Genome Biology, Andrés Bello University (UNAB), Life Science Foundation and Millennium Institute of Fundamental and Applied Biology, Santiago, Chile.

Susana Valenzuela Unit of Bacterial Cell Communication, Laboratory of Molecular Microbiology and Biotechnology, Department of Biology, Faculty of Sciences, University of Chile, Santiago, Chile.

Marisa Viera Cindefi (CONICET-UNLP), Universidad Nacional de La Plata, La Plata, Argentina.

PREFACE

In the last years, the application of microbiological methods to the extraction of metals from minerals has definitely gained a prominent role supported by the several bioleaching and biooxidation processes operating in different sites over the world. This may be an important reason why fundamental research has received a new powerful stimulus with fascinating discoveries and in addition it surely will become the cause of future development in the field. In 1997 Springer published an excellent book entitled 'Biomining' (edited by D. E. Rawlings) which not only provided critical discussion of microbial and physicochemical aspects of bioleaching processes (written by prestigious experts in biohydrometallurgical field) but also it concentrated many contributions by people employed in industries. During the last years since the publication of 'Biomining', the advances in molecular biology methods have been applied extensively to the study of microorganisms involved in bioleaching processes. In addition recent studies about proteomic and bioinformatics are bringing a new perspective on the microbial processes. Furthermore, there is a growing agreement about the mechanisms of bioleaching and the role played by exopolymers substances in the interfacial degradation of metal sulfides. Additionally new evidence has been supplied by new techniques (electrochemical techniques, atomic force microscope). Due to the growing literature in these and other aspects, we think a new book could be opportune to organize partially this new information. However, it is clear that covering the range of subject areas in depth would require several volumes of specialist text. That is why, it has been necessary to be selective. Since the book 'Biomining' is still a large reference to the applied technology, we hope that this new book could go some way towards introducing undergraduate and postgraduate students as well as interested industrialists to the main subjects of microbial processing with special emphasis to the last contributions of the chemical and microbial aspects of bioleaching process and use of microorganisms in the treatment of complex ores and concentrates.

Our sincere thanks all the contributors whose efforts made this book possible and the editorial staff of Springer for their help in readying the manuscript for publication.

We wish to express our grateful recognition to *Giovanni Rossi*, *Corale Brierley*, *James Brierley*, *Violaine Bonnefoy*, *Franz Glombitza*, *David Holmes*, *Jacco Huisman*, *Barrie Johnson*, *Douglas Rawlings*, *Erika Kalman*, *Halit Kuyumcu* and *Axel Schippers* for their generous contribution with the peer review of the chapters.

Wolfgang Sand and Edgardo Donati

SECTION I

FUNDAMENTALS, MICROORGANISMS AND MECHANISMS

CHAPTER 1

MICROORGANISMS INVOLVED IN BIOLEACHING AND NUCLEIC ACID-BASED MOLECULAR METHODS FOR THEIR IDENTIFICATION AND QUANTIFICATION

AXEL SCHIPPERS

Geomicrobiology, Federal Institute for Geosciences and Natural Resources, Stilleweg 2, D-30655 Hannover, Germany. E-mail: a.schippers@bgr.de

1. INTRODUCTION

Bioleaching is the biological conversion of an insoluble metal compound into a water soluble form. In case of metal sulfide bioleaching, metal sulfides are oxidized to metal ions and sulfate by aerobic, acidophilic Fe(II) and/or sulfur-compound oxidizing Bacteria or Archaea. Bioleaching involves chemical and biological reactions. Despite molecular oxygen being the final electron acceptor for the overall metal sulfide bioleaching process, Fe(III) ions are the relevant oxidizing agent for the metal sulfides. The metal sulfide oxidation itself is a chemical process in which Fe(III) ions are reduced to Fe(II) ions and the sulfur moiety of the metal sulfide is oxidized to sulfate, and various intermediate sulfur compounds, e.g. elemental sulfur, polysulfide, thiosulfate, and polythionates. For example the oxidation of sphalerite (ZnS) to elemental sulfur is given in the following equation:

$$(1) \qquad ZnS + 2Fe^{3+} \rightarrow Zn^{2+} + 0.125S_8 + 2Fe^{2+}$$

Because of two different groups of metal sulfides exist, two different metal sulfide oxidation mechanisms have been proposed, namely the thiosulfate mechanism (for acid-insoluble metal sulfides, such as pyrite) and the polysulfide mechanism (for acid-soluble metal sulfides, e.g. sphalerite or chalcopyrite, $CuFeS_2$). These mechanisms explain the occurrence of all inorganic sulfur compounds which have been detected in the course of metal sulfide oxidation (for review see: Sand *et al.*, 2001; Rohwerder *et al.*, 2003; Schippers, 2004; Chapter 2).

The role of the microorganisms in the bioleaching process is to oxidize the products of the chemical metal sulfide oxidation (Fe(II) ions and sulfur- compounds) in order to provide Fe(III) and protons, the metal sulfide attacking agents. In addition, proton production keeps the pH low and thus, the Fe ions in solution.

E.R. Donati and W. Sand (eds.), Microbial Processing of Metal Sulfides, 3–33.

Aerobic, acidophilic Fe(II) oxidizing Bacteria or Archaea provide Fe(III) by the following equation:

(2) $$2Fe^{2+} + 0.5O_2 + 2H^+ \rightarrow 2Fe^{3+} + H_2O$$

Aerobic, acidophilic sulfur-compound oxidizing Bacteria or Archaea oxidize intermediate sulfur compounds to sulfate and protons (sulfuric acid). Most relevant is the oxidation of elemental sulfur to sulfuric acid since elemental sulfur can only be biologically oxidized under bioleaching conditions:

(3) $$0.125S_8 + 1.5O_2 + H_2O \rightarrow SO_4^{2-} + 2H^+$$

The sulfur-compound oxidizing Bacteria or Archaea produce protons which dissolve metal sulfides besides pyrite which is not acid-soluble. Pyrite is only attacked by Fe(III) ions (not by protons) and therefore only dissolved by Fe(II) oxidizing Bacteria or Archaea.

This book chapter gives an update of previous excellent reviews on microorganisms involved in bioleaching (e.g. Harrison, 1984; Rossi, 1990; Rawlings, 1997, 2002; Johnson, 1998; Hallberg & Johnson, 2001). In the first part of this chapter, the metal sulfide oxidizing microorganisms are described. In the second part, acidophilic microorganisms which do not oxidize metal sulfides and their importance for bioleaching are reviewed. In the third part, nucleic-acid based methods for the identification and quantification of these microorganisms are introduced.

2. METAL SULFIDE OXIDIZING MICROORGANISMS

The most described acidophilic metal sulfide oxidizing microorganisms belong to the mesophilic and moderately thermophilic Bacteria. The Archaea are usually extremely thermophilic (besides the genus *Ferroplasma*). Most industrial heap and tank bioleaching operations run below 40 °C but operations at higher temperatures promise higher reaction rates (Olson *et al.*, 2003; Batty & Rorke, 2006). All acidophilic metal sulfide oxidizing microorganisms oxidize Fe(II) and/or sulfur compounds. Most of these microorganisms fix CO_2 and grow chemolithoautotrophically. A list of the metal sulfide oxidizing Bacteria or Archaea, their phylogeny and some of their physiological properties is given in the Tables 1–3.

The organisms can be separated in three groups according to their temperature optimum for growth: Mesophiles up to ~ 40 °C, moderate themophiles between ~40 – ~55 °C, and extreme thermophiles between ~55 – ~80 °C.

2.1. Mesophilic and Moderately Thermophilic Bacteria

2.1.1. Proteobacteria

***Acidithiobacillus* spp.** The genus *Acidithiobacillus* was proposed by Kelly & Wood (2000) after reclassification of some species of the genus *Thiobacillus*. The affiliation of the genus *Acidithiobacillus* to the β- or γ-Proteobacteria is not clearly

Table 1. Phylogeny of metal sulfide oxidizing, acidophilic microorganisms

Species[#]	Phylum	G+C (mol%)
Mesophilic and moderately thermophilic Bacteria		
Acidimicrobium ferrooxidans	Actinobacteria	67-69
Acidithiobacillus albertensis	Proteobacteria	61.5
Acidithiobacillus caldus	Proteobacteria	63-64
Acidithiobacillus ferrooxidans	Proteobacteria	58-59
Acidithiobacillus thiooxidans	Proteobacteria	52
Alicyclobacillus disulfidooxidans	Firmicutes	53
Alicyclobacillus tolerans	Firmicutes	49
"*Caldibacillus ferrivorus*"	Firmicutes	51
"*Ferrimicrobium acidiphilum*"	Actinobacteria	51-55
Leptospirillum ferriphilum	Nitrospira	55-58
"*Leptospirillum ferrodiazotrophum*"	Nitrospira	na
Leptospirillum ferrooxidans	Nitrospira	52
Sulfobacillus acidophilus	Firmicutes	55-57
"*Sulfobacillus montserratensis*"	Firmicutes	52
Sulfobacillus sibiricus	Firmicutes	48
Sulfobacillus thermosulfidooxidans	Firmicutes	48-50
Sulfobacillus thermotolerans	Firmicutes	48
"*Thiobacillus plumbophilus*"	Proteobacteria	66
"*Thiobacillus prosperus*"	Proteobacteria	64
Thiomonas cuprina	Proteobacteria	66-69
Mesophilic and moderately thermophilic Archaea		
"*Ferroplasma acidarmanus*"	Euryarchaeota	37
Ferroplasma acidiphilum	Euryarchaeota	36.5
"*Ferroplasma cupricumulans*"	Euryarchaeota	na
Extremely thermophilic Archaea		
Acidianus brierleyi	Crenarchaeota	31
Acidianus infernus	Crenarchaeota	31
Metallosphaera hakonensis	Crenarchaeota	46
Metallosphaera prunae	Crenarchaeota	46
Metallosphaera sedula	Crenarchaeota	45
Sulfolobus metallicus	Crenarchaeota	38
Sulfolobus yangmingensis	Crenarchaeota	42
Sulfurococcus mirabilis	Crenarchaeota	~44
Sulfurococcus yellowstonensis	Crenarchaeota	45

[#]Listed in alphabetical order; G + C = mole% guanine+cytosine content of genomic DNA; na = data not available; species without standing in nomenclature (http://www.bacterio.cict.fr/) are given in quotation marks

defined in the literature (Lane *et al.*, 1992; McDonald *et al.*, 1997; Kelly & Wood, 2000; Hallberg & Johnson, 2001). Species of the genus *Acidithiobacillus* are obligately acidophilic (pH < 4.0), Gram-negative, motile rods. CO_2 is fixed by means of the Benson-Calvin Cycle. The genus comprises the following species: *At. ferrooxidans, At. thiooxidans, At. caldus,* and *At. albertensis*.

Table 2. Optimum and range of growth for pH and temperature of metal sulfide oxidizing, acidophilic microorganisms

Species[#]	pH optimum	pH minimum-maximum	Temperature optimum (°C)	Temperature minimum-maximum (°C)
Mesophilic and moderately thermophilic Bacteria				
Acidimicrobium ferrooxidans	~2	na	45-50	<30-55
Acidithiobacillus albertensis	3.5-4.0	2.0-4.5	25-30	na
Acidithiobacillus caldus	2.0-2.5	1.0-3.5	45	32-52
Acidithiobacillus ferrooxidans	2.5	1.3-4.5	30-35	10-37
Acidithiobacillus thiooxidans	2.0-3.0	0.5-5.5	28-30	10-37
Alicyclobacillus disulfidooxidans	1.5-2.5	0.5-6.0	35	4-40
Alicyclobacillus tolerans	2.5-2.7	1.5-5	37-42	<20-55
"*Caldibacillus ferrivorus*"	1.8	na	45	<35->55
"*Ferrimicrobium acidiphilum*"	2-2.5	1.3-4.8	37	<10-45
Leptospirillum ferriphilum	1.3-1.8	na	30-37	na-45
"*Leptospirillum ferrodiazotrophum*"	na	<1.2<	na	<37<
Leptospirillum ferrooxidans	1.5-3.0	1.3-4.0	28-30	na
Sulfobacillus acidophilus	~2	na	45-50	<30-55
"*Sulfobacillus montserratensis*"	1.6	0.7->2	37	<30-43
Sulfobacillus sibiricus	2.2-2.5	1.1-3.5	55	17-60
Sulfobacillus thermosulfidooxidans	~2	1.5-5.5	45-48	20-60
Sulfobacillus thermotolerans	2-2.5	1.2-5	40	20-60
"*Thiobacillus plumbophilus*"	na	4.0-6.5	27	9-41
"*Thiobacillus prosperus*"	~2	1.0-4.5	33-37	23-41
Thiomonas cuprina	3.5-4	1.5-7.2	30-36	20-45
Mesophilic and moderately thermophilic Archaea				
"*Ferroplasma acidarmanus*"	1.2	<0-1.5	42	23-46
Ferroplasma acidiphilum	1.7	1.3-2.2	35	15-45
"*Ferroplasma cupricumulans*"	1-1.2	0.4-1.8	54	22-63
Extremely thermophilic Archaea				
Acidianus brierleyi	1.5-2.0	1-6	~70	45-75
Acidianus infernus	~2	1-5.5	~90	65-96
Metallosphaera hakonensis	3	1-4	70	50-80
Metallosphaera prunae	2-3	1-4.5	~75	55-80
Metallosphaera sedula	2-3	1-4.5	75	50-80
Sulfolobus metallicus	2-3	1-4.5	65	50-75
Sulfolobus yangmingensis	4	2-6	80	65-95
Sulfurococcus mirabilis	2-2.6	1-5.8	70-75	50-86
Sulfurococcus yellowstonensis	2-2.6	1-5.5	60	40-80

[#]Listed in alphabetical order; na = data not available; species without standing in nomenclature (http://www.bacterio.cict.fr/) are given in quotation marks

Table 3. Physiological properties of metal sulfide oxidizing, acidophilic microorganisms

Species[#]	Oxidation of Pyrite	Oxidation of other *MS	Oxidation of Fe(II) ions	Oxidation of Sulfur	Growth
Mesophilic and moderately thermophilic Bacteria					
Acidimicrobium ferrooxidans	+	na	+	-	F
Acidithiobacillus albertensis	-	+	-	+	A
Acidithiobacillus caldus	-	+	-	+	F
Acidithiobacillus ferrooxidans	+	+	+	+	A
Acidithiobacillus thiooxidans	-	+	-	+	A
Alicyclobacillus disulfidooxidans	+	na	+	+	F
Alicyclobacillus tolerans	+	+	+	+	F
"*Caldibacillus ferrivorus*"	+	na	+	+	F
"*Ferrimicrobium acidiphilum*"	+	na	+	-	H
Leptospirillum ferriphilum	+	+	+	-	A
"*Leptospirillum ferrodiazotrophum*"	na	na	+	na	A
Leptospirillum ferrooxidans	+	+	+	-	A
Sulfobacillus acidophilus	+	+	+	+	F
"*Sulfobacillus montserratensis*"	+	na	+	+	F
Sulfobacillus sibiricus	+	+	+	+	F
Sulfobacillus thermosulfidooxidans	+	+	+	+	F
Sulfobacillus thermotolerans	+	+	+	+	F
"*Thiobacillus plumbophilus*"	-	+	-	+	A
"*Thiobacillus prosperus*"	+	+	+	+	A
Thiomonas cuprina	-	+	-	+	F
Mesophilic and moderately thermophilic Archaea					
"*Ferroplasma acidarmanus*"	+	na	+	-	F
Ferroplasma acidiphilum	+	na	+	-	F
"*Ferroplasma cupricumulans*"	na	+	+	+	F
Extremely thermophilic Archaea					
Acidianus brierleyi	+	+	+	+	F
Acidianus infernus	+	+	+	+	A
Metallosphaera hakonensis	na	+	na	+	F
Metallosphaera prunae	+	+	+	+	F
Metallosphaera sedula	+	+	+	+	F
Sulfolobus metallicus	+	+	+	+	A
Sulfolobus yangmingensis	na	+	na	+	F
Sulfurococcus mirabilis	+	+	+	+	F
Sulfurococcus yellowstonensis	+	+	+	+	F

[#]Listed in alphabetical order; *MS = Metal sulfides other than pyrite; A = autotroph; F = facultative autotroph and/or mixotroph; H = heterotroph; na = data not available ; species without standing in nomenclature (http://www.bacterio.cict.fr/) are given in quotation marks

Acidithiobacillus ferrooxidans (type strain ATCC 23270 = CIP 104768 = DSM 14882). *At. ferrooxidans* (formerly *Thiobacillus ferrooxidans*) was the first described (Colmer & Hinkle, 1947; Temple & Colmer, 1951: Kelly & Wood, 2000) and is the most studied metal sulfide oxidizing organism. *At. ferrooxidans* is an obligate autotrophic, and derives energy from the oxidation of Fe(II) ions, various sulfur compounds, e.g. elemental sulfur, thiosulfate, trithionate, tetrathionate and sulfide, and hydrogen. It has been reported to oxidize arsenopyrite (AsFeS), bornite (Cu_5FeS_4), chalcocite (Cu_2S), chalcopyrite ($CuFeS_2$), covellite (CuS), enargite ($3Cu_2S.As_2S_5$), galena (PbS), millerite (NiS), orpiment (As_2S_3), pyrite (FeS_2), marcasite (FeS_2), sphalerite (ZnS), wurtzite (ZnS), stibnite (Sb_2S_3), pyrrhotite ($Fe_{1-x}S$ (x = 0 to x = 0.2)), and tetrahedrite ($Cu_8Sb_2S_7$), as well as gallium sulfide and the synthetic metal sulfides CdS, CoS, CuS, Cu_2S, FeS, MnS, MoS_2, NiS, PbS, SnS and ZnS (Silver & Torma, 1971; 1974; Sakaguchi *et al.*, 1976; Torma & Gabra, 1977; Torma, 1978; Torma & Sakaguchi, 1978; Tributsch & Bennett, 1981; Lizama & Suzuki, 1991; Bhatti *et al.*, 1993; Pistorio *et al.*, 1994; Tuovinen *et al.*, 1994; Garcia *et al.*, 1995a; 1995b; Schippers *et al.*, 1996; Fowler & Crundwell, 1998; Fowler *et al.*, 1999; Ehrlich, 2002). The ability to oxidize Fe(II) ions (and pyrite) is the key characteristic of this *Acidithiobacillus* species. During growth on sulfur compounds, *At. ferrooxidans* has been observed to accumulate fine sulfur deposits, which are predominantly associated with the cell wall (Hazeu *et al.*, 1988). Models of the electron transport pathways of Fe(II) and sulfur oxidation have recently been updated (Rohwerder *et al.*, 2003; Rawlings, 2005). The organism also grows anaerobically with sulfur compounds or hydrogen as electron donor and Fe(III) ions as electron acceptor (Brock & Gustafson, 1976). N_2-fixation has been demonstrated (Mackintosh, 1978; Norris *et al.*, 1995). Since *At. ferrooxidans* lives in an environment with high metal ion concentrations, it is resistent to many cationic metals, though metals which occur as oxyanions (such as molybdate) tend to be highly toxic (Dopson *et al.*, 2003; Rawlings, 2005). For example, metal or arsenic concentrations whereby metabolic activity of *At. ferrooxidans* still occurs are 84 mM As(III), 800 mM Cu(II), 1071 mM Zn(II), 500 mM Cd(II), and 1000 mM Ni(II). *At. ferrooxidans* strains have been often isolated from different acidic environments. The organism seems to be more abundant in heap leaching environments than in column or tank leaching operations (Groudev & Groudeva, 1993; Schippers *et al.*, 1995; Brierley, 1997; Olson *et al.*, 2003; Rawlings, 2005).

At. ferrooxidans is phylogenetically heterogene. The phenotypically indistinguishable strains of *At. ferrooxidans* are divided into at least four genomovars, and it may turn out that these represent separate species. The acidophilic Fe(II) oxidizing strains SPIII/3 and m-1 (DSM 2392) have shown not be related to the genus *Acidithiobacillus* (Harrison, 1982; Karavaiko *et al.*, 2003; Karsten *et al.*, 2005). Early reports of facultatively autotrophic *At. ferrooxidans* strains are now believed to have been due to the presence of facultatively autotrophic (see *Acidiphilium acidophilum*) or heterotrophic contaminants (*Acidiphilium* sp.), some of which have been difficult to remove, as their presence can stimulate the growth of *At. ferrooxidans* (Harrison, 1984).

Acidithiobacillus thiooxidans (type strain ATCC 19377 = CIP 104597 = DSM 14887 = JCM 3867 = NCIMB 8343) was described by Waksman and Joffe (1922). It grows obligately autotrophically with various sulfur compounds, e.g. elemental sulfur, thiosulfate, and tetrathionate. Growth on the following metal sulfides has been reported: covellite, galena, sphalerite, wurtzite (Lizama & Suzuki, 1991; Pistorio *et al.*, 1994; Curutchet *et al.*, 1995; Garcia *et al.*, 1995a; 1995b; Schippers & Sand, 1999; Pogliani & Donati, 2000). *At. thiooxidans* does not oxidize pyrite (Bacelar-Nicolau & Johnson, 1999). A scheme of the sulfur compound oxidation pathways has recently been presented by Kamimura *et al.* 2005. *At. thiooxidans* strains have been isolated from different acidic environments such as soil, sulfur deposits, and mine waste. The organism occurs in tank bioleaching operations (together with Fe(II) oxidizers, Battaglia *et al.*, 1994; Goebel & Stackebrandt, 1994; Rawlings, 2002, 2005; Rzhepishevska *et al.*, 2005).

Acidithiobacillus caldus (type strain: strain KU = DSM 8584 = ATCC 51756). The species was described by Hallberg and Lindström (1994). Like *At. thiooxidans, At. caldus* grows autotrophically with various sulfur compounds (elemental sulfur, sulfide, sulfite, thiosulfate, and tetrathionate, Hallberg *et al.*, 1996), and not with Fe(II) ions), but has a higher temperature growth optimum at 45 °C, and can also grow mixotrophically with yeast extract or glucose. *At. caldus* has shown to oxidize arsenopyrite (Dopson & Lindström, 1990). *At. caldus* has been found in tank bioleaching operations (together with Fe(II) oxidizers, Rawlings, 2002, 2005; Okibe *et al.*, 2003), and other acidic environments suchs as geothermal sites and acid mine drainage. One *Acidithiobacillus* strain has proposed to be a new species (*"Acidithiobacillus cuprithermicus"*) despite phylogenetically closely related to *At. caldus* according to 16S rRNA gene sequence analysis (Karavaiko *et al.*, 2003).

Acidithiobacillus albertensis (type strain ATCC 35403 = DSM 14366) was described by Bryant *et al.* (1983). Although physiologically similar to *At. thiooxidans*, it must be considered as a separate species. The cells have a condensed glycocalyx and, like *At. ferrooxidans*, accumulate internal sulfur deposits. *At. albertensis* was isolated from an acidic soil adjacent to a sulfur stockpile. Unfortunately, the original isolate has been lost (Kelly & Wood, 2000).

"Thiobacillus prosperus" (strain DSM 5130). The species was proposed by Huber and Stetter (1989). Cells are acidophilic, obligately chemolithoautotrophic, Gram-negative, motile rods. They grow aerobically on metal sulfides such as pyrite, sphalerite, chalcopyrite, arsenopyrite and galena, and on Fe(II) ions, elemental sulfur, and H_2S. Thiosulfate, tetrathionate, and the synthetic sulfides Ag_2S, CuS, MoS_2, Sb_2S, SnS and ZnS do not serve as substrate. The halotolerant not validly described species grows at seawater salt concentrations (up to 3.5% NaCl). Concentrations of 0.9 mM Ag, 1.3 mM As, 0.009 mM Cd, 170 mM Co, 16 mM Cu, 0.05 mM Hg, 1 mM Mo, 850 mM Ni, 8 mM Sb, 0.04 mM U, and 1500 mM Zn are tolerated. No significant homology with *Acidithiobacillus* was detected by DNA-DNA hybridization. The organism was originally isolated from a geothermal marine area off Italy.

"Thiobacillus plumbophilus" (strain DSM 6690) was isolated from an uranium mine in Germany by Drobner *et al.* (1992). Cells are strictly aerobic, mesoacidophilic, rod-shaped, obligately chemolithoautotrophic bacteria which are able to grow on H_2S, galena (PbS) or H_2, but not on Fe(II) ions, sulfur, thiosulfate or many other metal sulfides. No significant homology with *Acidithiobacillus* or *Thiomonas* was detected by DNA-DNA hybridization.

Thiomonas cuprina (type strain DSM 5495). The species was originally described by Huber and Stetter (1990) as *Thiobacillus cuprinus (ex),* and reclassified by Moreira and Amils (1997). The newly described genus *Thiomonas* contains many of the facultatively autotrophic members of the previously classified as *Thiobacillus* spp. Strains of *Thiomonas cuprina* are motile, Gram-negative, facultatively chemolithotrophic and mixotrophic, moderately acidophilic rods capable of growth on some metal sulfides including arsenopyrite, galena, sphalerite, chalcopyrite, and synthetic CdS and FeS, as well as on H_2S and elemental sulfur. No growth was obtained on thiosulfate, tetrathionate, Fe(II) ions, pitch blend or the following metal sulfides: bornite, chalcocite, covellite, pyrite, cinnabar, and synthetic Ag_2S, CuS, MoS_2, Sb_2S, SnS and ZnS. Organotrophic growth occurs e.g. on yeast extract, peptone and pyruvate. Resistance to arsenic and heavy metals during growth on ore was observed up to 1.3 mM As, 0.09 mM Cd, 170 mM Co, 7.9 mM Cu, 1 mM Mo, 170 mM Ni, 0.04 mM U, and 150 mM Zn. *Thiomonas cuprina* was isolated from solfatara fields in Iceland and an uranium mine waste heap in Germany.

2.1.2. *Nitrospira*

***Leptospirillum* spp.** The genus *Leptospirillum* was validly described by Hippe in 2000, though the first bacteria of this genus were isolated and described by Markosyan in 1972. Phylogentically, the genus belongs to the phylum Nitrospira. *Leptospirillum* spp. are obligate acidophilic (pH < 4.0) and aerobic. Cells are Gram-negative, motile vibroid to spirilla-shaped, but cocci or pseudococci can be formed. They grow obligately chemolihoautotrophically and derive energy only from the oxidation of Fe(II) ions but not from sulfur compounds. CO_2 is fixed by means of the Benson-Calvin Cycle (Balashova *et al.*, 1974; Tyson *et al.*, 2004). The oxidation of pyrite, sphalerite and chalcopyrite in pure culture has been reported (Norris, 1983; Sand *et al.*, 1992; Schippers *et al.*, 1996; Okibe & Johnson, 2004). *Leptospirillum* spp. are regularly found in bioleaching operations where they oxidize metal sulfides in co-culture with sulfur-oxidizers (Battaglia *et al.*, 1994; Goebel & Stackebrandt, 1994; Rawlings, 2002, 2005; Okibe *et al.*, 2003; Kinnunen & Puhakka, 2004; Rzhepishevska *et al.*, 2005; Hawkes *et al.*, 2006a, 2006b). They also occur in acid mine drainage environments (Schrenk *et al.*, 1998; Bond *et al.*, 2000a, 2000b; González-Toril *et al.*, 2003, 2005). The genus *Leptospirillum* comprises the following species: *L. ferrooxidans, L. ferriphilum, "L. ferrodiazotrophum"* and *L. thermoferrooxidans. L. thermoferrooxidans* was described as a thermophilic organism (Golovacheva *et al.*, 1992; Hippe, 2000). Since *L. thermoferrooxidans* could oxidize Fe(II) ions but not metal sulfides (pyrite and chalcopyrite) in pure culture, and the culture has been lost, the species is not further considered here.

Leptospirillum ferrooxidans (type strain L15 = ATCC 29047 = DSM 2705 = VKM B-1339). The mesophilic species was isolated by Markosyan (1972) and validly described by Hippe 2000. N_2-fixation has been shown (Norris *et al.*, 1995; Parro *et al.*, 2003). The biochemistry of the Fe(II) ion oxidation is different to that of *Acidithiobacillus ferrooxidans* (Rohwerder *et al.*, 2003; Rawlings, 2005) which allows *L. ferrooxidans* to oxidize Fe(II) ions at a high redox potential (Rawlings *et al.*, 1999). The following metal ion concentrations were tolerated: 7.5 g/L Al, 1 g/L Co, 25 g/L Cu, >30 g/L Mn, 7.5 g/L Ni, and 30 g/L Zn (Hallmann *et al.*, 1992).

Leptospirillum ferriphilum (type strain P3a = ATCC 49881 = DSM 14647). *L. ferriphilum* was described by Coram and Rawlings (2002) as a thermotolerant (up to 45°C) mesophilic Fe(II) ion oxidizer. The organism seems to be dominant in tank bioleaching operations at 35 – 50 °C. It is able to oxidize Fe(II) in high rate at pH < 1 (Kinnunen & Puhakka, 2005). *L. ferriphilum* does not have the nitrogen fixing (*nif*) operon (Tyson *et al.*, 2004) in contrast to *L. ferrooxidans* and "*L. ferrodiazotrophum*".

"*Leptospirillum ferrodiazotrophum*" (strain ATCC BAA-1181) was recently isolated from a subsurface acid mine drainage biofilm (Iron Mountain, California, USA, Tyson *et al.*, 2005). Due to its ability to oxidize Fe(II) ions and to fix nitrogen, the name "*L. ferrodiazotrophum*" has been proposed (although *L. ferrooxidans* also fixes nitrogen). The ability to oxidize metal sulfides has not yet been demonstrated for this recently proposed species.

2.1.3. Actinobacteria

Acidimicrobium ferrooxidans (type strain ICP = DSM 10331). *Am. ferrooxidans* was described by Clark & Norris (1996) as a species of the single genus *Acidimicrobium* of the family Acidimicrobiaceae. The strains recognized so far were isolated from a copper leaching dump in the USA (strain TH3; Brierley, 1978; Norris *et al.*, 1996) and a pyrite enrichment from an Icelandic geothermal site (type strain DSM 10331; Norris *et al.*, 1996). The two strains differ in their morphology in that strain TH3 grows in filaments in liquid media. The rod-shaped Gram-positive *Am. ferrooxidans* cells are acidophilic and moderately thermophilic. Autotrophic growth occurs on Fe(II) ions (not on sulfur compounds), heterotrophic growth on yeast extract during which cells are motile. Facultative anaerobic growth via reduction of Fe(III) has been reported (Bridge & Johnson, 1998). Mixed cultures of *Am. ferrooxidans* and *Sulfobacillus* oxidize Fe(II) ions more extensively than either strain does in pure culture (Clark & Norris, 1996).

"*Ferrimicrobium acidiphilum*" (strain T23). The proposed type strain of this acidophile was isolated from an abandoned mine in north Wales, and similar isolates were obtained from mines in the USA and elsewhere (Johnson *et al.*, 1995, 2001; Johnson & Roberto, 1997). It is closely related to *Am. ferrooxidans* but does not fix CO_2, thus, only grows heterotrophically (e.g. with yeast extract). Cells are motile rods but may form filaments. The mesophilic, acidophilic, Fe(II) ion and pyrite oxidizing actinobacterium does not grow on sulfur compounds. Mixed cultures of the Fe(II) ion oxidizer *"Ferrimicrobium acidiphilum"* with a sulfur oxidizer (*Acidithiobacillus thiooxidans* or *Acidiphilium acidophilum*) oxidize pyrite in higher rates than pure cultures of *"Ferrimicrobium acidiphilum"* (Bacelar-Nicolau & Johnson, 1999).

2.1.4. Firmicutes

Metal sulfide oxidizing Gram-positive bacteria that have low G+C mol% in their chromosomal DNA belong to the genera *Alicyclobacillus, Sulfobacillus,* and the as yet non-validated genus *"Caldibacillus"*. Physiologically, these bacteria are versatile, being able to grow lithotrophically with Fe(II) ions and/or sulfur compounds and/or organotrophically with different organic substances. Growth may be autotrophic (CO_2-fixation), heterotrophic (e.g yeast extract) or mixotrophic (CO_2 + yeast extract). Endospores may be formed. Mesophilic and moderately thermophilic species have been isolated from sulfidic heaps or thermal springs, many of them are not validly described yet.

Alicyclobacillus disulfidooxidans (type strain SD-11 = ATCC 51911 = DSM 12064). The facultativ anaerobic, mesophilic, physiologically versatile species was originally described as *Sulfobacillus disulfidooxidans* (*ex*) by Dufresne *et al.* (1996) and recently reclassified by Karavaiko *et al.* (2005). It was isolated from an enrichment of wastewater sludge. Cells are rod-shaped and non-motile. Yeast extract is necessary as growth factor for elemental sulfur oxidation. It has been doubted that the original culture was pure. Thus, Fe(II) ions and metal sulfides are probably not oxidized by *Alicyclobacillus disulfidooxidans.*

Alicyclobacillus tolerans (type strain K1 = VKM B-2304 = DSM 16297). The thermotolerant mesophilic, physiologically versatile species was originally described as *"Sulfobacillus thermosulfidooxidans* subsp. *thermotolerans"* by Kovalenko and Malakhova (1983) and recently reclassified by Karavaiko *et al.* (2005). It was isolated from lead-zinc ores of a deposit in Uzbekistan. Cells are rod-shaped and non-motile. Facultative anaerobic growth via reduction of Fe(III) is described.

Sulfobacillus acidophilus (type strain NAL = ATCC 700253 = DSM 10332). *Sb. acidophilus* was described by Norris *et al.* (1996) as a rod-shaped moderately thermophilic organism, isolated from a coal spoil heap, UK. Limited motility was observed. Facultative anaerobic growth via reduction of Fe(III) has been described (Bridge & Johnson, 1998).

"Sulfobacillus montserratensis" (strain L15) was isolated from a geothermal area of the Caribian island Montserrat (Yahya *et al.*, 1999; Johnson, 2001). The mesophilic, motile, rod-shaped bacterium tolerates 500 mM Fe^{2+}, 100 mM Fe^{3+}, 100 mM Cu^{2+}, >300 mM Zn^{2+}, 0.2 mM MoO_4^{2-} and an extremely low pH of 1. Growth on Fe(II) ions and sulfur compounds was enhanced in presence of yeast extract.

Sulfobacillus sibiricus (strain N1 = DSM 17363 = VKM B-2280) was isolated from an ore deposit in east Siberia and described by Melamud *et al.* (2003). The rod-shaped, moderately thermophilic bacterium oxidizes metal sulfides, Fe(II) ions and sulfur in presence of yeast extract.

Sulfobacillus thermosulfidooxidans (type strain AT-1 = DSM 9293 = VKM B-1269). The moderately thermophilic, facultative anaerobic (reduction of Fe(III)) strain was isolated from dumps in Russia and described as the first species of the genus *Sulfobacillus* (Golovacheva & Karavaiko, 1978). Some strains grow as coryneforms. Limited motility was observed. Metal concentrations whereby metabolic activity of *S. thermosulfidooxidans* still occurs are 6 mM Cu(II), 43 mM Zn(II), and 5 mM Ni(II) (Dopson *et al.*, 2003).

Sulfobacillus thermotolerans (type strain Kr1 = VKM B-2339 = DSM17362). Cells are straight to slightly curved rods. The thermotolerant, Gram-positive, aerobic, endospore-forming, acidophilic bacterium was isolated from a gold-recovery plant (Siberia). Growth is mixotrophic by oxidizing Fe(II) ions and sulfur compounds in the presense of yeast extract or other organic substrates (Bogdanova *et al.*, 2006).

"Caldibacillus ferrivorus" (strain GSM) was isolated from mine spoil material, Montana, USA (Johnson *et al.*, 2001). The moderate thermophile strain grows autotrophically and mixotrophically, and is facultative anaerobic (reduction of Fe(III)). As substrates, Fe(II) ions, sulfur compounds, and various organic compounds are used. The strain is phylogenetically closely related to the genus *Alicyclobacillus* (Karavaiko *et al.*, 2005). It appears likely that the genus *"Caldibacillus"* will not therefore be validated.

2.2. Mesophilic and Moderately Thermophilic Archaea

2.2.1. Euryarchaeota

All metal sulfide oxidizing, mesophilic and moderately thermophilic Archaea belong to the genus *Ferroplasma* within the family Ferroplasmaceae of the order Thermoplasmatales of the archaeal phylum Euryarchaeota.

***Ferroplasma* spp.** The genus *Ferroplasma* was first described by Golyshina *et al.*, 2000. Species of the genus *Ferroplasma* are acidophilic Archaea that oxidize Fe(II) ion, pyrite and other metal sulfides. Cells lack a cell wall and are pleomorphic (irregular cocci, varying from spherical to filamentous, forming duplex and triplex forms). In contrast to the original description as aerobic and obligately chemolithoautotrophic cells, growth may also be mixotrophic or organotrophic, and facultatively anaerobic via Fe(III) reduction (Dopson *et al.*, 2004). *Ferroplasma* species are widespread in very acidic mining environments (Bond *et al.*, 2000a, 2000b; Edwards *et al.*, 2000; Burton and Norris, 2000; Johnson & Hallberg, 2003; González-Toril *et al.*, 2003; Golyshina and Timmis, 2005). *Ferroplasma* species are regularly found in bioleaching operations as well (Okibe *et al.*, 2003; Okibe & Johnson, 2004; Rawlings, 2005; Rzhepishevska *et al.*, 2005; Hawkes *et al.*, 2006a).

The genus *Ferroplasma* comprises the following species: *Ferroplasma acidiphilum*, *"Ferroplasma acidarmanus"* and *"Ferroplasma cupricumulans"*.

Ferroplasma acidiphilum (type strain Y = DSM 12658 = JCM 10970) was isolated from a bioleaching bioreactor in Russia and described by Golyshina *et al.* (2000). Aerobic growth may be lithoautotrophic (Fe(II) ions + CO_2), organoheterotrophic (e.g. on yeast extract) or mixotrophic (Fe(II) and an organic carbon source). Anaerobic growth occurs on Fe(III) in the presence of yeast extract as electron donor (Dopson *et al.*, 2004).

"Ferroplasma acidarmanus" (type strain Fer1) was isolated from the AMD site Iron Mountain (California, USA) by Edwards *et al.* (2000) and described by Dopson *et al.* 2004. The type strain has been deposited in the American Type Culture Collection (ATTC) in the patent pending collection. Cells grow aerobically either organoheterotrophically (e.g. on yeast extract) or mixotrophically on Fe(II) and an organic carbon source, and anaerobically on Fe(III) in the presence of yeast extract as electron donor. Metal or arsenic concentrations whereby metabolic activity of *"F. acidarmanus"*still occurs are 13 mM As(III), 16 mM Cu(II), and 9 mM Cd(II) (Dopson *et al.*, 2003).

"Ferroplasma cupricumulans" (strain BH2 = DSM 16651) was isolated from a chalcocite heap bioleaching operation in Myanmar (Hawkes *et al.*, 2006a, 2006b). Mixotrophic growth with Fe(II) and yeast extract was observed. Cells grow anaerobically on Fe(III) in the presence of tetrathionate and yeast extract as electron donors.

2.3. Extremely Thermophilic Archaea

2.3.1. Crenarchaeota

All extremely thermophilic metal sulfide oxidizing Archaea belong to the family Sulfolobaceae within the order Sulfolobales of the archaeal phylum Crenarchaeota. Members of the Sulfolobales have optimal growth temperatures between 65 and 90°C and pH optima of around pH 2. Cells are motile or non-motile cocci, which occur usually singly or in pairs. Characteristically, they are highly irregular in shape, and often strongly lobed or edged, and stain Gram negatively. They grow either aerobically, facultatively anaerobically, or anaerobically. Under autotrophic conditions they gain energy by oxidation of elemental sulfur, thiosulfate, metal sulfides, or molecular hydrogen (H_2). Alternatively, heterotrophic growth occurs by aerobic respiration or anaerobic sulfur respiration or by fermentation of organic substrates (Segerer & Stetter, 1992; Huber & Stetter, 2001; Huber & Prangishvili, 2004).

Several isolates of Sulfolobales are able to extract metal ions from sulfidic ores. Dissolution of the metal sulfides chalcocite, chalcopyrite, molybdenite, pentlandite, pyrite, pyrrhotite and sphalerite has been reported (Brierley & Murr, 1973; Brierley & Lockwood, 1977; Brierley & Brierley, 1986; Huber *et al.*, 1986; Norris & Parrott, 1986; Norris *et al.*, 1988; Huber *et al.*, 1989; Huber & Stetter, 1991; Tobita *et al.*, 1994; Vitaya *et al.*, 1994; Fuchs *et al.*, 1995; Konishi *et al.*, 1995; Norris *et al.*, 2000).

The metal sulfide oxidizing species belong to the following genera of Sulfolobales: *Acidianus, Metallosphaera, Sulfolobus* and *Sulfurococcus*. Whereas some strains only oxidize metal sulfides weakly, others, e.g. *Acidianus brierleyi, Sulfolobus metallicus* and *Metallosphaera* species are very efficient ore leachers. *Sulfolobus acidocaldarius* and *Sulfolobus solfataricus* do not oxidize sulfur compounds including metal sulfides in contrast to several reports (Kargi & Robinson, 1985; Vitaya *et al.*, 1994; Tobita *et al.*, 1994). Either the cultures were not pure or they lost their ability to oxidize sulfur compounds (Marsh *et al.*, 1983; Huber *et al.*, 1989; Grogan, 1989, 1991; Hallberg & Johnson, 2001; Huber & Prangishvili, 2004).

Acidianus brierleyi (type strain DSM 1651 = IFO (now NBRC) 15269 = JCM 8954) was isolated from a solfataric spring in Yellowstone National Park, USA (Brierley & Brierley, 1973), one year after the genus *Sulfolobus* was established by Brock *et al.* (1972). The species was originally described as *Sulfolobus brierleyi* (*ex*) by Zillig *et al.* (1980) and reclassified as *Acidianus brierleyi* by Segerer *et al.* (1986). The facultative anaerobic, facultative chemolithoautotrophic organism uses metal sulfides, elemental sulfur, H_2, and organic compounds as substrates.

Acidianus infernus (type strain So4a = DSM 3191 = IFO (now NBRC) 15270 = JCM 8955) was isolated from a solfataric spring in Italy and described by Segerer *et al.* 1986. The facultative anaerobic, obligate chemolithoautotrophic organism uses metal sulfides, elemental sulfur and H_2 as substrates.

Metallosphaera hakonensis (type strain HO1-1 = IAM 14250 = JCM 8857 = DSM 7519 = ATCC 51241) was originally described as *Sulfolobus hakonensis (ex)* by Takayanagi *et al.* (1996) and reclassified as *Metallosphaera hakonensis* by Kurosawa *et al.* (2003). The species was isolated from hot springs in Hakone, Japan. The aerobic, facultative chemolithoautotrophic organism uses metal sulfides, elemental sulfur, tetrathionate, H_2S, and organic compounds as substrates.

Metallosphaera prunae (type strain Ron 12/II = DSM 10039) was isolated from a smoldering slag heap of an uranium mine in Germany and described by Fuchs *et al.* (1995). The aerobic, facultative chemolithoautotrophic organism uses metal sulfides, elemental sulfur, H_2 and organic compounds as substrates.

Metallosphaera sedula (type strain TH2 = ATCC 51363 = DSM 5348 = IFO (now NBRC) 15509 = JCM 9185) was isolated from a solfataric field in Italy and described by Huber *et al.* (1989). The aerobic, facultative chemolithoautotrophic organism uses metal sulfides (pyrite, chalcopyrite, sphalerite and the synthetic sulfides CdS, SnS, ZnS), elemental sulfur, and organic compounds as substrates. Metal or arsenic concentrations whereby metabolic activity of *M. sedula* still occurs are 1.3 mM As, 0.9 mM Cd, 0.85 mM Co, 16 mM Cu, 0.0005 mM Hg, 0.1 mM Mo, 0.8 mM Sb, 0.4 mM U, and 150 mM Zn.

Sulfolobus metallicus (type strain Kra 23 = DSM 6482 = IFO (now NBRC) 15436 = JCM 9184) was isolated from an Icelandic solfataric field and described by Huber & Stetter (1991). The aerobic, obligate chemolithoautotrophic organism uses metal sulfides (pyrite, chalcopyrite, sphalerite and the synthetic sulfides CdS, ZnS), and elemental sulfur as substrates.

Sulfolobus yangmingensis (type strain YM1) was isolated from a geothermal vent in Yang-Ming National Park in northern Taiwan and described by Jan *et al.* (1999). The aerobic, facultative chemolithoautotrophic organism uses FeS, elemental sulfur, tetrathionate and organic compounds as substrates.

Sulfurococcus mirabilis (type strain INMI AT-59) was isolated from a crater of the Uzon volcano in Kamchatka, Russia. The strain has no deposition number and is preserved at the Institute of Microbiology, Russian Academy of Sciences, Moscow, Russia (Golovacheva *et al.*, 1987a; Golovacheva *et al.*, 1987b). The aerobic, facultative chemolithoautotrophic organism uses metal sulfides, elemental sulfur and organic compounds as substrates.

Sulfurococcus yellowstonensis (type strain Str6kar) was isolated from a thermal spring in Yellowstone National Park, USA. The strain has no deposition number and is preserved at the Institute of Microbiology, Russian Academy of Sciences, Moscow, Russia (Karavaiko *et al.*, 1994). The aerobic, facultative chemolithoautotrophic organism uses metal sulfides, Fe(II) ions, elemental sulfur and organic compounds as substrates.

3. ACIDOPHILIC MICROORGANISMS THAT DO NOT OXIDIZE METAL SULFIDES

Most of the metal sulfide oxidizers described in the previous chapter are litho-autotrophic microorganisms. Some metal sulfide oxidizers grow facultatively autotrophically, mixotrophically, or heterotrophically, such as *Acidithiobacillus caldus*, *Thiomonas cuprina*, *Acidimicrobium ferrooxidans*,*"Ferrimicrobium acidiphilum"*, Firmicutes, *Ferroplasma*, and several extremely thermophilic Archaea. In this chapter further acidophilic microorganisms are introduced. Most of them grow heterotrophically. Some of them oxidize sulfur compounds, thus, have the potential to oxidize acid-soluble metal sulfides. However, since the ability to oxidize metal sulfides has not been demonstrated for these species, they are introduced in this chapter. The phylogenetic affiliation and physiological properties of the non-metal sulfide oxidizing, mainly heterotrophic, acidophilic Bacteria and Archaea are listed in Table 4 and 5 .

3.1. Mesophilic Bacteria

These bacteria are heterotrophs and belong to the genera *Acidiphilium* (Harrison, 1981; Kishimoto *et al.*, 1996), *Acidocella* (Kishimoto *et al.*, 1996), *Acidomonas* (Urakami *et al.*, 1989; Yamashita *et al.*, 2004), *Acidisphaera* (Hiraishi *et al.*, 2000), and *Acidobacterium* (Kishimoto *et al.*, 1991). The species have mostly been isolated from acidic mining environments or cultures of *Acidithiobacillus ferrooxidans.*

The mesoacidophilic, heterotrophic Bacteria are Gram-negative, and grow as motile or non-motile rod-shaped cells. They oxidize various organic substrates but not Fe(II) ions. Some of them also oxidize sulfur compounds such as *Acidiphilium*

Table 4. Phylogeny of acidophilic microorganisms that do not oxidize metal sulfides (mainly heterotrophs)

Species[#]	Phylum	G+C (mol%)	Sulfur oxidation
Mesophilic Bacteria			
Acidiphilium acidophilum	Proteobacteria	63-64	+
Acidiphilium angustum	Proteobacteria	67	-
Acidiphilium cryptum	Proteobacteria	64-70	+
Acidiphilium multivorum	Proteobacteria	66-68	-
Acidiphilium organovorum	Proteobacteria	64	-
Acidiphilium rubrum	Proteobacteria	63	+
Acidisphaera rubrifaciens	Proteobacteria	69-70	-
Acidobacterium capsulatum	Acidobacteria	60	-
Acidocella aminolytica	Proteobacteria	59	-
Acidocella facilis	Proteobacteria	65	-
Acidomonas methanolica	Proteobacteria	63-65	-
Moderately thermophilic Bacteria			
Acidicaldus organivorans	Proteobacteria	72	+
Alicyclobacillus acidiphilus	Firmicutes	54	na
Alicyclobacillus acidocaldarius	Firmicutes	61-63	-
Alicyclobacillus acidoterrestris	Firmicutes	53	-
Alicyclobacillus cycloheptanicus	Firmicutes	54-57	-
Alicyclobacillus herbarius	Firmicutes	56	na
Alicyclobacillus hesperidum	Firmicutes	53-54	na
Alicyclobacillus pomorum	Firmicutes	53	na
Alicyclobacillus sendaiensis	Firmicutes	62	na
Alicyclobacillus vulcanalis	Firmicutes	62	-
Hydrogenobaculum acidophilum (autotroph)	Aquificae	35	+
Archaea			
Acidianus ambivalens (autotroph)	Crenarchaeota	33	+
Picrophilus oshimae	Euryarchaeota	36	-
Picrophilus torridus	Euryarchaeota	na	-
Sulfolobus acidocaldarius	Crenarchaeota	37	-
Sulfolobus shibatae	Crenarchaeota	35	+
Sulfolobus solfataricus	Crenarchaeota	35	-
Sulfolobus tokodaii	Crenarchaeota	33	+
Sulfolobus yangmingensis	Crenarchaeota	42	+
Sulfurisphaera ohwakuensis	Crenarchaeota	33	+
Thermoplasma acidophilum	Euryarchaeota	46	-
Thermoplasma volcanicum	Euryarchaeota	38-40	-

[#]Listed in alphabetical order; G + C = mole% guanine+cytosine content of genomic DNA; na = data not available

acidophilum (formerly *Thiobacillus acidophilus* (*ex*), reclassified by Hiraishi *et al.* (1998)), *Acidiphilium crytum*, *Acidiphilium rubrum* and *Acidiphilium* sp. strain SJH (Hallberg *et al.*, 2001).

Table 5. Optimum and range of growth for pH and temperature of acidophilic microorganisms that do not oxidize metal sulfides (mainly heterotrophs)

Species[#]	pH optimum	pH minimum-maximum	Temperature optimum (°C)	Temperature minimum-maximum (°C)
Mesophilic Bacteria				
Acidiphilium acidophilum	2.5-3	1.5-6.5	27-30	<25-37
Acidiphilium angustum	na	2.5-6	na	na
Acidiphilium cryptum	3	1.9-5.9	35-41	20-41
Acidiphilium multivorum	~ 3.5	1.9-5.6	27-35	17-42
Acidiphilium organovorum	3	2-5.5	37	20-45
Acidiphilium rubrum	na	2.5-6	na	na
Acidisphaera rubrifaciens	4.5-5	3.5-6	30-35	20-40
Acidobacterium capsulatum	na	3-6	na	20-37
Acidocella aminolytica	na	3-6	na	20-37
Acidocella facilis	na	2.5-6	na	25-37
Acidomonas methanolica	na	2-5.5	na	<30-42
Moderately thermophilic Bacteria				
Acidicaldus organivorans	2.5-3	1.8-3	50-55	na-65
Alicyclobacillus acidiphilus	3	2.5-5.5	50	20-55
Alicyclobacillus acidocaldarius	3-4	2-6	60-65	45-70
Alicyclobacillus acidoterrestris	na	2.2-5.8	42-53	<35-55
Alicyclobacillus cycloheptanicus	na	3-5.5	48	40-53
Alicyclobacillus herbarius	4.5-5	3.5-6	55-60	35-65
Alicyclobacillus hesperidum	3.5-4	>2-<6	50-53	>35-<60
Alicyclobacillus pomorum	4.5-5	>2.5-<6	45-50	30-60
Alicyclobacillus sendaiensis	5.5	2.5-6.5	55	40-65
Alicyclobacillus vulcanalis	4	2-6	55	35-65
Hydrogenobaculum acidophilum (autotroph)	3-4	2-na	65	na-~70
Archaea				
Acidianus ambivalens (autotroph)	2.5	1-3.5	80	na-87
Picrophilus oshimae	0.7	0-3.5	60	47-65
Picrophilus torridus	0.7	0-3.5	60	47-65
Sulfolobus acidocaldarius	2-3	1-6	70-75	55-85
Sulfolobus shibatae	3	nd	81	na-86
Sulfolobus solfataricus	3-4.5	2-5.5	85	50-87
Sulfolobus tokodaii	2.5-3	2-5	80	70-85
Sulfolobus yangmingensis	4	2-6	80	65-95
Sulfurisphaera ohwakuensis	2	1-5	84	63-92
Thermoplasma acidophilum	1-2	0.5-4	59	45-63
Thermoplasma volcanicum	2	1-4	59-60	33-67

[#]Listed in alphabetical order; na = data not available

The mesoacidophilic, heterotrophic Bacteria live closely associated with autotrophic metal sulfide oxidizers such as *Acidithiobacillus ferrooxidans*, and are often found in bioleaching operations or sulfidic mining environments together with the autotrophs. Presumably, the heterotrophs promote the bioleaching activity of the autotrophic metal sulfide oxidizers by consuming inhibitory organic compounds which are produced by the autotrophs. Furthermore, *Acidiphilium* species are able to reduce Fe(III), and thereby deliver Fe(II) as substrate for autotrophic Fe(II) oxidizers. Metal or arsenic concentrations at which metabolic activity of still occurs are up to 30 mM As(III), 30 mM Cu(II), 900 mM Zn(II), 700 mM Cd(II), and 350 mM Ni(II) (Belly & Brock, 1974; Harrison, 1981, 1984; Wichlacz & Thomson, 1988; Johnson & McGinness, 1991; Pronk & Johnson, 1993; Goebel & Stackebrandt, 1994; Schippers *et al.*, 1995; Johnson & Roberto, 1997; Hiraishi *et al.*, 1998; Küsel *et al.*, 1999; Bridge & Johnson, 2000; Hallberg & Johnson, 2001; Dopson *et al.*, 2003; Johnson & Hallberg, 2003; Wenderoth & Abraham, 2005).

3.2. Moderately Thermophilic Bacteria

Acidicaldus organivorans (Johnson *et al.*, 2006) grows heterotrophically on a range of organic substrates, and oxidizes sulfur but not autotrophically. Facultativ anaerobic growth with Fe(III) was observed. The Gram-negative bacterium was isolated from geothermal sites in Yellowstone National Park, USA.

Alicyclobacillus disulfidooxidans and *Alicyclobacillus tolerans* were described as facultatively anaerobic, metal sulfide-, Fe(II) ions- and sulfur compound-oxidizing bacteria (Karavaiko *et al.*, 2005). As far as known, the other *Alicyclobacillus* species (isolated from various environments) only grow aerobically and heterotrophically with organic substrates. Cells are Gram-positive, spore-forming rods.

The Gram-negative, motile *Hydrogenobaculum acidophilum* has been isolated from a solfataric field in Japan, and reclassified (formerly *Hydrogenobacter acidophilus* (*ex*), Stöhr *et al.*, 2001). This organism grows autotrophically (CO_2 fixation via reductive tricarboxylic acid cycle) with H_2 and sulfur compounds as substrates.

3.3. Archaea

Several species of the phylum Crenarchaeota (order Sulfolobales) have been described as metal sulfide-, Fe(II) ions- and sulfur compound-oxidizing extremely thermoacidophilic Archaea. Most of the species of the genera *Acidianus*, *Sulfolobus* and *Sulfurisphaera* listed in Table 4 and 5 oxidize sulfur but metal sulfide oxidation has not been demonstrated. *Acidianus ambivalens* does not grow heterotrophically (obligate autotroph). In addition to the Crenarchaeota, moderately thermophilic Euryarchaeota of the genera *Thermoplasma* and *Picrophilus* are listed.

Thermoplasma was firstly isolated from a coal refuse pile in Indiana, USA (Darland *et al.*, 1970). Further *Thermoplasma* isolates were also obtained from self-heated smoldering coal refuse piles (Belly *et al.*, 1973; Brock, 1978), and from different solfatara fields by Segerer *et al.* 1988, who described the species. *Thermoplasma* species lack cell walls and are mostly motile, pleomorphic cells. They are facultative anaerobic, obligate heterotrophs and do not oxidize Fe(II) ions and sulfur compounds.

Picrophilus strains were isolated from geothermal solfataric soils and springs in Hokkaido, northern Japan, and described by Schleper *et al.*. (1995, 1996). Cells are irregular cocci. They are strict aerobic, obligate heterotrophs and do not oxidize Fe(II) ions and sulfur compounds.

4. NUCLEIC ACID-BASED MOLECULAR METHODS

To control and optimize metal bioleaching, quick and reliable methods to identify and quantify single species in complex bioleaching communities are needed. Microbial communities can be analyzed using microscopic techniques, cultivation techniques, immunological techniques and nucleic-acid based molecular techniques. The latter are introduced here.

Total cell numbers can be determined by counting cells under a fluorescence microscope after application of nucleic acid-staining fluorochromes (e.g. SybrGreen, acridine orange, DAPI). The drawback of this technique is that these fluorochromes bind unspecifically to nucleic acids and thus, do not provide information on the viability of the cells. Potentially, a major part of the counted cells could be dormant or even dead and yet retain stainable DNA (Kepner & Pratt, 1994; Morita, 1997).

Using classical cultivation techniques, i.e., the most-probable-number (MPN) cultivation method (Schippers & Bosecker, 2005) or the dual-layer agarose plate technique (Johnson *et al.*, 1995), acidophilic autotrophic Fe(II) and sulfur oxidizing bioleaching microorganisms have been enriched from bioleaching communities. By cultivation techniques, however, only a subset of the whole microbial community can be detected, though media have been designed to select different groups on the basis of their physiologies. Furthermore, cultivation techniques are labor-intensive and results are only available after incubation times of several days or even weeks which does not allow a monitoring of bioleaching operations.

Immunoassays with specific antibodies have been applied to enumerate *At. ferrooxidans* and other bioleaching microorganisms (Jerez, 1997; Dziurla *et al.*, 1998), however, their application is time-consuming and requires thorough knowlegde of the microbes occurring in the bioleaching operation.

Over the last years, nucleic-acid based molecular techniques have been increasingly used to identify and quantify microorganisms in the environment and technical applications. Most of them are based on the extraction of DNA from a culture, a bioreactor or an environmental sample, followed by the amplification of DNA using the Polymerase Chain Reaction (PCR), and finally an analysis of the DNA amplification products. In most of the cases, the 16S ribosomal RNA gene (16S rDNA) of prokaryotes (Bacteria and Archaea) is targeted, but also functional genes coding for key enzymes of particular metabolic interest have been analyzed (e.g. the *rus* gene coding for rusticyanin in *At. ferrooxidans*). In the following, several PCR-based techniques for the identification of microorganisms are briefly introduced.

Techniques for the quantification of microorganisms are quantitative, real-time PCR and Fluorescence In Situ Hybridization (FISH) or its modification Catalyzed Reporter Deposition – Fluorescence In Situ Hybridization (CARD – FISH). These quantitative techniques and their application are described afterwards.

4.1. PCR-based Molecular Methods for Identification of Microorganisms

Usually the 16S rRNA gene is analyzed for the PCR-based identification of microorganisms. To address the biodiversity and to identify new species, PCR products can be cloned and the 16S rRNA gene of the various clones in the clone library can be sequenced. The similarities of the sequences can then be shown in a phylogenetic tree, to address the phylogenetic affiliation of the microorganisms in the sample. This approach has been choosen to analyze the microbial communities in natural, acidic environments and bioleaching operations (Goebel & Stackebrandt, 1994, 1995; Bruneel *et al.*, 2005; González-Toril *et al.*, 2005).

Alternatively to clone libraries, the PCR products can be separated in denaturing gradient gel electrophoresis (DGGE) which allows a separation of DNA fragments of the same length but different base-pair sequences. Bands in the DGGE can be excised and the 16S rRNA gene sequenced to address the phylogenetic affiliation of the organisms. This method has been applied to bioleaching communities as well (González-Toril *et al.*, 2003; Dopson & Lindstrom, 2004; Kinnunen & Puhakka, 2004; Demergasso *et al.*, 2005b; Rzhepishevska *et al.*, 2005).

Less labour intensive are DNA fingerprinting techniques which allow only the identification of known organisms. The DNA fingerprinting techniques RFLP (restriction fragment length polymorphism) and ARDREA (amplified ribosomal DNA restriction enzyme analysis) have been applied to identify bioleaching organism (Rawlings, 1995; Selenska-Pobell *et al.*, 1998; 2001; Bond *et al.*, 2000a; Asmah *et al.*, 2001, Bergamo *et al.*, 2004; Coupland & Johnson, 2004; Bruneel *et al.*, 2005; Diaby *et al.*, 2006; Johnson *et al.*, 2005; Bryan *et al.*, 2006a). These techniques include the digestion of the PCR product with one or more restriction enzymes to produce fragments of varying sizes that are resolved on appropriate gels.

Further PCR-based techniques for the identification of bioleaching organisms are: RAPD (randomly amplified polymorphic DNA) and rep-APD (Novo *et al.*, 1996; Selenska-Pobell *et al.*, 1998; 2001; Waltenbury *et al.*, 2005), SSCP (single stranded conformation polymorphism, Battaglia-Brunet *et al.*, 2002), analysis of the PCR-amplified 16S-23S rRNA gene intergenic spacer (Pizarro *et al.*, 1996; Vásquez & Espejo, 1997; Bergamo *et al.*, 2004), and the use of microbe-specific PCR primers (Wulf-Durand *et al.*, 1997).

4.2. Real-time PCR for Quantification of Microorganisms

Real-time PCR (quantitative PCR) is a technique with high sensitivity used in environmental microbiology to quantify different phylogenetic groups and genera, e.g. Bacteria, Archaea, the Fe(III) reducer *Geobacter*, methanogens or cyanobacteria (Suzuki *et al.*, 2000; Takai & Horikoshi, 2000; Stults *et al.*, 2001; Beller *et al.*, 2002; Nadkarni *et al.*, 2002; Harms *et al.*, 2003; Kolb *et al.*, 2003; Sharkey *et al.*, 2004; Schippers *et al.*, 2005). The technique is based on the online fluorescence detection of PCR products and allows the rapid detection and quantification of gene sequences without the need for labor-intensive post-PCR processing (Heid *et al.*, 1996). DNA is quantitatively extracted from samples, purified, and specifically amplified with a thermocycler using sequence-specific fluorescently labeled probes. There are different chemical assasys for real-time PCR, but the most common are sequence-specific TaqMan probes (Heid *et al.*, 1996) and the intercalating non specific SybrGreen dye (Wittwer *et al.*, 1997). The detection limit of the method depends on the target of interest, sample purity, PCR conditions and other factors, but theoretically allows the detection of a single DNA molecule (Lockey *et al.*, 1998; Smith *et al.*, 2005). Real-time PCR has been applied to quantify Bacteria and Archaea oxidizing metal sulfides in mine heaps (Kock & Schippers, 2006). Furthermore, SybrGreen based protocols have been developed to quantify single species in bioleaching communities such as *Acidianus brierleyi, Sulfolobus* sp., *Sulfobacillus thermosulfidooxidans, Sulfobacillus acidophilus, Acidithiobacillus caldus*, and *Leptospirillum ferrooxidans* (Liu *et al.*, 2006).

In addition to DNA, RNA can be quantified after application of an additional reverse transcription step (real-time RT-PCR), which allows quantification of gene expression in environmental microbiology (Wilson *et al.*, 1999; Wawrik *et al.*, 2002; Sharkey *et al.*, 2004). Concerning bioleaching organisms, gene expression has been studied in pure cultures of *Acidithiobacillus ferrooxidans.* The expression of the *rus* gene coding for rusticyanin, a protein involved in Fe(II) oxidation, and of several additional genes important for Fe(II) and sulfur oxidation as well as for CO_2-fixation has been measured by real-time PCR (Yarzábal *et al.*, 2003, 2004; Quatrini *et al.*, 2006; Appia-Ayme *et al.*, 2006). It has been shown that the biological Fe(II) oxidation is most relevant for metal sulfide oxidation during bioleaching (Sand *et al.*, 2001; Schippers, 2004) and that the *rus* gene in *At. ferrooxidans* is particularly expressed during Fe(II) and metal sulfide oxidation rather than during sulfur oxidation (Yarzábal *et al.*, 2003, 2004; Ramírez *et al.*, 2004). Rusticyanin may not be expressed in all strains classified as *At. ferrooxidans*. However, the *rus* gene should be a good target to monitor the abundance and activity of *At. ferrooxidans* in biomining.

4.3. Fluorescence In Situ Hybridization (FISH) for Quantification of Microorganisms

A powerful technique to quantify microbial cells in environmental samples is FISH (Amann *et al.*, 1990, 1995). Since FISH targets ribosomal RNA (rRNA), which is indicative of actively metabolizing bacteria, FISH can provide quantitative information on living bacteria in an environmental sample. FISH has been successfully applied to quantify acidophilic Fe(II) oxidizing *Acidithiobacillus*,

Leptospirillum, *Ferroplasma* and other microorganisms in acid mine drainage environments (Schrenk *et al.*, 1998; Bond *et al.*, 2000a; Edwards *et al.*, 1999a, 2000; González-Toril *et al.*, 2003; Bernier & Warren, 2005; Mahmoud *et al.*, 2005) and in bioleaching operations (Peccia *et al.*, 2000; Okibe & Johnson, 2004; Ebrahimi *et al.*, 2005; Kimura *et al.*, 2005; Coram-Uliana *et al.*, 2006). A drawback of the technique is that a sufficient content of cellular ribosomes is prerequisite for its successful application (Amann *et al.*, 1995; Ludwig & Schleifer, 2000; Schippers *et al.*, 2005). Recently, modified FISH protocols (CARD-FISH = Catalyzed Reporter Deposition - Fluorescence In Situ Hybridization) have been published which allow the detection of less active cells in environmental samples as well (Pernthaler *et al.*, 2002; Sekar *et al.*, 2003; Teira *et al.*, 2004). So far, these protocols have successfully been applied to quantify Bacteria and Archaea oxidizing metal sulfides in mine heaps (Demergasso *et al.*, 2005a; Kock & Schippers, 2006).

5. REFERENCES

Amann RI, Krumholz L, Stahl DA. 1990. Fluorescent-oligonucleotide probing of whole cells for determinative, phylogenetic, and environmental studies in microbiology. J Bacteriol 172: 762-770.

Amann RI, Ludwig W, Schleifer K-H. 1995. Phylogenetic identification and in situ detection of individual microbial cells without cultivation. Microbiol Rev 59: 143-169.

Appia-Ayme C, Quatrini R, Denis Y, Denizot F, Silver S, Roberto F, Veloso F, Valdes J, Cárdenas JP, Esparza M, Orellana O, Jedlicki E, Bonnefoy V, Holmes D. 2006. Microarray and bioinformatic analyses suggest models for carbon metabolism in the autotroph *Acidithiobacillus ferrooxidans*. Hydrometallurgy 83: 273–280.

Asmah RH, Clement C, Bosompem KM, Wilson MD, Brown CA, Osei YD, Addy ME. 2001. Molecular characterization of mineral leaching bacteria from a gold mine in Ghana using PCR and RFLP. In: Ciminelli VST, Garcia Jr. O, eds. Biohydrometallurgy: Fundamentals, Technology and Sustainable Development. Elsevier Science BV, Amsterdam, Part A, 317-324.

Bacelar-Nicolau P, Johnson DB. 1999. Leaching of pyrite by acidophilic heterotrophic iron-oxidizing bacteria in pure and mixed cultures. Appl Environ Microbiol 65: 585-590.

Balashova VV, Vedenina IYa, Markosyan GE, Zavarzin GA. 1974. The autotrophic growth of *Leptospirillum ferrooxidans*. Microbiology 43: 491-494.

Battaglia F, Morin D, Garcia J-L, Ollivier P. 1994. Isolation and study of two strains of *Leptospirillum*-like bacteria from a natural mixed population cultured on a cobaltiferrous pyrite substrate. Antonie van Leeuwenhoek 66: 295-302.

Battaglia-Brunet F, Clarens M, d'Hugues P, Godon JJ, Foucher S, Morin D. 2002. Monitoring of a pyrite-oxidising bacterial population using DNA single-strand conformation polymorphism and microscopic techniques. Appl Microbiol Biotechnol 60: 206-211.

Batty JD, Rorke GV. 2006. Development and commercial demonstration of the BioCOP^{TM} thermophile process. Hydrometallurgy 83: 83–89.

Beller HR, Kane SR, Legler TC, Alvarez PJJ. 2002. A real-time polymerase chain reaction method for monitoring anaerobic, hydrocarbon-degrading bacteria based on a catabolic gene. Environ Sci Technol 36: 3977-3984.

Belly RT, Bohlool BB, Brock TD. 1973. The genus*Thermoplasma*. Ann. NY Acad. Sci. 225: 94-107.

Bergamo RF, Novo MTM, Verissimo RV, Paulino LC, Stoppe NC, Sato MIZ, Manfio GP, Prado PI, Garcia O, Ottoboni LMM. 2004. Differentiation of *Acidithiobacillus ferrooxidans* and *A. thiooxidans* strains based on 16S-23S rDNA spacer polymorphism analysis. Res Microbiol 155: 559-567.

Bernier L, Warren LA. 2005. Microbially driven acidity generation in a tailings lake. Geobiology 5: 115-133.

Bhatti TM, Bigham JM, Carlson L, Tuovinen OH. 1993. Mineral products of pyrrhotite oxidation by *Thiobacillus ferrooxidans*. Appl Environ Microbiol 59: 1984-1990.

Bogdanova TI, Tsaplina IA, Kondrat'eva TF, Duda VI, Suzina NE, Melamud VS, Tourova1 TP, Karavaiko GI. 2006. *Sulfobacillus thermotolerans* sp. nov., a thermotolerant, chemolithotrophic bacterium. Int J Syst Evol Microbiol 56: 1039-1042.

Bond PL, Druschel GK, Banfield JF. 2000a. Comparison of acid mine drainage microbial communities in physically and geochemically distinct ecosystems. Appl Environ Microbiol 66: 4962-4971.

Bond PL, Smriga SP, Banfield JF. 2000b. Phylogeny of microorganisms populating a thick, subaerial, predominantly lithotrophic biofilm at an extreme acid mine drainage site. Appl Environ Microbiol 66: 3842-3849.

Bridge TAM, Johnson DB. 1998. Reduction of soluble iron and reductive dissolution of ferric iron-containing minerals by moderately thermophilic iron-oxidizing bacteria. Appl Environ Microbiol 64: 2181-2186.

Bridge TAM, Johnson DB. 2000. Reductive dissolution of ferric iron minerals by *Acidiphilium* SJH. Geomicrobiol J 17: 193-206.

Brierley CL. 1978. Bacterial leaching. CRC Crit Rev Microbiol 6: 207-262.

Brierley CL. 1997. Mining Biotechnology: Research to commercial development and beyond. In: Rawlings DE, ed. Biomining: Theory, Microbes and Industrial Processes. Springer-Verlag, Berlin, 3-17.

Brierley CL, Brierley JA. 1973. A chemoautotrophic and thermophilic microorganism isolated from an acid hot spring. Can J Microbiol 19: 183-188.

Brierley CL, Brierley JA. 1986. Microbial mining using thermophilic microorganisms. In: Brock TD, ed. Thermophiles: General, Molecular, and Applied Microbiology. Wiley, New York, 279-305.

Brierley JA, Lockwood AS. 1977. The occurrence of thermophilic iron-oxidizing bacteria in a copper leaching system. FEMS Microbiol Lett 2: 163-165.

Brierley CL, Murr LE. 1973. Leaching: use of a thermophilic and chemoautotrophic microbe. Science 179: 488-490.

Brock TD. 1978. Thermophilic Microorganisms and Life at High Temperatures. Springer-Verlag, New York.

Brock TD, Gustafson J. 1976. Ferric iron reduction by sulfur- and iron-oxidizing bacteria. Appl Environ Microbiol 32: 567-571.

Brock TD, Brock KM, Belly RT, Weiss RL. 1972. *Sulfolobus*: a new genus of sulfur-oxidizing bacteria living at low pH and high temperature. Arch Mikrobiol 84: 54-68.

Bruneel O, Duran R, Koffi K, Casiot C, Fourcans A, Elbaz-Poulichet F, Personne JC. 2005. Microbial diversity in a pyrite-rich tailings impoundment (Carnoules, France). Geomicrobiol J 22: 249-257.

Bryan CG, Hallberg KB, Johnson DB. 2006. Mobilisation of metals in mineral tailings at the abandoned São Domingos copper mine (Portugal) by indigenous acidophilic bacteria. Hydrometallurgy 83: 184–194.

Bryant RD, McGroarty KM, Costerion JW, Laishley EJ. 1983. Isolation and characterization of a new acidophilic *Thiobacillus* species (*T. albertis*). Can J Microbiol 29: 1159-1170.

Burton NP, Norris PR. 2000. Microbiology of acidic, geothermal springs of Montserrat: environmental rDNA analysis. Extremophiles 4: 315-320.

Clark DA, Norris PR. 1996. *Acidimicrobium ferrooxidans* gen. nov., sp. nov.: mixed-culture ferrous iron oxidation with *Sulfobacillus* species. Microbiology 142: 785-790.

Colmer AR, Hinkle ME. 1947. The role of microorganisms in acid mine drainage. Science 106: 253-256.

Coram NJ, Rawlings DE. 2002. Molecular relationship between two groups of the genus *Leptospirillum* and the finding that *Leptospirillum ferriphilum* sp. nov. dominates South African commercial biooxidation tanks that operate at 40 °C. Appl Environ Microbiol 68: 838-845.

Coram-Uliana JM, van Hille RP, Kohr WJ, Harrison STL. 2006. Development of a method to assay the microbial population in heap bioleaching operations. Hydrometallurgy 83: 237–244.

Coupland K, Johnson DB. 2004. Geochemistry and microbiology of an impounded subterranean acidic water body at Mynydd Parys, Anglesey, Wales. Geobiology 2:77-86.

Curutchet G, Pogliani C, Donati E. 1995. Indirect bioleaching of covellite by *Thiobacillus thooxidans* with an oxidant agent. Biotechnol Lett 17: 1251-1256.

Darland G, Brock TD, Samsonoff W, Conti SF. 1970. A thermophilic acidophilic *Mycoplasma* isolated from a coal refuse pile. Science 170: 1416-1418.

Demergasso C, Echeverría A, Escudero L, Gallguillos P, Zepeda V, Castillo D. 2005a. Comparison of fluorescent in situ hybridization (FISH) and catalyzed reporter deposition (CARD-FISH) for visualization and enumeration of archaea and bacteria ratio in industrial heap bioleaching operations. In: Harrison STL, Rawlings DE, Petersen J, eds. Proceedings of the 16th International Biohydrometallurgy Symposium, September 25–29, Cape Town, South Africa. Produced by Compress www.compress.co.za, 843-851.

Demergasso CS, Pedro A, Galleguillos P, Lorena V, Escudero G, Víctor J, Zepeda A, Castillo D, Casamayor EO. 2005b. Molecular characterization of microbial populations in a low-grade copper ore bioleaching test heap. Hydrometallurgy 80: 241-253.

Diaby N, Dold B, Pfeifer HR, Holliger C, Johnson DB, Hallberg KB. 2006. Microbial communities in a porphyry copper tailings impoundment and their impact on the geochemical dynamics of the mine waste. Environ Microbiol, online.

Dopson M, Lindström EB. 2004. Analysis of community composition during moderately thermophilic bioleaching of pyrite, arsenical pyrite, and chalcopyrite. Microb Ecol 48: 19-28.

Dopson M, Baker-Austin C, Koppineedi PR, Bond PL. 2003. Growth in sulfidic mineral environments: metal resistance mechanisms in acidophilic micro-organisms. Microbiology 149:1959-1970.

Dopson M, Baker-Austin C, Hind A, Bowman JP, Bond PL. 2004. Characterization of *Ferroplasma* isolates and *Ferroplasma acidarmanus* sp nov., extreme acidophiles from acid mine drainage and industrial bioleaching environments. Appl Environ Microbiol 70: 2079-2088.

Drobner E, Huber H, Rachel R, Stetter KO. 1992. *Thiobacillus plumbophilus* spec. nov., a novel galena and hydrogen oxidizer. Arch Microbiol 157: 213-217.

Dufresne S, Bousquet J, Boissinot M, Guay R. 1996. *Sulfobacillus disulfidooxidans* sp. nov., a new acidophilic, disulfide-oxidizing, gram-positiv, spore-forming bacterium. Int J Syst Bacteriol 46: 1056-1064.

Dziurla MA, Achouak W, Lam BT, Heulin T, Berthelin J. 1998. Enzyme-linked immunofiltration assay to estimate attachment of thiobacilli to pyrite. Appl Environ Microbiol 64: 2937-2942.

Ebrahimi S, Morales FJF, Kleerebezem R, Heijnen JJ, van Loosdrecht MCM. 2005. High-rate acidophilic ferrous iron oxidation in a biofilm airlift reactor and the role of the carrier material. Biotechnol Bioeng 90: 462-472.

Edwards KJ, Gihiring TM, Banfield JF. 1999a. Seasonal variations in microbial populations and environmental conditions in an extreme acid mine drainage environment. Appl Environ Microbiol 65: 3627-3632.

Edwards KJ, Goebel BM, Rodgers TM, Schrenk MO, Gihiring TM, Cardona MM, Hu B, McGuire MM, Hamers RJ, Pace NR. 1999b. Geomicrobiology of pyrite (FeS_2) dissolution: case study at Iron Mountain, California. Geomicrobiol J 16: 155-179.

Edwards KJ, Bond PL, Gihring TM, Banfield JF. 2000a. An archaeal iron-oxidizing extreme acidophile important in acid mine drainage. Science 287: 1796-1799.

Edwards KJ, Bond PL, Druschel GK, McGuire MM, Hamers RJ, Banfield JF. 2000b. Geochemical and biological aspects of sulfide mineral dissolution: lessons from Iron Mountain, California. Chem Geol 169: 383-397.

Ehrlich HL. 2002. Geomicrobiology. Marcel Dekker Inc., New York.

Fowler TA, Crundwell FK. 1999. Leaching of zinc sulfide by *Thiobacillus ferrooxidans*: bacterial oxidation of the sulfur product layer increases the rate of zinc sulfide dissolution at high concentrations of ferrous ions. Appl Environ Microbiol 65: 5285-5292.

Fowler TA, Holmes PR, Crundwell FK. 1999. Mechanism of pyrite dissolution in the presence of *Thiobacillus ferrooxidans*. Appl Environ Microbiol 65: 2987-2993.

Fuchs T, Huber H, Teiner K Burggraf S, Stetter KO. 1995. *Metallosphaera prunae*, sp. nov., a novel metal-mobilizing, thermoacidophilic *Archaeum*, isolated from a uranium mine in Germany. Syst Appl Microbiol 18: 560-566.

Garcia O, Bigham JM, Tuovinen OH. 1995a. Oxidation of galena by *Thiobacillus ferrooxidans* and *Thiobacillus thiooxidans*. Can J Microbiol 41: 508-514.

Garcia O, Bigham JM, Tuovinen OH. 1995b. Sphalerite oxidation by *Thiobacillus ferrooxidans* and *Thiobacillus thiooxidans*. Can J Microbiol 41: 578-584.

Ghauri MA, Johnson DB. 1991. Physiological diversity amongst some moderately thermophilic iron-oxidizing bacteria. FEMS Microbiol Ecol 85: 327-334.

Goebel BM, Stackebrandt E. 1994. Cultural and phylogenetic analysis of mixed microbial populations found in natural and commercial bioleaching environments. Appl Environ Microbiol 60: 1614-1621.

Golovacheva RS, Karavaiko GI. 1978. A new genus of thermophilic spore-forming bacteria, *Sulfobacillus*. Microbiology 47: 658-665.

Golovacheva RS, Valieho-Roman KM, Troitskii AV. 1987a. *Sulfurococcus mirabilis* gen. nov., sp. nov., a new thermophilic archaebacterium with the ability to oxidize sulfur. Mikrobiologiya 56: 100-107.

Golovacheva RS, Zhukova IG, Nikultseva TP, Ostrovinskii DN. 1987b. Some properties of*Sulfurococcus mirabilis*, a new thermoacidophilic archaebacterium. Mikrobiologiya 56: 281-287.

Golovacheva RS, Golyshina OV, Karavaiko GI, Dorofeev AG, Pivovarova TA, Chernykh NA. 1992. A new ironoxidizing bacterium, *Leptospirillum thermoferrooxidans* sp. nov. Microbiology 61: 744-750.

Golyshina OV. Timmis KN. 2005. *Ferroplasma* and relatives, recently discovered cell wall-lacking archaea making a living in extremely acid, heavy metal-rich environments. Environ Microbiol 7: 1277-1288.

Golyshina OV, Pivovarova TA, Karavaiko GI, Kondrat'eva TF, Moore ERB, Abraham WR, Lunsdorf H, Timmis KN, Yakimov MM, Golyshin PN. 2000. *Ferroplasma acidiphilum* gen. nov., sp nov., an acidophilic, autotrophic, ferrous-iron-oxidizing, cell-wall-lacking, mesophilic member of the Ferroplasmaceae fam. nov., comprising a distinct lineage of the Archaea. Int J Syst Evol Microbiol 50: 997-1006.

González-Toril E, Llobet-Brossa E, Casamayor EO, Amann R, Amils R. 2003. Microbial ecology of an extreme acidic environment, the Tinto River. Appl Environ Microbiol 69: 4853-4865.

González-Toril E, Garcia-Moyano A, Amils R. 2005. Phylogeny of prokaryotic microorganisms from the Tinto River. In: Harrison STL, Rawlings DE, Petersen J, eds. Proceedings of the 16th International Biohydrometallurgy Symposium, September 25–29, Cape Town, South Africa. Produced by Compress www.compress.co.za, 737-749.

Grogan DW. 1989. Phenotypic characterization of the archaebacterial genus *Sulfolobus*: comparison of five wild-type strains. J Bacteriol 171: 6710-6719.

Grogan D. 1991. Selectable mutant phenotypes of the extremely thermophilic archaea-bacterium *Sulfolobus acidocaldarius*. J Bacteriol 173: 7725-7727.

Groudev SN, Groudeva VI. 1993. Microbial communities in four industrial copper dump leaching operations in Bulgaria. FEMS Microbiol Rev 11: 261-268.

Hallberg KB, Johnson DB. 2001. Biodiversity of acidophilic prokaryotes. Adv Appl Microbiol 49: 37-84.

Hallberg KB, Lindström EB. 1994. Characterization of *Thiobacillus caldus* sp. nov., a moderately thermophilic acidophile. Microbiol 140: 3451-3456.

Hallberg KB, Dopson M, Lindstrom EB. 1996. Reduced sulfur compound oxidation by *Thiobacillus caldus*. J Bacteriol 178: 6-11.

Hallberg KB, Thomson HEC, Boeselt I, Johnson DB. 2001. Aerobic and anaerobic sulfur metabolism by acidophilic bacteria. In: Ciminelli VST, Garcia Jr. O. Biohydrometallurgy: Fundamentals, Technology and Sustainable Development. Elsevier Science BV, Amsterdam, Part A, 423-431.

Hallmann R, Friedrich A, Koops H-P, Pommerening-Röser A, Rohde K, Zenneck C, Sand W. 1993. Physiological characteristics of *Thiobacillus ferrooxidans* and *Leptospirillum ferrooxidans* and physicochemical factors influence microbial metal leaching. Geomicrobiol J 10: 193-206.

Harms G, Layton AC, Dionisi HM, Gregory IR, Garrett VM, Hawkins SA, Robinson KG, Sayler GS. 2003. Real-time PCR quantification of nitrifying bacteria in a municipal wastewater treatment plant. Environ Sci Technol 37: 343-351.

Harrison Jr AP. 1981. *Acidiphilium cryptum* gen. nov., sp. nov., heterotrophic bacterium from acidic mineral environments. Int J Syst Bacteriol 31: 327-332.

Harrison Jr AP. 1982. Genomic and physiological diversity amongst strains of *Thiobacillus ferrooxidans*, and genomic comparison with *Thiobacillus thiooxidans*. Arch Micobiol 131: 68-76.

Harrison Jr AP. 1983. Genomic and physiological comparisons between heterotrophic Thiobacilli and *Acidiphilium cryptum: Thiobacillus versutus* sp. nov., and*Thiobacillus acidophilus* nom. rev. Int J Syst Bacteriol 33: 211-217.

Harrison Jr AP. 1984. The acidophilic Thiobacilli and other acidophilic bacteria that share their habitat. Ann Rev Microbiol 38: 265-292.

Hawkes RB, Franzmann PD, Plumb JJ. 2006a. Moderate thermophiles including "*Ferroplasma cupricumulans*" sp. nov. dominate an industrial-scale chalcocite heap bioleaching operation. Hydrometallurgy 83: 229–236.

Hawkes RB, Franzmann PD, O'hara G, Plumb JJ. 2006b. *Ferroplasma cupricumulans* sp. nov., a novel moderately thermophilic, acidophilic archaeon isolated from an industrial-scale chalcocite bioleach heap. Extremophiles, online.

Hazeu W, Batenburg-van der Vegte WH, Bos P, van der Pas RK, Kuenen JG. 1988. The production and utilization of intermediary elemental sulfur during the oxidation of reduced sulfur compounds by *Thiobacillus ferrooxidans*. Arch Microbiol 150: 574-579.

Heid CA, Stevens J, Livak KJ, Williams PM. 1996. Real time quantitative PCR. Genome Res 6: 986-994.

Hippe H. 2000. *Leptospirillum* gen. nov. (ex Markosyan 1972), nom. rev., including *Leptospirillum ferrooxidans* sp. nov. (ex Markosyan 1972), nom. rev. and *Leptospirillum thermoferrooxidans* sp. nov. (Golovacheva *et al.* 1992). Int J Syst Evol Microbiol 50: 501-503.

Hiraishi A, Nagashima KVP, Matsuura K, Shimada K, Takaichi S, Wakao N, Katayama Y. 1998. Phylogeny and photosynthetic features of *Thiobacillus acidophilus* and related acidophilic bacteria: its transfer to the genus *Acidiphilium* as *Acidiphilium acidophilum* comb. nov. Int J Syst Bacteriol 48: 1389-1398.

Huber H, Prangishvili D. 2004. Sulfolobales, Release 3.18. In: The prokaryotes, electronic version, 1994-2004. Springer-Verlag, New York.

Huber H, Stetter KO. 1989. *Thiobacillus prosperus* sp. nov., represents a new group of halotolerant metal-mobilizing bacteria isolated from a marine geothermal field. Arch Microbiol 151: 479-485.

Huber H, Stetter KO. 1990. *Thiobacillus cuprinus* sp. nov., a novel facultativ organotrophic metal-mobilizing bacterium. Appl Environ Microbiol 56: 315-322.

Huber H, Stetter KO. 1991. *Sulfolobus metallicus* sp. nov., a novel strictly chemolithotrophic thermophilic archaeal species of metal-mobilizers. Syst Appl Microbiol 14: 372-378.

Huber H, Stetter KO. 2001. Order III: Sulfolobales. In: Garrity G, ed. Bergey's Manual of Systematic Bacteriology, 2nd ed. Springer-Verlag, New York, Vol. 1, 198.

Huber G, Huber H, Stetter KO. 1986. Isolation and characterization of new metal-mobilizing bacteria. Biotech Bioeng Symp 16: 239-251.

Huber G, Spinnler C, Gambacorta A, Stetter KO. 1989. *Metallosphaera sedula* gen. and sp. nov. represents a new genus of aerobic, metal-mobilizing, thermoacidophilic archaebacteria. Syst Appl Microbiol 12: 38-47.

Jan R-J, Wu J, Chaw S-M, Tsai C-W, Tsen S-D. 1999. A novel species of thermoacidophilic archaeon, *Sulfolobus yangmingensis* sp. nov. Int J Syst Bacteriol 49: 1809-1816.

Jerez CA. 1997. Molecular methods for the identification and enumeration of bioleaching microorganisms. In: Rawlings DE, ed. Biomining: Theory, Microbes and Industrial Processes. Springer-Verlag, Berlin, 281-297.

Johnson DB. 1995. Selective solid media for isolating and enumerating acidophilic bacteria. J Microbiol Meth 23: 205-218.

Johnson DB. 1998. Biodiversity and ecology of acidophilic microorganisms. FEMS Microbiol Ecol 27: 307-317.

Johnson DB, Hallberg KB. 2003. The microbiology of acidic mine waters. Res Microbiol 154: 466-473.

Johnson DB, McGinness S. 1991. Ferric ion reduction by acidophilic heterotrophic bacteria. Appl Environ Microbiol 57: 207-211.

Johnson DB, Roberto FF. 1997. Heterotrophic acidophiles and their roles in bioleaching of sulfide minerals. In: Rawlings DE, ed. Biomining: Theory, Microbes and Industrial Processes. Springer-Verlag, Berlin, 259-279.

Johnson DB, Bacelar-Nicolau P, Okibe N, Yahya A, Hallberg K. 2001. Role of pure and mixed culures of Gram-positive eubacteria in mineral leaching. In: Ciminelli VST, Garcia Jr. O. Biohydrometallurgy: Fundamentals, Technology and Sustainable Development. Elsevier Science BV, Amsterdam, Part A, 461-470.

Johnson DB, Okibe N, Hallberg KB. 2005. Differentiation and identification of iron-oxidizing acidophilic bacteria using cultivation techniques and amplified ribosomal DNA restriction enzyme analysis. J Microbiol Meth 60: 299-313. Erratum: J Microbiol Meth 63: 216-217.

Johnson DB, Stallwood B, Kimura S, Hallberg KB. 2006. Characteristics of *Acidicaldus organovorus*, gen. nov., sp. nov.; a novel thermo-acidophilic heterotrophic proteobacterium. Arch Microbiol 185: 212-221.

Kamimura K, Higashino E, Kanao T, Sugio T. 2005. Effects of inhibitors and NaCl on the oxidation of reduced inorganic sulfur compounds by a marine acidophilic, sulfur-oxidizing bacterium, *Acidithiobacillus thiooxidans* strain SH. Extremophiles 9: 45-51.

Karavaiko GI, Golyshina OV, Troitskii AV, Valieho-Roman KM, Golovacheva RS, Pivovarova TA. 1994. *Sulfurococcus yellowstonii* sp. nov., a new species of iron- and sulphur-oxidizing thermoacidophilic archaebacteria. Microbiology 63: 379-387.

Karavaiko GI, Turova TP, Kondrat'eva TF, Lysenko AM, Kolganova TV, Ageeva SN, Muntyan LN, Pivovarova TA. 2003. Phylogenetic heterogeneity of the species *Acidithiobacillus ferrooxidans*. Int J Syst Evol Microbiol 53: 113-119.

Karavaiko GI, Bogdanova TI, Tourova TP, Kondrat'eva TF, Tsaplina IA, Egorova MA, Krasil'nikova EN, Zakharchuk LM. 2005. Reclassification of '*Sulfobacillus thermosulfidooxidans* subsp *thermotolerans*' strain K1 as *Alicyclobacillus tolerans* sp nov and *Sulfobacillus disulfidooxidans* Dufresne *et al* 1996 as *Alicyclobacillus disulfidooxidans* comb. nov., and emended description of the genus *Alicyclobacillus*. Int J Syst Evol Microbiol 55: 941-947.

Kargi F, Robinson JM. 1985. Biological removal of pyritic sulfur from coal by the thermophilic organism *Sulfolobus acidocaldarius*. Biotechnol Bioeng 27: 41-49.

Karsten C, Harneit K, Sand W, Stackebrandt, E, Schumann P. 2005. Characterization of novel iron-oxidizing bacteria. In: Harrison STL, Rawlings DE, Petersen J, eds. Proceedings of the 16th International Biohydrometallurgy Symposium, September 25–29, Cape Town, South Africa. Produced by Compress www.compress.co.za, 729-735.

Kelly DP, Wood AP. 2000. Reclassification of some species of *Thiobacillus* to the newly designated genera *Acidithiobacillus* gen. nov., *Halothiobacillus* gen. nov. and *Thermithiobacillus* gen. nov. Int J Syst Evol Microbiol 50: 511-516.

Kepner Jr RL, Pratt JR. 1994. Use of fluorochromes for direct enumeration of total bacteria in environmental samples: past and present. Microbiol Rev 58: 603-615.

Kimura S, Hallberg KB, Johnson DB. 2005. Biodiversity of microbial populations in macroscopic "acid streamer" growths at an abandoned pyrite mine, elucidated using a combined cultivation-based and cultivation-independent approach. In: Harrison STL, Rawlings DE, Petersen J, eds. Proceedings of the 16th International Biohydrometallurgy Symposium, September 25–29, Cape Town, South Africa. Produced by Compress www.compress. co.za, 687-696.

Kinnunen PHM, Puhakka JA. 2004. High-rate ferric sulfate generation by a *Leptospirillum ferriphilum*-dominated biofilm and the role of jarosite in biomass retainment in a fluidized-bed reactor. Biotechnol Bioeng 85: 697-705.

Kinnunen PHM, Puhakka JA. 2005. High-rate iron oxidation at below pH 1 and at elevated iron and copper concentrations by a *Leptospirillum ferriphilum* dominated biofilm. Process Biochem 40: 3536-3541.

Kishimoto N, Kosako Y, Tano T. 1993. *Acidiphilium aminolytica* sp. nov.: an acidophilic chemoorganotrophic bacterium isolated from acidic mineral environment. Curr Microbiol 27: 131-136.

Kishimoto N, Kosako Y, Wakao N, Tano T, Hiraishi A. 1995. Transfer of *Acidiphilium facilis* and *Acidiphilium aminolytica* to the genus *Acidocella* gen. nov., and emendation of the genus *Acidiphilium*. Syst Appl Microbiol 18: 85-91.

Kock D, Schippers A. 2006. Geomicrobiological investigation of two different mine waste tailings generating acid mine drainage. Hydrometallurgy 83: 167–175.

Kolb S, Knief C, Stubner S, Conrad R. 2003. Quantitative detection of methanotrophs in soil by novel *pmoA*-targeted real-time PCR assays. Appl Environ Microbiol 69: 2423-2429.

Konishi Y, Yoshida S, Asai S. 1995. Bioleaching of pyrite by acidophilic thermophile *Acidianus brierleyi*. Biotechnol Bioeng 48: 592-600.

Kovalenko EV, Malakhova PT. 1983. The spore-forming iron-oxidizing bacterium *Sulfobacillus thermosulfidooxidans*. Mikrobiologiya 52: 19-25 (in Russian).

Kurosawa N, Itoh YH, Itoh T. 2003. Reclassification of *Sulfolobus hakonensis* Takayanagi *et al.* 1996 as *Metallosphaera hakonensis* comb. nov. based on phylogenetic evidence and DNA G+C content. Int J Syst Evol Microbiol 53: 1607-1608.

Küsel K, Dorsch T, Acker G, Stackebrandt E. 1999. Microbial reduction of Fe(III) in acidic sediments: isolation of *Acidiphilium cryptum* JF-5 capable of coupling the reduction of Fe(III) to the oxidation of glucose. Appl Environ Microbiol 65: 3633-3640.

Lane DJ, Harrison Jr AP, Stahl D, Pace B, Giovannoni SJ, Olsen GJ, Pace NR. 1992. Evolutionary relationship among sulfur- and iron-oxidizing eubacteria. J Bacteriol 174: 269-278.

Liu C-Q, Plumb J, Hendry P. 2006. Rapid specific detection and quantification of Bacteria and Archaea involved in mineral sulfide bioleaching using real-time PCR. Biotechnol Bioeng 94: 330-336.

Lizama HM, Suzuki I. 1991. Interaction of chalcopyrite and sphalerite with pyrite during leaching by *Thiobacillus ferrooxidans* and *Thiobacillus thiooxidans*. Can J Microbiol 37: 304-311.

Lockey C, Otto E, Long Z. 1998. Real-time fluorescence detection of a single DNA molecule. BioTechniques 24: 744-746.

Ludwig W, Schleifer K-H. 2000. How quantitative is quantitative PCR with respect to cell counts? Syst Appl Microbiol 23: 556-562.

Mackintosh ME. 1978. Nitrogen fixation by*Thiobacillus ferrooxidans*. J Gen Microbiol 105: 215-218.

Mahmoud KK, Leduc LG, Ferroni GD. 2005. Detection of *Acidithiobacillus ferrooxidans* in acid mine drainage environments using fluorescent in situ hybridization (FISH). J Microbiol Meth 61: 33-45.

Markosyan GE. 1972. A new iron-oxidizing bacterium *Leptospirillum ferrooxidans* nov. gen. nov. sp. Biol J Armenia 25: 26-29 (in Russian).

Marsh RM, Norris PR, LeRoux NW. 1983. Growth and mineral oxidation studies with *Sulfolobus*. In: Rossi G, Torma AE, eds. Recent Progress in Biohydrometallurgy. Associazione Mineraria Sarda. Iglesias, Italy, 71-81.

McDonald IR, Kelly DP, Murrell JC, Wood AP. 1997. Taxonomic relationship of *Thiobacillus halophilus*, *T. aquaesulis*, and other species of *Thiobacillus*, as determined using 16S rDNA sequencing. Arch Microbiol 166: 394-398.

Melamud VS, Pivovarova TA, Tourova TP, Kolganova TV, Osipov GA, Lysenko AM, Kondrat'eva TF, Karavaiko GI. 2003. *Sulfobacillus sibiricus* sp nov., a new moderately thermophilic bacterium. Microbiology 72: 605-612.

Moreira D, Amils R. 1997. Phylogeny of *Thiobacillus cuprinus* and other mixotrophic thiobacilli proposal for *Thiomonas* gen. nov. Int J Syst Bacteriol 47: 522-528.

Morita RY, ed. 1997. Bacteria in Oligotrophic Environments. Chapman & Hall, New York.

Nadkarni M, Martin FE, Jacques NA, Hunter N. 2002. Determination of bacterial load by real-time PCR using a broad range (universal) probe and primer set. Microbiol 148: 257-266.

Norris PR. 1983. Iron and mineral oxidation studies with leptospirillum-like bacteria. In: Rossi G, Torma AE, eds. Recent Progress in Biohydrometallurgy. Associazione Mineraria Sarda, Iglesias, Italy, 83-96.

Norris PR, Parrott L. 1986. High temperature, mineral concentrate dissolution with *Sulfolobus*. In: Lawrence RW, Branion RMR, Ebner HG, eds. Fundamental and Applied Biohydrometallurgy. Elsevier, Amsterdam, 355-365.

Norris PR, Barr DW, Hinson D. 1988. Iron and mineral oxidation by acidophilic bacteria: affinities for iron and attachment to pyrite. In: Norris PR, Kelly DP, eds. Biohydrometallurgy, Proceedings of the International Symposium, Warwick, Antony Rowe Ltd. Chippenham, Wiltshire, UK, 43-59.

Norris PR, Murrell JC, Hinson D. 1995. The potential for diazotrophy in iron- and sulfur oxidizing acidophilic bacteria. Arch Microbiol 164: 294-300.

Norris PR, Clark DA, Owen JP, Waterhouse S. 1996. Characteristics of *Sulfobacillus acidophilus* sp. nov. and other moderately thermophilic mineral-sulphide-oxidizing bacteria. Microbiol 142: 775-783.

Norris PR, Burton NP, Foulis NAM. 2000. Acidophiles in bioreactor mineral processing. Extremophiles 4: 71-76.

Novo MTM, De Souza AP, Garcia O, Ottoboni LMM. 1996. RAPD genomic fingerprinting differentiates *Thiobacillus ferrooxidans* strains. Syst Appl Microbiol 19: 91-95.

Okibe N, Johnson DB. 2004. Biooxidation of pyrite by defined mixed cultures of moderately thermophilic acidophiles in pH-controlled bioreactors: significance of microbial interactions. Biotechnol Bioeng 87: 574-583.

Okibe N, Gericke M, Hallberg KB, Johnson DB. 2003. Enumeration and characterization of acidophilic microorganisms isolated from a pilot plant stirred-tank bioleaching operation Appl Environ Microbiol 69: 1936-1943.

Olson GJ, Brierley JA, Brierley CL. 2003. Bioleaching review part B: Progress in bioleaching: applications of microbial processes by the minerals industries. Appl Microbiol Biotechnol 63: 249-257.

Parro V, Moreno-Paz M. 2003. Gene function analysis in environmental isolates: the nif regulon of the strict iron oxidizing bacterium *Leptospirillum ferrooxidans*. PNAS 100: 7883-7888.

Pernthaler A, Pernthaler J, Amann R. 2002. Fluorescence in situ hybridization and catalyzed reporter deposition for the identification of marine bacteria. Appl Environ Microbiol 68: 3094-3101.

Pistorio M, Curutchet G, Donati R, Tedesco P. 1994. Direct zinc sulfide bioleaching by *Thiobacillus ferrooxidans* and *Thiobacillus thiooxidans*. Biotechnol Lett 16: 419-424.

Pizarro J, Jedlicki E, Orellana O, Romero J, Espejo RT. 1996. Bacterial populations in samples of bioleached copper ore as revealed by analysis of DNA obtained before and after cultivation. Appl Environ Microbiol 62: 1323-1328.

Pogliani C, Donati E. 2000. Immobilisation of *Thiobacillus ferrooxidans*: importance of jarosite precipitation. Process Biochem 35: 997-1004.

Pronk JT, Johnson DB. 1993. Oxidation and reduction of iron by acidophilic bacteria. Geomicrobiol J 10: 153-171.

Quatrini R, Appia-Ayme C, Denis Y, Ratouchniak J, Veloso F, Valdes J, Lefimil C, Silver S, Roberto F, Orellana O, Denizot F, Jedlicki E, Holmes D, Bonnefoy V. 2006. Insights into the iron and sulfur energetic metabolism of *Acidithiobacillus ferrooxidans* by microarray transcriptome profiling. Hydrometallurgy 83: 263–272.

Ramírez P, Guiliani N, Valenzuela L, Beard S, Jerez CA. 2004. Differential protein expression during growth of *Acidithiobacillus ferrooxidans* on ferrous iron, sulfur compounds, or metal sulfides. Appl Environ Microbiol 70: 4491-4498.

Rawlings DE. 1995. Restriction enzyme analysis of 16S rRNA genes for the rapid identification of *Thiobacillus ferrooxidans, Thiobacillus thiooxidans* and *Leptospirillum ferrooxidans* strains in leaching environments. In: Jerez JCA, Vargas T, Toledo H, Wiertz JV, eds. Biohydrometallurgical processing, Proceedings of the International Biohydrometallurgy Symposium IBS-95, Vina del Mar, Chile. University of Chile, Santiago de Chile, Vol. 2, 9-17.

Rawlings DE. 1997. Mesophilic, autotrophic bioleaching bacteria: description, physiology and role. In: Rawlings DE, ed. Biomining: Theory, Microbes and Industrial Processes. Springer-Verlag, Berlin, 229-245.

Rawlings DE. 2002. Heavy metal mining using microbes. Ann Rev Microbiol 56: 65-91.

Rawlings DE. 2005. Characteristics and adaptability of iron- and sulfur-oxidizing microorganisms used for the recovery of metals from minerals and their concentrates. Microbial Cell Factories (via www.uni-potsdam.de - PDF - 402 KB).

Rawlings DE, Coram NJ, Gardner MN, Deane SM. 1999. In: Amils R, Ballester A, eds. Biohydrometallurgy and the environment toward the mining of the 21st century, Proceedings of the International Biohydrometallurgy Symposium IBS-99, El Escorial, Spain. Elsevier, Amsterdam, Part A, 777-786.

Rohwerder T, Gehrke T, Kinzler K, Sand W. 2003. Bioleaching review part A: Progress in bioleaching: fundamentals and mechanisms of bacterial metal sulfide oxidation. Appl Microbiol Biotechnol 63: 239-248.

Rossi G. 1990. Biohydrometallurgy. McGraw-Hill, Hamburg.

Rzhepishevska OI, Lindstrom EB, Tuovinen OH, Dopson M. 2005. Bioleaching of sulfidic tailing samples with a novel, vacuum-positive pressure driven bioreactor. Biotechnol Bioeng 92: 559-567.

Sakaguchi H, Torma AE, Silver M. 1976. Microbiological oxidation of synthetic chalcocite and covellite by *Thiobacillus ferrooxidans*. Appl Environ Microbiol 31: 7-10.

Sand W, Rohde K, Sobotke B, Zenneck C. 1992. Evaluation of *Leptospirillum ferrooxidans* for leaching. Appl Environ Microb 58: 85-92.

Sand W, Gehrke T, Jozsa P-G, Schippers A. 2001.(Bio)chemistry of bacterial leaching - direct vs. indirect bioleaching. Hydrometallurgy 59: 159-175.

Schippers A, Sand W. 1999. Bacterial leaching of metal sulfides proceeds by two indirect mechanisms via thiosulfate or via polysulfides and sulfur. Appl Environ Microbiol 65: 319-321.

Schippers A, Hallmann R, Wentzien S, Sand W. 1995. Microbial diversity in uranium mine waste heaps. Appl Environ Microbiol 61: 2930-2935.

Schippers A, Jozsa P-G, Sand W. 1996. Sulfur chemistry in bacterial leaching of pyrite. Appl Environ Microbiol 62: 3424-3431.

Schippers A. 2004. Biogeochmistry of metal sulfide oxidation in mining environments, sediments and soils. In: Amend JP, Edwards KJ, Lyons TW, eds. Sulfur biogeochemistry - Past and present. Special Paper 379. Geological Society of America, Boulder, Colorado, 49-62.

Schippers A, Bosecker K. 2005. Bioleaching: Analysis of microbial communities dissolving metal sulfides. In: Barredo J-L, ed. Methods in Biotechnology, Vol. 18: Microbial Processes and Products. Humana Press Inc., Totowa, New York, 405-412.

Schippers A, Neretin LN, Kallmeyer J, Ferdelman TG, Cragg BA, Parkes RJ, Jørgensen BB. 2005. Prokaryotic cells of the deep sub-seafloor biosphere identified as living bacteria. Nature 433: 861-864.

Schleper C, Puehler G, Holz I, Gambacorta A, Janekovic D, Santarius U, Klenk H-P, Zillig W. 1995. *Picrophilus* gen. nov., fam. nov.: a novel aerobic, heterotrophic, thermo-acidophilic genus and family comprising archaea capable of growth around pH 0. Int J Syst Bacteriol 177: 7050-7059.

Schleper C, Pühler G, Klenk H-P, Zillig W. 1996. *Picrophilus oshimae* and *Picrophilus torridus* fam. nov., gen. nov., sp. nov., two species of hyperacidophilic, thermophilic, heterotrophic, aerobic archaea. Int J Syst Bacteriol 46: 814-816.

Schrenk MO, Edwards KJ, Goodman RM, Hamers RJ, Banfield JF. 1998. Distribution of *Thiobacillus ferrooxidans* and *Leptospirillum ferrooxidans* for generation of acid mine drainage. Science 279: 1519-1522.

Segerer AH, Stetter KO. 1992. The order Sulfolobales. In: Balows A, Trüper HG, Dworkin M, Harder W, Schleifer K-H, eds. The Prokaryotes, 2nd ed., Springer-Verlag, New York, 684-701.

Segerer A, Neuner A, Kristjansson JK, Stetter KO. 1986. *Acidianus infernus* gen. nov., sp. nov., and *Acidianus* brierleyi comb. nov.: facultatively aerobic, extremely acidophilic thermophilic sulfur-metabolizing archaebacteria. Int J Syst Bacteriol 36: 559-564.

Segerer A, Langworthy TA, Stetter KO. 1988. *Thermoplasma acidophilum* and*Thermoplasma volcanium* sp. nov. from solfatara fields. Syst Appl Microbiol 10: 161-171.

Sekar R, Pernthaler A, Pernthaler J, Warnecke F, Posch T, Amann R. 2003. An improved protocol for quantification of freshwater actinobacteria by fluorescence in situ hybridization. Appl Environ Microbiol 69: 2928-2935.

Selenska-Pobell S, Otto A, Kutschke S. 1998. Identification and discrimination of thiobacilli using ARDREA, RAPD and rep-APD. J Appl Microbiol 84:1085-1091.

Selenska-Pobell S, Kampf G, Flemming K, Radeva G, Satchanska G. 2001. Bacterial diversity in soil samples from two uranium waste piles as determined by rep-APD, RISA and 16S rDNA retrieval. Ant Leeuwenh 79: 149-161.

Sharkey FH, Banat IM, Marchant R. 2004. Detection and quantification of gene expression in environmental bacteriology. Appl Environ Microbiol 70: 3795-3806.

Silver S, Torma AE. 1974. Oxidation of metal sulfides by *Thiobacillus ferrooxidans* grown on different substrates. Can J. Microbiol 20: 141-147.

Smith CJ, Nedwell DB, Dong LF, Osborn AM. 2006. Evaluation of quantitative polymerase chain reaction-based approaches for determining gene copy and gene transcript numbers in environmental samples. Environ Microbiol 8: 804-815.

Stöhr R, Waberski A, Volker H, Tindall BJ, Thomm M. 2001. *Hydrogenothermus marinus* gen. nov., sp. nov., a novel thermophilic hydrogen-oxidizing bacterium, recognition of *Calderobacterium hydrogenophilum* as a member of the genus *Hydrogenobacter* and proposal of the reclassification of *Hydrogenobacter acidophilus* as *Hydrogenobaculum acidophilum* gen. nov., comb. nov., in the phylum '*Hydrogenobacter/Aquifex*'. Int J Syst Evol Microbiol 51: 1853-1862.

Stults JR, Snoeyenbos-West O, Methe B, Lovley DR, Chandler DP. 2001. Application of the 5'fluorogenic exonulease assay (TaqMan) for quantitative ribosomal DNA and rRNA analysis in sediments. Appl Environ Microbiol 67: 2781-2789.

Suzuki MT, Taylor LT, DeLong EF. 2000. Quantitative analysis of small-subunit rRNA genes in mixed microbial populations via 5'-nuclease assays. Appl Environ Microbiol 66: 4605-4614.

Takai K, Horikoshi K. 2000. Rapid detection and quantification of members of the archaeal community by quantitative PCR using fluorogenic probes. Appl Environ Microbiol 66: 5066-5072.

Takayanagi S, Kawasaki H, Sugimori K, Yamada T, Sugai A, Ito T, Yamasato K, Shioda M. 1996. *Sulfolobus hakonensis* sp. nov., a novel species of acidothermophilic archaeon. Int J Syst Bacteriol 46: 377–382.

Teira E, Reinthaler T, Pernthaler A, Pernthaler J, Herndl GJ. 2004. Combining catalyzed reporter deposition-fluorescence in situ hybridization and microautoradiography to detect substrate utilization by Bacteria and Archaea in the deep ocean. Appl Environ Microbiol 70: 4411-4414.

Temple KL, Colmer AR. 1951. The autotrophic oxidation of iron by a new bacterium, *Thiobacillus ferrooxidans*. J Bacteriol 62: 605-611.

Tobita M, Yokozeki M, Nishikawa N, Kawakami Y. 1994. Pyrite oxidation by *Sulfolobus acidocaldarius*. Biosci Biotech Biochem 58: 771-772.

Torma AE. 1971. Microbial oxidation of synthetic cobalt, nickel and zinc sulfides by *Thiobacillus ferrooxidans*. Rev Can Biol 30: 209-216.

Torma AE. 1978. Oxidation of gallium sulfides by *Thiobacillus ferrooxidans*. Can J Microbiol 24: 888-891.

Torma AE, Gabra GG. 1977. Oxidation of stibnite by *Thiobacillus ferrooxidans*. Ant Leeuwenh 43: 1-6.

Torma AE, Sakaguchi H. 1978. Relation between the solubility product and the rate of metal sulfide oxidation by *Thiobacillus ferrooxidans*. J Ferment Technol 56: 173-178.

Tributsch H, Bennett JC. 1981. Semiconductor-electrochemical aspects of bacterial leaching. Part 2. Survey of rate-controlling sulphide properties. J Chem Tech Biotechnol 31: 627-635.

Tuovinen OH, Bhatti TM, Bigham JM, Hallberg KB, Garcia Jr O, Lindström EB. 1994. Oxidative dissolution of arsenopyrite by mesophilic and moderately acidophilic thermophiles. Appl Environ Microbiol 60: 3268-3274.

Tyson GW, Chapman J, Hugenholtz P, Allen EE, Ram RJ, Richardson PM, Solovyev VV, Rubin EM, Rokhsar DS, Banfield JF. 2004. Community structure and metabolism through reconstruction of microbial genomes from the environment. Nature 428: 37-43.

Tyson GW, Lo I, Baker BJ, Allen EE, Hugenholtz P, Banfield JF. 2005. Genome-directed isolation of the key nitrogen fixer *Leptospirillum ferrodiazotrophum* sp nov from an acidophilic microbial community. Appl Environ Microbiol 71: 6319-6324.

Urakami T, Tamaoka J, Suzuki KI, Komagata K. 1989. *Acidomonas* gen. nov., incorporating *Acetobacter methanolicus* as *Acidomonas methanolica* comb. nov. Int J Syst Bacteriol 39: 50-55.

Vásquez M, Espejo RT. 1997. Chemolithotrophic bacteria in copper ores leached at high sulfuric acid concentrations. Appl Environ Microbiol 63: 332-334.

Vitaya VB, Koizumi J-I, Toda K. 1994. A kinetic assessment of substantial oxidation by *Sulfolobus acidocaldarius* in pyrite oxidation. J Ferment Bioeng 77: 528-534.

Waksman SA, Joffe JS. 1922. Microorganisms concerned in the oxidation of sulfur in the soil. II. *Thiobacillus thiooxidans*, a new sulfur-oxidizing organism isolated from the soil. J Bacteriol 7: 239-256.

Waltenbury DR, Leduc LG, Ferroni GD. 2005. The use of RAPD genomic fingerprinting to study relatedness in strains of *Acidithiobacillus ferrooxidans*. J Microbiol Meth 62: 103-112.

Wawrik B, Paul JH, Tabita FR. 2002. Real-time PCR quantification of *rbcL* (ribulose-1,5-bisphosphate carboxylase/oxygenase) mRNA in diatoms and pelagophytes. Appl Environ Microbiol 68: 3771-3779.

Wenderoth DF, Abraham W-R. 2005. Microbial indicator groups in acidic mining lakes. Environ Microbiol 7: 133-139.

Wilson M, Bakermans C, Madsen EL. 1999. In situ, real-time catabolic gene expression: extraction and characterization of naphtalene dioxygenase mRNA transcripts from groundwater. Appl Environ Microbiol 65: 80-87.

Wittwer CT, Herrmann MG, Moss AA, Rasmussen RP. 1997. Continuous fluorescence monitoring of rapid cycle DNA amplification. BioTechniques 22: 130-138.

Wulf-Durand P de, Bryant LJ, Sly LI. 1997. PCR-mediated detection of acidophilic, bioleaching-associated bacteria. Appl Environ Microbiol 63: 2944-2948.

Yahya A, Roberto FF, Johnson DB. 1999. Novel mineral-oxidizing bacteria from Montserrat (W.I.): physiological and phylogenetic characteristics. In: Amils R, Ballester A, eds. Biohydrometallurgy and the environment toward the mining of the 21st century, Proceedings of the International Biohydrometallurgy Symposium IBS-99, El Escorial, Spain. Elsevier, Amsterdam, Part A, 729-739.

Yamashita S, Uchimura T, Komagata K. 2004. Emendation of the genus Acidomonas Urakami, Tamaoka, Suzuki and Komagata 1989. Int J Syst Evol Microbiol 54: 865-870.

Yarzábal A, Duquesne K, Bonnefoy V. 2003. Rusticyanin gene expression of *Acidithiobacillus ferrooxidans* ATCC 33020 in sulfur- and in ferrous iron media. Hydrometallurgy 71: 107-114.

Yarzábal A, Appia-Ayme C, Ratouchniak J, Bonnefoy V. 2004. Regulation of the expression of the *Acidithiobacillus ferrooxidans rus* operon encoding two cytochromes c, a cytochrome oxidase and rusticyanin. Microbiology 150: 2113-2123.

Zillig W, Stetter KO, Wunderl S, Schulz W, Priess H, Scholz J. 1980. The Sulfolobus-"Caldariella" group: taxonomy on the basis of the structure of DNA-dependent RNA polymerases. Arch Microbiol 125: 259-269.

CHAPTER 2

MECHANISMS AND BIOCHEMICAL FUNDAMENTALS OF BACTERIAL METAL SULFIDE OXIDATION

THORE ROHWERDER AND WOLFGANG SAND

University of Duisburg-Essen, Biofilm Centre, Aquatic Biotechnology, Geibelstr. 41, D-47057 Duisburg, Germany. Tel/Fax 0049 203 379 4475/1941. E-mail: wolfgang.sand@uni-due.de

1. INTRODUCTION

The interest in the fundamentals of metal sulfide bioleaching has significantly increased in the course of the last decade. One driving force is the metal winning industry, as the recovery of heavy metals from sulfidic ores employing microorganisms is now an established biotechnology (Bosecker, 1997; Rawlings, 1997, 2002; Rohwerder *et al.*, 2002; Olson *et al.*, 2003; Rawlings *et al.*, 2003). On the other hand, the problem of acid rock/mine drainage (ARD/AMD) increasingly attracts attention, because environmental awareness and legislation force mining companies and state agencies into alliances for fighting effectively these problems or at least reducing their impact on the environment. Consequently, the underlying reaction mechanisms have been the subject of intensive research. In combination with the emerging possibilities resulting from exciting developments in the fields of chemical and surface analysis as well as biochemistry and molecular biology, our understanding of the responsible bioprocesses have been dramatically improved.

The first and decisive step for the renewed interest was the clarification of the sulfur chemistry behind metal sulfide oxidation (Schippers *et al.*, 1996; Schippers & Sand, 1999; Schippers, 2004), which brought a decades-long, often contradictory discussion to an end and, as a consequence, changed the scientific approach for studying these processes. It has been shown that by only two different pathways all relevant redox reactions and sulfur compound speciation may be explained. Most important was that the original concept of a "direct leaching mechanism" i.e. the direct enzymatic oxidation of the sulfur moiety of heavy metal sulfides (reviewed by Ehrlich, 2002 and Sand *et al.*, 1995), lost any justification. The "indirect mechanism", i.e. the non-enzymatic metal sulfide oxidation by iron(III) ions combined with the enzymatic (re)oxidation of the resulting iron(II) ions, is to be considered as the only relevant mechanism. It is now divided into

E.R. Donati and W. Sand (eds.), Microbial Processing of Metal Sulfides, 35–58.

two sub-mechanisms, the "contact" and the "non-contact" one (Sand *et al.*, 2001; Tributsch, 2001; Rawlings, 2002; Rohwerder *et al.*, 2003). In the non-contact mechanism, planktonic cells oxidize the iron(II) ions in the bulk solution and regenerate in this way the oxidizing agent which afterwards chemically reacts with metal sulfides (corresponding to the previously designated indirect mechanism; Sand *et al.*, 1995). The contact mechanism, on the other hand, takes into account that the majority of cells attaches to the surface of sulfide minerals. This means that the process of dissolution of metal sulfides, basically an electrochemical reaction between iron(III) ions and metal sulfide, takes place at the interface between bacterial cell and mineral surface. The space between the cell wall and the surface can be considered from our current point of view as a biogenic reaction space, which is filled with bacterial extracellular polymeric substances (EPS). Consequently, the composition and properties of the EPS as well as their biosynthesis are subjects of current research (Barreto *et al.*, 2005a, 2005b; Sand & Gehrke, 2006), and only preliminary results can be presented here. In contrast, the biochemical understanding of iron(II) ion and sulfur compound oxidation, which is important in both sub-mechanisms, has significantly developed. At least for the mesophilic *Acidithiobacillus ferrooxidans*, the redox chain from iron(II) ions to the final electron acceptor oxygen has been resolved (Lemesle-Meunier *et al.*, 2001). In addition, most of the enzymes involved in sulfur compound oxidation have been identified and models for the oxidation of hydrogen sulfide, elemental sulfur as well as other relevant sulfur compounds will be proposed here.

This chapter is based on our previous reviews (Sand *et al.*, 1995; Rohwerder *et al.*, 2003; Sand & Gehrke, 2006) and will summarize the current understanding of bioleaching fundamentals. It contains besides the well-known hypotheses new ideas on sulfur chemistry and attack mechanisms of bacterial metal sulfide oxidation, EPS properties and biosynthesis, and biochemistry of iron(II) and sulfur compound oxidation. Finally, conclusions will be drawn about the importance of these ideas for metal winning biotechnology as well as mitigation strategies against ARD/AMD processes.

2. THE TWO PATHWAYS

Due to the recent work of Schippers and colleagues we now know that only two different reaction mechanisms, the thiosulfate and the polysulfide pathway, control the dissolution of metal sulfides. Generally, leaching is achieved by a combination of proton attack and oxidation processes. Which one of the two pathways takes place is mineral specific (Schippers & Sand, 1999; Sand *et al.*, 2001; Schippers, 2004). However, the crystal structure (e.g. monosulfide or disulfide structure) as the most obvious criterion does not determine the pathway of dissolution. In fact, the acid solubility of metal sulfides is the relevant criterion. The latter property is determined by the electronic configuration (Tributsch & Bennett, 1981; Schippers & Sand, 1999; Edelbro *et al.*, 2003,). Metal sulfides with valence bands which are derived only from orbitals of the metal atoms cannot be attacked by protons (acid-insoluble metal sulfides). In contrast, metal sulfides with valence bands derived from both the metal and the sulfide atom orbitals are more or less soluble in acids (acid-soluble metal sulfides).

2.1. Thiosulfate Pathway for Acid-Insoluble Metal Sulfides

Representatives of the group of acid-insoluble metal sulfides such as pyrite, molybdenite, and tungstenite (FeS_2, MoS_2, and WS_2, respectively) need in any case an oxidation process to become dissolved. The chemical bonds between sulfur atom and metal atom do not break until a total of 6 successive one-electron oxidation steps have been conducted and thiosulfate is liberated (Luther, 1987; Moses *et al.*, 1987; Schippers *et al.*, 1996) (Figures 1A and 4B). Under bioleaching conditions, iron(III) ions are the only relevant oxidant for this reaction and, thus, are reduced to iron(II) ions by receiving the valence band electrons primarily from the metal atoms of the metal sulfide. According to its first free sulfur compound, this mechanism is called "thiosulfate pathway" (Schippers *et al.*, 1996). Thiosulfate is mainly oxidized via tetrathionate, the highly reactive disulfane monosulfonic acid, and other polythionates finally to sulfate (Figure 1A), but also significant amounts of elemental sulfur (10 to 20 %) may be produced in the absence of bacterial sulfur compound oxidation (purely chemical oxidation by iron(III) ions or by *Leptospirillum ferrooxidans*; Rohwerder *et al.*, 1998; Schippers *et al.*, 1999). Due to the requirement that the electron extraction is effected by iron(III) ions and the fact that around pH 2 iron(II) ions are not abiotically oxidized at significant rates (Singer & Stumm, 1970), only iron(II)-oxidizing bacteria such as *A. ferrooxidans* and *L. ferrooxidans* are able to leach acid-insoluble metal sulfides under acidic conditions. These bacteria are able to regenerate the iron(III) ions consumed in the initial oxidation processes. Consequently, leaching kinetics of metal sulfides dissolved via the thiosulfate pathway are determined mainly by bacterial iron(II) ion oxidation activity (Hansford, 1997). Under acidic conditions in the absence of bacteria and without detectable amounts of iron(III) ions, pyrite leaching rates are dramatically reduced (Rohwerder *et al.*, 1998; Schippers *et al.*, 1999). But also under these conditions the thiosulfate mechanism seems to be effective, although the further reactions of thiosulfate may deviate from the above mentioned ones (Descostes *et al.*, 2004).

2.2. Polysulfide Pathway for Acid-Soluble Metal Sulfides

Metal sulfides such as sphalerite, galena, arsenopyrite, chalcopyrite, and hauerite (ZnS, PbS, FeAsS, $CuFeS_2$, and MnS_2, respectively) are dissolved by a combined action of electron extraction by iron(III) ions and proton attack, i.e. the extraction of valence band electrons from the bonding metal atom-sulfur atom orbitals. Consequently, in this group of metal sulfides the chemical bonds between metal and sulfur moiety can be broken by iron(III) ion and, in addition, by proton attack. Therefore,

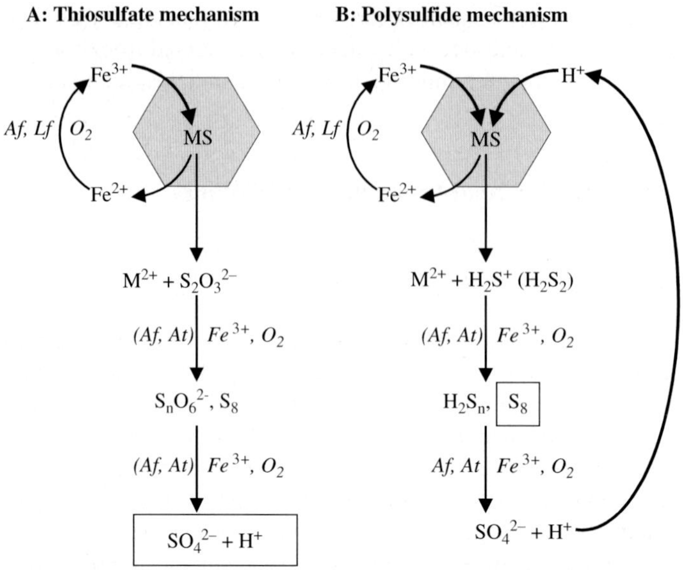

Figure 1. Scheme of the two metal sulfide (bio)leaching pathways, the thiosulfate (**A**) and polysulfide (**B**) one (Schippers *et al.*, 1996; Schippers & Sand, 1999). Iron(III) ions extract electrons from the metal sulfide (MS) and are thereby reduced to the iron(II) form. As a result, the metal sulfide crystal releases metal cations (M^{2+}) and water-soluble intermediary sulfur compounds. Iron(II)-oxidizing bacteria such as *A. ferrooxidans* and *L. ferrooxidans* (*Af, Lf*) catalyze the recycling of iron(III) ions in acidic solutions. In the case of acid-soluble metal sulfides (**B**), an additional attack is performed by protons. The liberated sulfur compounds are oxidized abiotically and also by sulfur compound-oxidizing bacteria such as *A. ferrooxidans* and *A. thiooxidans* (*Af, At*). Where the reactions are mainly abiotic the contribution of bacteria is put in parentheses. The main electron acceptors of oxidation reactions other then the initial iron(III) attack are given on the right site of the arrows. The main reaction products which accumulate in the absence of sulfur compound oxidizers are put in boxes, i.e. sulfuric acid in (**A**) and elemental sulfur in (**B**). (Reproduced from Rohwerder *et al.*, 2003 with permission of the publisher.)

one could assume that after binding of two protons hydrogen sulfide (H_2S) is liberated from the metal sulfide (which could occur in cases where the metal sulfide is absolutely iron ion-free, e.g. ZnS). However, in the presence of iron(III) ions (which are almost always present as an impurity of metal sulfides) the sulfur moiety should be oxidized in a one-electron step concomitantly with the proton attack. Therefore, the first free sulfur compound most likely is a sulfide cation (H_2S^+, i.e. H_2S minus one electron), which can spontaneously dimerize to a free disulfide (H_2S_2) and is further oxidized via higher polysulfides and polysulfide radicals to elemental sulfur (Steudel, 1996; Figure 1B). Consequently, this mechanism was named "polysulfide pathway" (Schippers & Sand, 1999). In the course of polysulfide oxidation, more than 90% of the sulfide moiety of a metal sulfide is transformed to elemental sulfur in the absence of sulfur-oxidizing bacteria (Schippers & Sand, 1999). Minor products formed are thiosulfate, polythionates, and sulfate (Steudel, 1996; Schippers & Sand, 1999). Because the oxidizing action of iron(III) ions is not an absolute requirement for the polysulfide pathway (because here chemical bonds between metal and sulfide can be broken by proton attack), acid-soluble metal sulfides may also be dissolved by the activity of sulfur-oxidizing bacteria. In the absence of iron(III) ions, these bacteria oxidize free sulfide (H_2S), resulting from the proton attack on the metal sulfide, via elemental sulfur to sulfuric acid and, thus, regenerate the protons previously consumed by the metal sulfide dissolution.

2.3. Implications of Sulfur Chemistry to Bioleaching Kinetics

It needs to be highlighted that in both pathways the main role of leaching bacteria is the regeneration of iron(III) ions, which are the most important oxidant in acidic biotopes (Figure 1). Thus, the acidophilic iron(II) ion oxidizers control the redox potential of their environment as it is determined mainly by the iron(III)/iron(II) ratio in leaching solutions. Besides, the acidophilic sulfur oxidizers contribute to these processes by the transformation of the intermediary sulfur compounds to sulfuric acid (Schippers & Sand, 1999; Schippers *et al.*, 1999). In the case of elemental sulfur, the oxidation is carried out exclusively by bacteria, because this sulfur species is almost inert at moderate temperatures and pressures against any abiotic oxidation in acidic environments (Figure 1B). Consequently, elemental sulfur may accumulate in the course of metal sulfide dissolution, if sulfur-oxidizing bacteria are missing or inhibited in their activity. Generally, the production of sulfuric acid from reduced sulfur compounds is needed to regenerate the protons which are consumed by the initial leaching process via the polysulfide pathway (Figure 1B). Besides, sulfur oxidizers can influence the leaching kinetics in a particular manner. Elemental sulfur may occur in leaching systems suspended as free aggregate and/or crystal (without influence on leaching rate) or can form a layer on the metal sulfide surface (with negative consequence; Mustin *et al.*, 1993; Fowler & Crundwell, 1999). In the latter case, the electrochemical properties of the metal sulfide surface might change and/or a barrier may be formed reducing diffusion rates for ions and oxygen. Both phenomena have a negative influence on leaching kinetics. Leaching rate-decreasing sulfur layers were observed on acid-soluble sphalerite at low redox potentials in the absence of sulfur oxidizers (Fowler & Crundwell, 1999). Similar problems are known for chalcopyrite (Bevilaqua *et al.*, 2002). In contrast, at high redox potentials (about 750 mV vs. SHE, i.e. against standard hydrogen electrode) no inhibiting sulfur layers were observed, neither with acid-soluble nor with acid-insoluble metal sulfides (Fowler & Crundwell, 1998; Fowler *et al.*, 1999). Although in the latter two cases elemental sulfur was also formed, it probably occurred only as free aggregates, which do not decrease leaching rates under pH-controlled conditions. The factors which influence the formation and properties of surfacial sulfur layers formed during metal sulfide oxidation are not yet fully understood. Therefore, the detailed surface chemistry of this process needs to be elucidated in future research for controlling (preventing) the formation of inhibition layers in leaching plant operations.

3. BACTERIAL ATTACHMENT TO THE METAL SULFIDE SURFACE

3.1. Role of EPS

Since it has become accepted in general that most bacteria grow attached to material surfaces in form of a biofilm, indicating that also leaching bacteria form biofilms on metal sulfide surfaces, the influence of the factors involved in this form of growth have become of high importance for a thorough understanding of the bioleaching process. When a non-limiting surface space is provided in leaching systems by addition of finely grained minerals, usually more than 80% of an inoculum disappeared from solution within a few hours and attached to the mineral surface (DiSpirito *et al.*, 1983; Bagdigian & Meyerson, 1986; Gehrke, 1998; Harneit *et al.*, 2005). In principle, bacterial attachment predominantly is mediated by EPS, which surround the cells and whose production is even stimulated by the presence of an appropriate substratum and by the attachment process itself (Vandevivere & Kirchman, 1993). In case of cells of *A. ferrooxidans* R1 growing attached on pyrite, we were able to demonstrate that these EPS consisted of the sugars glucose, rhamnose, fucose, xylose, mannose, $C_{12}$to C_{20} saturated fatty acids, glucuronic acid, and iron(III) ions (Gehrke *et al.*, 1998, 2001; Kinzler *et al.*, 2003). Due to a mainly electrostatic interaction of the positively charged cells (two moles of negatively charged glucuronic acid residues complex one mole of positively charged iron(III) ions, which for the whole is resulting in a net positive charge) with the negatively charged pyrite (at pH 2 in sulfuric acid solution; Solari *et al.*, 1992; Blake *et al.*, 1994; Grützner, 2001) attachment can occur. In contrast, for initial attachment to metal sulfide surfaces hydrophobic interactions seem not to contribute significantly (Gehrke *et al.*, 1998; Sampson *et al.*, 2000). However, when bacterial adhesion proceeds to stages of specific and tight bonds/surface interactions in the range of 10 to 20 nm between bacterial outer membrane and metal sulfide surface more intimate non-ionic interactions may significantly contribute to the bonding between bacteria and mineral (Lazar, 2004; Sand & Gehrke, 2006). Although data is only available for a few cases, it may be assumed that EPS production and composition are determined by the growth medium and in particular by the kind of substratum (iron(II) sulfate or metal sulfide) present during growth. This means that the bacteria are able to adapt their EPS according to the substrate/growth conditions. It could be demonstrated in agreement with this hypothesis that planktonic cells grown with soluble substrates, e.g. iron(II) sulfate, produce almost no EPS. In addition, cells grown on elemental sulfur produce significant amounts of EPS, but do not attach to pyrite. These EPS contain considerably less sugars and uronic acids but much more fatty acids than the pyrite-grown ones. The most important difference is the total lack of complexed iron(III) ions or other positively charged groups in the EPS of sulfur-grown cells. Consequently, for attachment to elemental sulfur hydrophobic interactions are the most relevant ones (Gehrke *et al.*, 1998).

3.2. Attachment Site

The site of attachment and the detection/sensing of this site are still open questions. There are indications from literature (Andrews, 1988; Ohmura *et al.*, 1993; Shrihari *et al.*, 1995; Dziurla *et al.*, 1998; Edwards *et al.*, 1998, 1999; Sanhueza *et al.*, 1999; Edwards & Rutenberg, 2001; Ndlovu & Monhemius, 2005) and own work (Telegdi *et al.*, 1998; Gehrke *et al.*, 1998, 2001; Kinzler *et al.*, 2003) that attachment to metal sulfides does not occur by random (Figure 2). For example, AFM images demonstrate that cells of *A. ferrooxidans* preferentially (>80%) attach to sites with visible surface imperfections (scratches etc). Furthermore, attachment to areas with low crystallization is favored and the sessile cells seem to orient themselves in the process of attachment along the crystallographic axes in which direction oxidation fronts propagate. Whereas the adhesion to scratches could be explained by imperfections in the crystal lattice causing a mere contact area enhancement, areas with low crystallization and crystallographic axes are often not related to changes in surface topography. Therefore, attachment to specific sites on the mineral surface is related on principle to different attractant forces, most likely caused by dissolution processes liberating metal ions and sulfur compounds. *A. ferrooxidans* as well as *L. ferrooxidans* have clearly been shown to possess a chemosensory system – chemotaxis – reacting positively to gradients of iron(II)/(III) ions, thiosulfate etc. (Acuña *et al.*, 1992; Meyer *et al.*, 2002). These compounds inevitably occur in the course of metal sulfide dissolution (Figure 1). The dissolution occurs (in an electrochemical sense) at local anodes (Figure 4B) (the place where in case of pyrite iron(II) ions and thiosulfate go into solution; a review on the anodic and cathodic reactions is given by Rimstidt & Vaughan, 2003). It may be speculated that these local anodes or cathodes are the sites towards which the cells are chemotactically attracted.

However, experiments with a Kelvin probe to detect local anodes and cathodes on a pyrite surface with a lateral resolution of the instrument of 10 μm remained without success (Gehrke *et al.*, 1998), although in case of biocorrosion of metals and alloys (e.g. steel) local anodes and cathodes are easily detectable with this instrument. This finding could not be explained at that time, however, recent findings using electron and atomic force microscopy show that the single crystals, which form the pyrite lattice, have an average unit size of approximately 25 nm (5 to 50 nm) in diameter (Rojas-Chapana & Tributsch, 2004; Sand & Telegdi, unpubl.; Figure 3).

Consequently, any local anode or cathode will remain below the detection limit of the Kelvin probe and, very important, already a single cell of a leaching bacterium will cover several crystal units. In a similar case, Little and White with coworkers (Little *et al.*, 2000), testing the attachment sites of sulfate-reducing bacteria on steel surfaces, detected that the bacteria were attached in the immediate vicinity (nm-range) of the anode. Consequently, also the cathode must be in the vicinity of the anode (because sulfate-reducing bacteria preferentially attach to cathodic sites). However, with about 20 μm average size for steel grains these are 3 orders of magnitude larger than the core units of pyrite. Under normal conditions on steel, local anodes and cathodes form and decay (repassivate). Only due to bacterial attachment and/or cell-free EPS, anode and cathode become permanent (manifest)

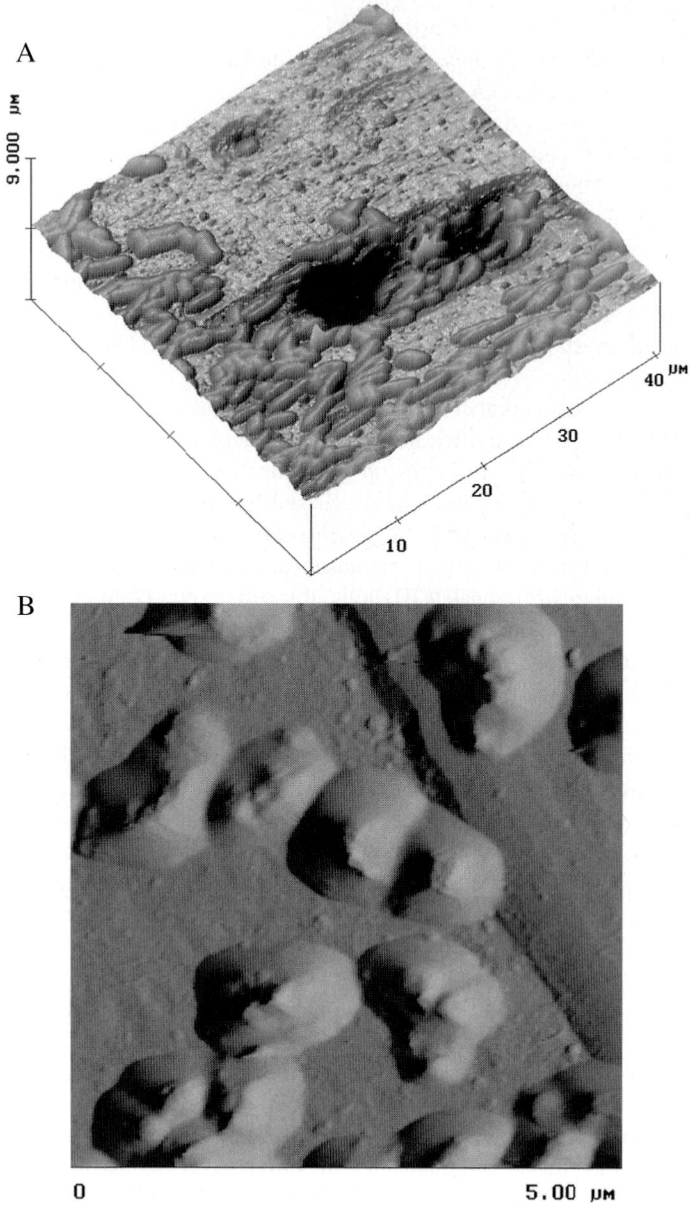

Figure 2. Atomic force microscopy (AFM) images of cells of *A. ferrooxidans* (**A**) and *L. ferrooxidans* (**B**) being specifically attached to dislocation areas (surface faults) on the pyrite surface and producing enormous amounts of extracellular polymeric substances. (Sand and Telegdi, unpubl.)

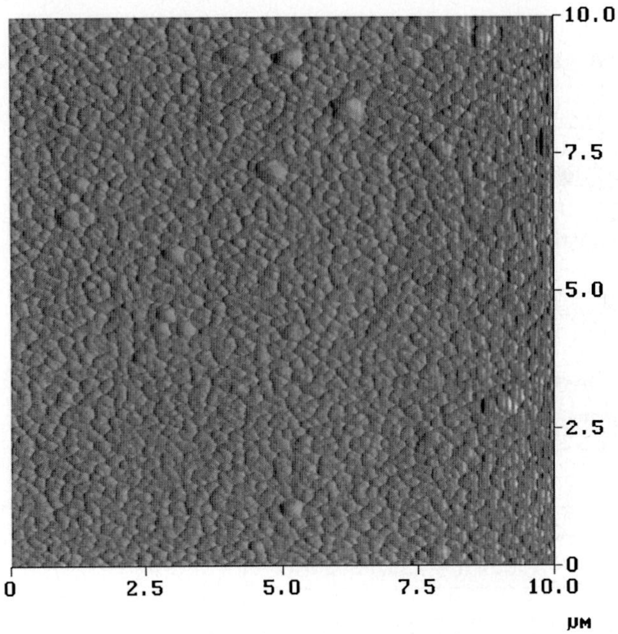

Figure 3. Atomic force microscopy image of the untreated pyrite surface, showing the unit size of single crystals forming the pyrite lattice (Sand and Telegdi, unpubl)

and steel corrosion gets started. This also seems to be fully applicable to bioleaching of metal sulfides. Only in the presence of bacteria or EPS a dissolution of the metal sulfide may occur.

Summarizing, cells of leaching bacteria are attracted to the dissolution sites by their chemotactic sensory system and determine the anodes and cathodes on the metal sulfide surface to become permanent (and start in this way the dissolution). The site of attachment, if similar to the mechanism of sulfate-reducing bacteria, is most likely the cathode. The latter is, due to the size of the crystal units, buried under the microbial cell.

3.3. Reaction Space Concept and Contact Mechanism

The EPS fill and structure the void between the outer membrane (of the cells) and the surface layer (of the metal sulfide) due to their composition (sugars, lipids, uronic acids and iron ions). Thus, they form the reaction space where the dissolution process occurs (Figure 4). Tributsch and coworkers in their pioneering work (Rodriguez-Leiva & Tributsch, 1988) demonstrated that this distance is 10 to 100 nm wide. Recent investigations performing SFM line scanning of single etch pits of bioleached pyrite also imply a reaction space within this 100 nm range (Telegdi *et al.*, 1998;

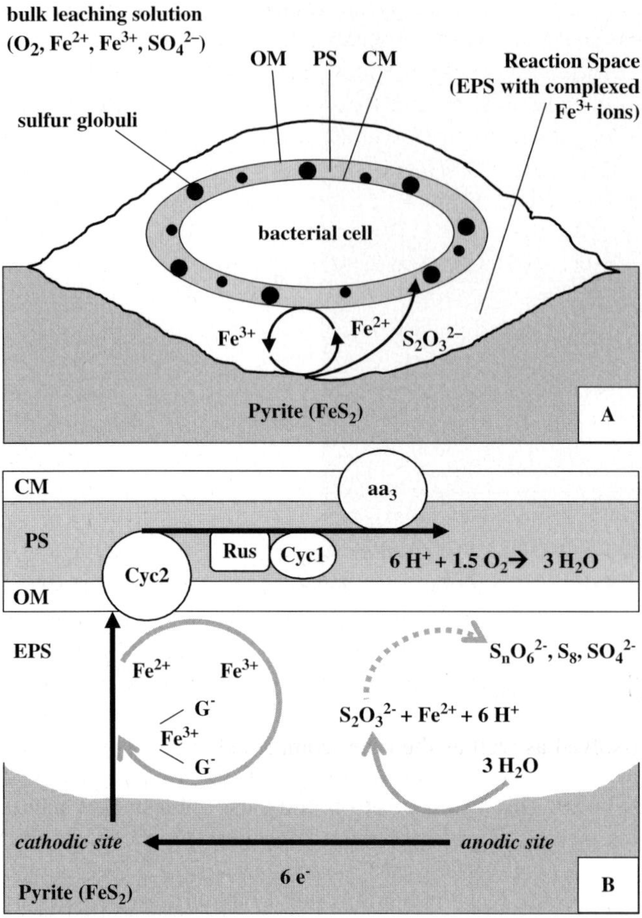

Figure 4. Model for the contact mechanism catalyzed by a cell of *A. ferrooxidans*. (**A**) Overview, showing the bacterial cell embedded in its extracellular polymeric substances (EPS). Positions of the cytoplasmic membrane (CM), the periplasmic space (PS), and the outer membrane (OM) are indicated. (**B**) Detail of the electron transport from pyrite to molecular oxygen (black arrows). Electrons are extracted by iron(III) ions complexed by glucuronic acid residues (G^-) located in the EPS. The resulting iron(II) ions are released from the complex and (re)oxidized by cytochrome Cyc2 (grey circle). The electrons then are transferred via rusticyanin (Rus) and/or cytochrome Cyc1 to the cytochrome c oxidase (aa_3). After 6 one-electron extractions the sulfur moiety of pyrite is released as thiosulfate which is oxidized to elemental sulfur and sulfate via polythionates (broken arrow; see also Fig. 1A and 5). Aggregates of elemental sulfur and higher polythionates can form in the periplasmic space during bacterial thiosulfate and tetrathionate oxidation (indicated as sulfur globuli in **A**). (Modified from Rohwerder *et al.*, 2003, with permission of the publisher.)

Pace *et al.*, 2005). However, precise data is not available. In case of acid- insoluble metal sulfides such as pyrite which need an oxidizing attack by iron(III) ions for dissolution, the EPS-complexed iron(III) ions must fulfill this function (Figure 4B). In addition, considering the utilization of iron(III) ions and the necessary iron(II) (re)oxidation, an electron transport from the pyrite surface to the iron(II)-oxidizing protein complexes of the bacterial outer membrane and further to the cytoplasmic membrane has to be postulated. Currently, the most likely explanation is based on two plausible assumptions. In order to reduce the iron(III) ions, we firstly assume that electrons coming from pyrite (donor) can tunnel to the nearest EPS-complexed iron(III) ion (acceptor), unless the iron(III) ion is already in physical contact with the pyrite surface and a direct reduction can occur. The maximum distance for tunneling of electrons and consequently a reduction of the complexed iron(III) ions amounts to about 2 nm. Such a long-range electron transfer takes place e.g. in the membrane-bound protein complexes of bacterial and mitochondrial respiratory chains. Here, the transfer between the distinct redox centers is mediated by the surrounding proteins (Farver & Pecht, 1997; Skourtis & Beratan, 1997; Winkler & Gray, 1997; Medvedev & Stuchebrukhov, 2001). Probably, a similar role can be played by the EPS chains forming a bridge between the complexed iron(III) ions and the pyrite surface. Consequently, the iron(III) ions have to be maximally within 2 nm from the pyrite surface (to be reducible by tunneling electrons). Considering the far longer distance between the initially reduced iron(III) ion complexes and the respiratory chain in the bacterial membrane (10 to 100 nm), it is secondly assumed that iron(II) ion-glucuronic acid complexes are less stable than the respective iron(III) ion ones. This has been demonstrated for various iron-carbonic acid complexes (NIST 2003). Consequently, the iron(II) ions produced by the cathodic electron transfer are released from their EPS chelators. The remaining uronic acid complex at the pyrite surface will recruit a new iron(III) ion out of solution as it stands in equilibrium with the dissolved as well as the other complexed iron(III) ions. If the mobile iron(II) ions diffuse towards the outer membrane, they will be (re)oxidized by the enzymatic system of the cells. Only these two assumptions are needed for explaining the electrochemical mechanisms/surface reactions taking place in the (bio)leaching of metal sulfides. The chemical reactions occur outside of the cells, respectively outside of the outer membrane, but still in the EPS-generated microenvironment (Figure 4B). Because in this reaction space concept bacterial EPS chains are in direct contact with the pyrite surface and the electron-extracting attack is performed not by free but EPS-complexed iron(III) ions, we use the term "contact mechanism" for this bioleaching sub-mechanism.

4. BIOCHEMICAL FUNDAMENTALS

Finally, we have to discuss the biochemical fundamentals underlying the leaching processes. Considering the sulfur chemistry of the two leaching pathways and the crucial role of EPS in the contact mechanism, the next section will consequently summarize the current knowledge on structural components of EPS and their biosynthesis in leaching bacteria, followed by proposals of electron transfer chains from iron(II) ions and reduced sulfur compounds to molecular oxygen. As a conclusion, biotechnological and ecological consequences directly resulting from the biochemical diversity found in leaching bacteria will be outlined.

4.1. Structure and Biosynthesis of EPS

Although the term EPS often incorrectly refers only to extracellular polysaccharides, we use it here in the currently accepted sense to describe the entirety of extracellular polymeric substances excreted by the cell into its environment (Flemming & Wingender, 2001). These polymers may be polysaccharides, lipids, and proteins or a combination of these different substance classes and can form discrete capsular structures on the cell surface or are only loosely associated to the cell surface. Thus far, little is known about the composition, structure, and synthesis of EPS in leaching bacteria. But, considering their crucial role in biofilm formation during the contact mechanism of bioleaching (Figure 4), it is clear that our knowledge on these EPS has to be improved in order to really understand the fundamentals of bacterial metal sulfide dissolution.

Recent investigations at least shed light on the synthesis pathways of the extracellular polysaccharides in *A. ferrooxidans*. In the genome of this leaching bacterium, several genes have been identified and proposed to encode enzymes involved in the production of extracellular sugar polymers via the Leloir pathway (Barreto *et al.*, 2005a, 2005b). An important role of the sugar monomer galactose is suggested. Surprisingly, it was demonstrated that supplementation of growth medium with this sugar influenced polymer production and increased biofilm formation of *A. ferrooxidans* on pyrite. Obviously, although usually classified as a strict autotrophic bacterium, *A. ferrooxidans* is able to import galactose and incorporate it in its extracellular sugar polymers. Whether the increased biofilm formation also enhances leaching rates, is not clear at the moment. Since EPS-complexed iron(III) ions are assumed to act as the metal sulfide oxidant attention should also be drawn to the enzymatic steps responsible for synthesis of the complexing uronic acids. Indeed, the gene of a putative UDP-glucose dehydrogenase transforming UDP-glucose to UDP-glucuronate has been identified in *A. ferrooxidans* (Barreto *et al.*, 2005b). Now, one can speculate if this finding may result in bacterial strains having an increased uronic acid content in their EPS and, consequently, show better leaching performance. Own work seems to indicate that the pyrite oxidation rates increase linearly with increased amounts of complexed iron(III) ions (Harneit & Sand, unpubl.).

Bioinformatics have also detected genes for extracellular polysaccharide synthesis in *Leptospirillum* spp. (Tyson *et al.*, 2004). Here, a cellulose synthase and numerous glycosyltransferases are proposed to be involved in biofilm production. On the other hand, no EPS-related genes have thus far been found in genomes of *Ferroplasma* spp. (Tyson *et al.*, 2004), suggesting that these species may profit from other EPS-producing bacteria, which share their biofilm habitat, by using the former for attachment. Similar phenomena and interdependencies certainly will be elucidated in further research and will allow to "fine-tune" leaching operations.

4.2. Iron(II) Oxidation

Actually, little is known about the details of the redox chains of aerobic iron(II)-oxidizing bacteria other than *A. ferrooxidans* (Barr *et al.*, 1990; Blake *et al.*, 1993a,

1993b; Tyson *et al.*, 2004; Dopson *et al.*, 2005 ; Ram *et al.*, 2005). Most of the biochemical and genetic research has been done with Gram-negative bacteria, where the electrons have to cross the periplasmic space, i.e. from the outer membrane to the cytoplasmic membrane-bound cytochrome c oxidase (Figure 4B). Although biochemical details are only known for *A. ferrooxidans*, it can be stated on the basis of spectroscopic and other analyses that the iron(II)-oxidizing systems of these bacteria must be totally different with respect to their redox components involved.

In the case of *A. ferrooxidans*, there are several models under discussion, but none has been proven unequivocally (Blake *et al.*, 1993a; Yamanaka & Fukumori, 1995; Appia-Ayme *et al.*, 1999). At least two c-type cytochromes (Cyc1 and Cyc2), the blue copper protein rusticyanin (Rus), and an aa_3-type cytochrome oxidase (CoxABCD) are involved (Figure 4B). This can be deduced from genetic and biochemical analyses. It was demonstrated that all 7 genes coding for the corresponding structural proteins of the 4 above-mentioned redox components are located in a single operon and are cotranscribed (Appia-Ayme *et al.*, 1999; Yarzábal *et al.*, 2004). Recent findings indicate that the 46 kDa Cyc 2 with a midpoint potential of +560 mV is located in the outer membrane and suggest that it functions as the primary electron acceptor in iron(II) ion oxidation (Yarzábal *et al.*, 2002b). The electron is then transferred to the periplasmic 21 kDa Cyc1 (cytochrome c_{552}, a dihemic c_4-type cytochrome) most likely via rusticyanin (Giudici-Orticoni *et al.*, 1999). The soluble Cyc1 hands over the electron to the aa_3-type cytochrome oxidase, whose orientation is not fully known. The catalytic site of oxygen reduction may be located at the cytoplasmic site, as it is observed in mitochondria and other neutrophilic species. On the other hand, from the pH optimum of the oxidase around 3.5 it is concluded that oxygen is reduced at the periplasmic site (Kai *et al.*, 1989; Yamanaka *et al.*, 1991; Lemesle-Meunier *et al.*, 2001). In contrast to this model employing Cyc 2 for iron(II) oxidation (Figure 4B), it has been suggested for a long time that a high-potential iron-sulfur protein, termed Iro, is the actual iron(II) oxidase in *A. ferrooxidans* (Fukumori *et al.*, 1988 ; Yamanaka & Fukumori, 1995). However, recent work on the genetic context of the *iro* gene suggests a role of Iro only in the electron transfer between a bc_1 complex and a terminal oxidase but not in iron(II) oxidation. Consequently, the renaming of Iro in Hip has been suggested (Bruscella *et al.*, 2005; Nouailler *et al.* 2006). The role of rusticyanin is still unresolved. Yamanaka and coworkers presented some evidence that the electron chain functions *in vitro* without its involvement (reviewed by Yamanaka & Fukumori, 1995). They considered rusticyanin as a "broadening" of the electron pathway. In addition, the other Gram-negative bacterium investigated thus far, *L. ferrooxidans*, does not have any rusticyanin and, nevertheless, oxidizes iron(II) ions. On the other hand, rusticyanin must be an essential component of the electron transport chain of *A. ferrooxidans*. An organism, who lives of a substrate with such little energy like iron(II) ions, may contain up to 5% of its whole cell protein in the form of rusticyanin (Cox & Boxer, 1978). With only a ΔG of about −30 kJ/mol available from iron(II) ion oxidation (with oxygen as electron acceptor at pH 2), *A. ferrooxidans* cannot afford to produce several percent of its biomass without important function.

This contradiction can most likely be explained, when assuming that rusticyanin functions as an electron reservoir in *A. ferrooxidans*. Rusticyanin has a midpoint redox potential of +680 mV (Ingledew & Cobley, 1980). As a consequence, it may take up electrons to become reduced up to potentials of around +800 mV. This agrees well with the data presented by Boon and coworkers (Boon *et al.*, 1998; see also section 4.4), which are also supported by other studies (Meruane *et al.*, 2002). Thus, it seems likely that rusticyanin, due to its large concentration, efficiently takes up every electron which becomes available at the outer membrane in the course of iron(II) ion oxidation and channels it downhill into the oxidation pathway. This electron reservoir hypothesis is in agreement with the electron flow schemes previously presented for *A. ferrooxidans* (Yamanaka & Fukumori, 1995; Rawlings, 1997; Appia-Ayme *et al.*, 1999; Lemesle-Meunier *et al.*, 2001). Short-term decreases in the electron flow (e.g. caused by fluctuations in the oxygen concentration) are balanced by the redox buffering function of rusticyanin. The primary electron acceptor (probably Cyc2) remains oxidized. Consequently, the driving force for iron(II) ion oxidation is at a maximum [i.e. for a certain iron(II)/iron(III) ratio the ΔG value of iron(II) ion oxidation is highest because the other redox partner, the electron acceptor Cyc2, is fully oxidized]. This has the advantage that most electrons available from iron(II) ions can be collected, however, only in the redox range of rusticyanin. Especially for the use of such a low-energy substrate, this seems to be highly beneficial.

4.3. Biochemistry of Sulfur Compound Oxidation

Considering the sulfur chemistry of the two metal sulfide oxidation pathways it is clear that various reduced sulfur compounds are occurring during bioleaching activity. As indicated in Figure 1, these sulfur compounds are oxidized abiotically as well as by enzymatic catalysis. In this connection, we have already stressed the importance of bacterial oxidation of elemental sulfur (section 2.3). In addition, the oxidation of hydrogen sulfide, thiosufate, sulfite, tetrathionate, and also higher polythionates might be relevant. The underlying biochemistry is very complex and has recently been reviewed (Friedrich, 1998; Suzuki, 1999; Brüser, 2000; Friedrich *et al.*, 2001, 2005; Kletzin *et al.*, 2004) and various models have been developed. In particular, for many neutrophilic bacteria the relevant reactions seem to be resolved in molecular detail and a common mechanism was proposed by Friedrich and coworkers (Friedrich *et al.*, 2001). In these bacteria, a suite of proteins, the so-called Sox system, should be responsible for the oxidation of sulfide, elemental sulfur, thiosulfate, and sulfite to sulfate accompanied by an electron transfer to the cytochrome c level. Contrary to this finding, in *A. ferrooxidans* and other acidophilic sulfur-oxidizing bacteria the Sox system is absent (Müller *et al.*, 2004; Ramírez *et al.*, 2004; Friedrich *et al.*, 2005; chapter 12 in this book). Nevertheless, although we do not know the exact route(s) of sulfur compound oxidation in the acidophilic bacteria, we are now able to draw an approximate picture and name the employed enzymatic steps. In addition to the substantial work published over more than 50 years, this has been made possible by recent studies reevaluating the involvement of the electron transfer from sulfur compound oxidation to the quinone pool (Brasseur *et al.*, 2004; Müller *et al.*, 2004; Wakai *et al.*, 2004).

In the mesophilic *A. ferrooxidans* as well as in related species such as *A. thiooxidans*, elemental sulfur is oxidized to sulfite by a thiol-dependent dioxygenase. This enzyme has been only poorly characterized thus far, but its activity has been demonstrated in various mesophilic bacteria. Now on the basis of the previous studies, the following pathway can be outlined (Figure 5). The sulfur dioxygenase (SDO) is located in the periplasmic space of Gram-negative bacteria and, consequently, cannot oxidize extracellular elemental sulfur directly. We propose the employment of thiol group-containing transport proteins (OMP) embedded in the outer membrane to move sulfur atoms into the periplasmic space. In this course, octameric sulfur reacts with thiol groups to form thiol-bound sulfane sulfur atoms. Only these highly reactive sulfur species but not sulfide can be oxidized by the SDO

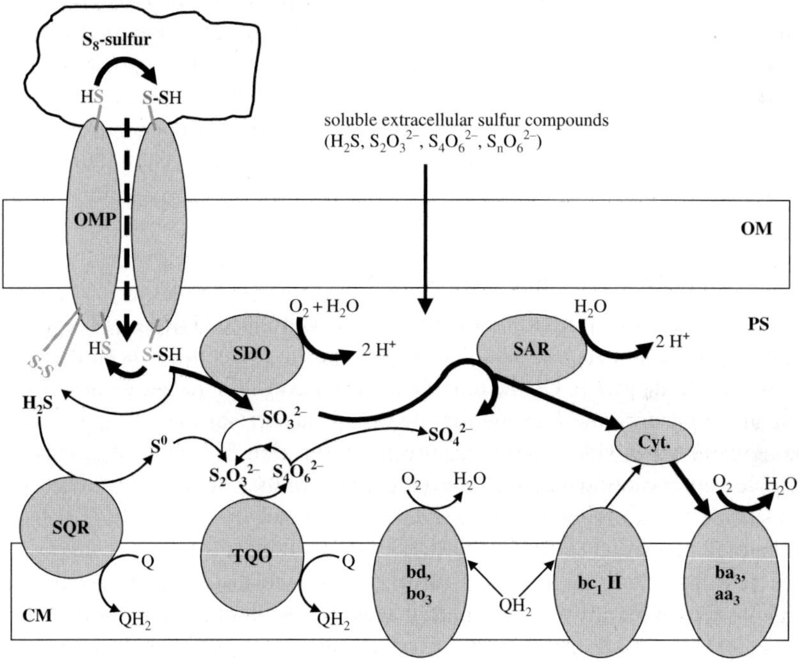

Figure 5. Model for the oxidation of elemental sulfur and other reduced sulfur compounds in *Acidithiobacillus* spp. and related Gram-negative sulfur-oxidizing bacteria. Elemental sulfur (S_8) is transported through the outer membrane (OM) into the periplasmic space (PS) by thiol-bearing outer membrane proteins (OMP). Then it is mainly oxidized by periplasmic sulfur dioxygenase (SDO) and sulfite:acceptor oxidoreductase (SAR) to sulfate. SAR transfers electrons to a periplasmic cytochrome (Cyt.) which is oxidized by terminal oxidases (ba_3, aa_3). Besides, elemental sulfur can also be reduced to sulfide which is oxidized by sulfide:quinone oxidoreductase (SQR) to free periplasmic sulfur (S^0). With sulfite coming from SDO activity S^0 can form thiosulfate which is oxidized to tetrathionate by thiosulfate:quinone oxidoreductase (TQO). SQR and TQO are located in the cytoplasmic membrane (CM) and interact with the quinone pool (Q, QH_2). Quinones are oxidized by quinone oxidases (bd, bo_3) or by a bc_1 complex (bc_1 II) (for details see text).

(Rohwerder & Sand, 2003). Details of the transport mechanism are not known, but it has been shown that some outer membrane proteins increasingly are expressed in sulfur- compared with iron(II)-grown cells and, thus, are candidates for the proposed thiol-bearing transport proteins (Buonfiglio *et al.*, 1999; Ramírez *et al.*, 2004). By handing over the sulfane sulfur to SDO, the thiol groups are set free and again can attack elemental sulfur. As a dioxygenase, SDO does not feed any electrons into the respiratory chain. The resulting sulfite is mainly oxidized further to sulfite by sulfite:acceptor oxidoreductase (SAR), most likely reducing soluble cytochromes, which then transfer the electrons to terminal cytochrome oxidase systems of the aa_3- or ba_3-type. Contrary to this sulfur oxidation scheme, enzymes catalyzing the quinone-reducing oxidation of sulfide (sulfide:quinone oxidoreductase, SQR) and thiosulfate (thiosulfate:quinone oxidoreductase, TQO) have been demonstrated to be active during sulfur oxidation in *A. ferrooxidans*. (Brasseur *et al.*, 2004; Wakai *et al.*, 2004). In addition, the genes encoding SQR (Bronstein *et al.*, 2000) and TQO (Müller *et al.*, 2004) have been identified in the genome of *A. ferrooxidans*. The involvement of electron transfer via quinones to oxygen is also supported by the finding that bd-type and bo_3-type quinone oxidases (Brasseur *et al.*, 2004; Wakai *et al.*, 2004) as well as a downhill electron transport via a bc_1 complex (Brasseur *et al.*, 2004) seem to be active during sulfur oxidation. A quinone-dependent electron route has been already postulated by Corbett and Ingledew (1987). However, the formation of sulfide and thiosulfate during elemental sulfur oxidation is not obvious at first sight. Interestingly, Bacon and Ingledew (1989) demonstrated sulfide production by *A. ferrooxidans* cells in the presence of elemental sulfur and, consequently, proposed a reductive initial step in the sulfur oxidation pathway. Alternatively, elemental sulfur disproportionation has to be postulated. The latter seems to be an unlikely process in *Acidithiobacillus* spp., because it has never been demonstrated in this bacterial group. On the other hand, if we consider the above-described activation and transport mechanism for elemental sulfur, sulfide formation can easily be explained as a side reaction occurring by reaction of free thiol groups of the OMP with sulfane sulfur-bearing ones (Figure 5). In the course of this reaction, hydrogen sulfide and a protein disulfide bridge are formed. The free sulfide is then oxidized back to elemental sulfur by SQR. This zero-valent sulfur (S^0) is supposed to be free and may form aggregates (sulfur globuli; see also Figure 4A) in the periplasmic space. In part, it will be captured again by the thiol groups of the OMP, whereas the excess S^0 can react with sulfite coming from SDO-catalyzed sulfur oxidation to form thiosulfate. The latter can be oxidized further to tetrathionate by TQO. Tetrathionate and other polythionates are thought to be degraded in a cyclic pathway employing hydrolyzing and oxidizing steps and resulting in the formation of sulfate and thiosulfate (Pronk *et al.*, 1990). It is not clear to which portion the route via hydrogen sulfide and thiosulfate contribute to sulfur oxidation. Considering the high reactivity of the sulfane sulfur in close neighborhood to thiol groups (Figure 5) it is a likely process and could occur to a significant extent besides the SDO- and SAR-catalyzed main route. However, it has to be stressed that in the pathway via sulfide two quinone electrons per sulfur atom have to be employed for the reactivation of the OMP by reducing the disulfide bridge. Consequently, not all electrons can be transferred via quinone oxidases or bc_1 complex to oxygen. Soluble sulfur compounds extracellularly formed during bioleaching of metal sulfides, comprising hydrogen sulfide, thiosulfate, polythionates, and sulfite, may be imported by porins or other transport mechanisms to the periplasmic space and can then be degraded by the above described routes (Figure 5).

In acidophilic *Archaea*, the sulfur oxidation biochemistry deviates a little from the above scheme proposed for *Acidithiobacillus* spp. and related Gram-negative bacteria. Leaching bacteria relevant at high temperatures such as *Acidianus* spp. and other *Sulfolobales* lack an outer membrane and a periplasmic space (which is also true for Gram-positive leaching bacteria). Therefore, most enzymatic reactions are proposed to occur in the cytoplasm. In addition, SDO for elemental sulfur oxidation is thiol-independent (Emmel *et al.*, 1986), possibly due to the increased reactivity of elemental sulfur at elevated temperatures, and a second enzyme concomitantly catalyzing both sulfur oxygenation and reduction has been demonstrated (sulfur oxygenase reductase, SOR) (Kletzin, 1989, 1992; Urich *et al.*, 2006). Kletzin and coworkers propose that all electrons available from sulfur reduce quinones and that a cytochrome c reduction seems not to be involved in sulfur oxidation (Kletzin *et al.*, 2004). This hypothesis is based on the identification of TQO besides a membrane-bound SAR in *Acidianus* spp. (Zimmermann *et al.*, 1999; Kletzin *et al.*, 2004; Müller *et al.*, 2004; Chen *et al.*, 2005).

4.4. Biotechnological and Ecological Consequences

The above described EPS metabolism of leaching bacteria and also their biochemistry of iron(II) ion and sulfur compound oxidation together with physical and chemical factors besides biological interactions determine leaching performance in biomining processes as well as bacterial activity in AMD/ARD cases. Generally, at industrial scale bioleaching processes employ more or less stable consortia, because the non-sterile incubation conditions in stirred tank reactors or other biomining facilities do not support the maintenance of pure cultures. Obviously, this is also valid for natural habitats such as abandoned (open pit) mines or waste heaps. Consequently, even in these extreme ecosystems a diverse microbial community can be found (see chapter 1 in this book). It has to be stressed that each species of these consortia has a unique environmental niche due to its unique biochemical background and, consequently, an important but often unrecognized function in the dissolution process. Which of these biochemical concepts prevail over other bacteria mainly depends on the site-specific conditions, comprising ore and salt composition, pH, temperature, and many other abiotic, physical and chemical parameters of varying importance, but also on up to now often unrecognized biotic interactions.

In this connection on the basis of metagenome and metaproteome studies (Tyson *et al.*, 2004; Ram *et al.*, 2005) it has been proposed that among other features EPS production for biopolymer/biofilm generation is partitioned among members of bacterial leaching consortia. In the case of *Ferroplasma* spp., which have been attributed an important role in some leaching habitats (Edwards *et al.*, 2000; Golyshina & Timmis, 2005), genes involved in EPS production could not be detected thus far. Possibly, these species profit from other EPS-producing bacteria which share their biofilm habitat. Likely candidates are polymer-producing *Leptospirillum* spp., whose genomes contain many EPS biosynthesis genes and who are known from own experience to produce copious amounts of EPS. Consequently, if for a special ore the presence of *Ferroplasma* spp. in the biofilm community is required, conditions may be chosen in bioreactors which allow the establishment of these bacteria in the presence of suitable biofilm/biocoenosis partners.

A second interesting point is the finding that in laboratory experiments, although the majority of cells attaches rapidly to the metal sulfide surface, some cells always remain in the planktonic state (Sand *et al.*, 1998; Harneit *et al.*, 2005). The reason for this is unknown, but the surface properties of these cells are likely to be the decisive factor. Currently, we assume that either limitations caused by availability or non-availability of anodes respective cathodes or the existence of genomically diverse planktonic and sessile subpopulations might be the reason for this finding. Subpopulations with defects in their EPS biosynthesis pathways may produce lower amounts or altered polymers, which might not allow attachment to metal sulfides or to other cells any more. Consequently in order to achieve optimal leaching performance via the contact mechanism, the reasons for the development of such non-attaching subpopulations should be analyzed and measures been taken to minimize or even exclude such cells.

Considering the diversity in iron(II) ion oxidation pathways, two major strategies can be observed, i.e. employment of the electron buffer system rusticyanin or related blue copper proteins (e.g. *Acidithiobacillus* and *Ferroplasma* spp.) or, alternatively, the use of cytochromes c with very high potentials at the beginning of the respiratory chain (e.g. *Leptospirillum* spp.). By comparing biochemical and ecological data, it can be concluded that these differences determine which species become the dominant bacteria in mining habitats such as leach dumps, (underground) ore bodies, or bioreactors (Sand *et al.*, 1992; Schrenk *et al.*, 1998; Rawlings *et al.*, 1999; Bond 2000a, 2000b; Rawlings, 2002). Consequently, the clarification of the impact of the iron(II) ion oxidation pathways is of industrial importance due to possible improvements in the use of the microbes and/or the design of bioleaching plants. Boon and Hansford with coworkers (Hansford, 1997; Boon *et al.*, 1998) described the phenomenon that in case of *A. ferrooxidans* iron(II) oxidation activity was optimal at low redox potentials and ceased completely at values above +850 mV, whereas with *L. ferrooxidans* iron(II) oxidation occurred at redox potentials of up to +950 mV. In a recent study, a *Leptospirillum*-like iron(II) oxidation behavior was also demonstrated with the thermophilc *Sulfolobus metallicus* (Petersen & Dixon, 2005). Consequently, the employment of either *Leptospirillum* spp. at mesophilic or *Sulfolobus* spp. at thermophilic conditions is recommendable in leaching operations where ores are oxidized mainly via the thiosulfate pathway. The species-specific variations deciding at which redox potentials iron(II) ions are still oxidizable by leaching bacteria must be related to the fact described by Norris *et al.*, (1988): the inhibitory concentration of iron(III) ions is much lower for *A. ferrooxidans* (3.1 mM) than for *L. ferrooxidans* (42.8 mM). Both phenomena can be explained by the biochemical differences in the composition of the energy-conserving electron transport chain from iron(II) ion to molecular oxygen. Most likely, the capability of iron(II) ion oxidation at extremely high redox potentials is achieved by applying electron acceptor components with appropriate high potentials already at the entrance to the respiratory chain, i.e. the primary electron acceptor of iron(II) oxidation. This adaptation to high redox potentials is rather inefficient with respect to energy conservation. Consequently, *L. ferrooxidans* has considerable lower oxidation and growth rates with iron(II) ions than *A. ferrooxidans* (Sand *et al.*, 1992; Hallmann *et al.*, 1993). This is one of the reasons, why often *L. ferrooxidans* in standard tests is not detected: it is overgrown by *A. ferrooxidans*. Thus, at an appropriate low redox potential the possession of rusticyanin or other redox buffers confers a considerable growth advantage.

5. ACKNOWLEDGMENTS

Part of the presented work has been supported by the BioMinE project of the EU. The authors are indebted to Axel Schippers and Carlos Jerez for providing manuscripts on diversity and proteomics of leaching bacteria prior to publication. In addition, we appreciate very much the cooperation with Judit Telegdi for investigating bioleaching processes by AFM. Thanks also to Gaël Brasseur and Kazuo Kamimura for their helpful discussion on aspects of the sulfur biochemistry of *Acidithiobacillus* spp.

6. REFERENCES

Acuña J, Rojas J, Amaro AM, Toledo H, Jerez CA. 1992. Chemotaxis of *Leptospirillum ferrooxidans* and other acidophilic chemolithotrophs: comparison with the *Escherichia coli* chemosensory system. FEMS Microbiol Lett 96: 37-42.

Andrews GF. 1988. The selective adsorption of thiobacilli to dislocation sites on pyrite surfaces. Biotechnol Bioeng 31: 378-381.

Appia-Ayme C, Guiliani N, Ratouchniak J, Bonnefoy V. 1999. Characterization of an operon encoding two c-type cytochromes, an aa_3-type cytochrome oxidase, and rusticyanin in *Thiobacillus ferrooxidans* ATCC 33020. Appl Environ Microbiol 65: 4781-4787.

Bacon M, Ingledew WJ. 1989. The reductive reactions of *Thiobacillus ferrooxidans* on sulphur and selenium. FEMS Microbiol Lett 58: 189-194.

Bagdigian RM, Meyerson AS. 1986. The adsorption of *Thiobacillus ferrooxidans* on coal surfaces. Biotechnol Bioeng 28: 467-479.

Barr DW, Ingledew WJ, Norris PR. 1990. Respiratory chain components of iron-oxidizing, acidophilic bacteria. FEMS Microbiol Lett 70: 85-90.

Barreto M, Gehrke T, Harneit K, Sand W, Jedlicki E, Holmes D. 2005b. Unexpected insights into biofilm formation by *Acidithiobacillus ferrooxidans* revealed by genome analysis and experimental approaches. In: Harrison STL, Rawlings DE, Petersen J, eds. Proceedings of the 16th International Biohydrometallurgy Symposium, September 25–29, Cape Town, South Africa. Produced by Compress www.compress.co.za, 817-825.

Barreto M, Jedlicki E, Holmes DS. 2005a. Identification of a gene cluster for the formation of extracellular polysaccharide precursors in the chemolithoautotroph *Acidithiobacillus ferrooxidans*. Appl Environ Microbiol 71: 2902-2909.

Bevilaqua D, Leite ALLC, Garcia O Jr., Tuovinen OH. 2002. Oxidation of chalcopyrite by *Acidithiobacillus ferrooxidans* and *Acidithiobacillus thiooxidans* in shake flasks. Process Biochem 38: 587-592.

Blake RC II, Shute EA, Greenwood MM, Spencer GH, Ingledew WJ. 1993a. Enzymes of aerobic respiration on iron. FEMS Microbiol Rev 11: 9-18.

Blake RC II, Shute EA, Howard GT. 1994. Solubilization of minerals by bacteria: electrophoretic mobility of *Thiobacillus ferrooxidans* in the presence of iron, pyrite, and sulfur. Appl Environ Microbiol 60: 3349-3357.

Blake RC II, Shute EA, Waskovsky J, Harrison AP Jr. 1993b. Respiratory components in acidophilic bacteria that respire on iron. Geomicrobiol J 10: 173-192.

Bond PL, Druschel GK, Banfield JF. 2000a. Comparison of acid mine drainage microbial communities in physically and geochemically distinct ecosystems. Appl Environ Microbiol 66: 4962-4971.

Bond PL, Smriga SP, Banfield JF. 2000b. Phylogeny of microorganisms populating a thick, subaerial, predominantly lithotrophic biofilm at an extreme acid mine drainage site. Appl Environ Microbiol 66: 3842-3849.

Boon M, Heijnen JJ, Hansford GS. 1998. The mechanism and kinetics of bioleaching sulphide minerals. Min Pro Ext Met Rev 19: 107-115.

Bosecker K. 1997. Bioleaching: metal solubilization by microorganisms. FEMS Microbiol Rev 20: 591-604.

Brasseur G, Levican G, Bonnefoy V, Holmes D, Jedlicki E, Lemesle-Meunier D. 2004. Apparent redundancy of electron transfer pathways via bc_1 complexes and terminal oxidases in the extremophilic chemolithoautotrophic *Acidithiobacillus ferrooxidans*. Biochim Biophys Acta 1656: 114-126.

Bronstein M, Schütz M, Hauska G, Padan E, Shahak Y. 2000. Cyanobacterial sulfide-quinone reductase: cloning and heterologous expression. J Bacteriol 182: 3336-3344.

Bruscella P, Cassagnaud L, Ratouchniak J, Brasseur G, Lojou E, Amils R, Bonnefoy V. 2005. The HiPIP from the acidophilic *Acidithiobacillus ferrooxidans* is correctly processed and translocated in *Escherichia coli*, in spite of the periplasm pH difference between these two micro-organisms. Microbiology 151: 1421-1431.

Brüser T, Lens PNL, Trüper HG. 2000. The biological sulfur cycle. In: Lens PNL, Hulshoff Pol L, eds. Environmental technologies to treat sulfur pollution. IWA Publishing, London, 47-85.

Buonfiglio V, Polidoro M, Soyer F, Valenti P, Shively J. 1999. A novel gene encoding a sulfur-regulated outer membrane protein in *Thiobacillus ferrooxidans*. J Biotechnol 72: 85-93.

Chen Z-W, Jiang C-Y, She Q, Liu S-J, Zhou P-J. 2005. Key role of cysteine residues in catalysis and subcellular localization of sulfur oxygenase-reductase of *Acidianus tengchongensis*. Appl Environ Microbiol 71: 621-628.

Corbett CM, Ingledew WJ. 1987. Is $Fe^{3+/2+}$ cycling an intermediate in sulfur oxidation by Fe^{2+}-grown *Thiobacillus ferrooxidans*? FEMS Microbiol Lett 41: 1-6.

Cox JC, Boxer DH. 1978. The purification and some properties of rusticyanin, a blue copper protein involved in iron(II) oxidation from *Thiobacillus ferrooxidans*. Biochem J 174: 497-502.

Descostes M, Vitorge P, Beaucaire C. 2004. Pyrite dissolution in acidic media. Geochim Cosmochim Acta 68: 4559-4569.

DiSpirito AA, Dugan PR, Tuovinen OH. 1983. Sorption of *Thiobacillus ferrooxidans* to particulate material. Biotechnol Bioeng 25: 1163-1168.

Dopson M, Baker-Austin C, Bond PL. 2005. Analysis of differential protein expression during growth states of *Ferroplasma* strains and insights into electron transport for iron oxidation. Microbiology 151: 4127-4137.

Dziurla MA, Achouak W, Lam BT, Heulin T, Berthelin J. 1998. Enzyme-linked immunofiltration assay to estimate attachment of thiobacilli to pyrite. Appl Environ Microbiol 64: 2937-2942.

Edelbro R, Sandström A, Paul J. 2003. Full potential calculations on the electron bandstructures of spalerite, pyrite and chalcopyrite. Appl Surf Sci 206: 300-313.

Edwards KJ, Bond PL, Gihring TM, Banfield JF. 2000. An archaeal iron-oxidizing extreme acidophile important in acid mine drainage. Science 287: 1796-1799.

Edwards KJ, Goebel BM, Rodgers TM, Schrenk MO, Gihring TM, Cardona MM, Mcguire MM, Hamers RJ, Pace NR, Banfield JF. 1999. Geomicrobiology of pyrite (FeS_2) dissolution: case study at Iron Mountain, California. Geomicrobiol J 16: 155-179.

Edwards KJ, Rutenberg AD. 2001. Microbial response to surface microtopography: the role of metabolism in localized mineral dissolution. Chem Geol 180: 19-32.

Edwards KJ, Schrenk MO, Hamers R, Banfield JF. 1998. Microbial oxidation of pyrite: Experiments using microorganisms from an extreme acidic environment. Am Mineral 83: 1444-1453.

Ehrlich HL. 2002. Geomicrobiology, 3rd ed. Marcel Dekker Inc., New York.

Emmel T, Sand W, König W, Bock E. 1986. Evidence for the existence of a sulphur oxygenase in *Sulfolobus brierleyi*. J Gen Microbiol 132: 3415-3420.

Farver O, Pecht I. 1997. The role of the medium in long-range electron transfer. J. Biol Inorg Chem 2: 387-392.

Flemming HC, Wingender J. 2001. Relevance of microbial extracellular polymeric substances (EPSs) – Part I: Structural and ecological aspects. Wat Sci Tech 43: 1-8.

Fowler TA, Crundwell FK. 1998. Leaching of zinc sulfide by *Thiobacillus ferrooxidans*: experiments with a controlled redox potential indicate no direct bacterial mechanism. Appl Environ Microbiol 64: 3570-3575.

Fowler TA, Crundwell FK. 1999. Leaching of zinc sulfide by *Thiobacillus ferrooxidans*: bacterial oxidation of the sulfur product layer increases the rate of zinc sulfide dissolution at high concentrations of ferrous ions. Appl Environ Microbiol 65: 5285-5292.

Fowler TA, Holmes PR, Crundwell FK. 1999. Mechanism of pyrite dissolution in the presence of *Thiobacillus ferrooxidans*. Appl Environ Microbiol 65: 2987-2993.

Friedrich CG, Bradischewsky F, Rother D, Quentmeier A, Fischer J. 2005. Prokaryotic sulfur oxidation. Curr Opin Microbiol 8: 253-259.

Friedrich CG, Rother D, Bradischewsky F, Quentmeier A, Fischer J. 2001. Oxidation of reduced inorganic sulfur compounds by bacteria: emergence of a common mechanism? Appl Environ Microbiol 67: 2873-2882.

Friedrich CG. 1998. Physiology and genetics of sulphur-oxidizing bacteria. Adv Microbiol Physiol 39: 235-289.

Fukumori, Y, Yano, T, Sato, A, Yamanaka, T. 1988. Fe(II)-oxidizing enzyme purified from *Thiobacillus ferrooxidans*. FEMS Microbiol Lett 50: 169-172.

Gehrke T, Hallmann R, Kinzler K, Sand W. 2001. The EPS of *Acidithiobacillus ferrooxidans* - a model for structure-function relationships of attached bacteria and their physiology. Water Sci Technol 43: 159-167.

Gehrke T, Telegdi J, Thierry D, Sand W. 1998. Importance of extracellular polymeric substances from *Thiobacillus ferrooxidans* for bioleaching. Appl Environ Microbiol 64: 2743-2747.

Gehrke T. 1998. Bedeutung extrazellulärer polymerer Substanzen von *Thiobacillus ferrooxidans* für die mikrobielle Besiedelung und Laugung von Pyrit und Schwefel. PhD thesis, University of Hamburg, Hamburg.

Giudici-Orticoni MT, Guerlesquin F, Bruschi M, Nitschke W. 1999. Interaction-induced redox switch in the electron transfer complex rusticyanin-cytochrome c_4. J Biol Chem 274: 30365-30369.

Golyshina OV, Timmis KN. 2005. *Ferroplasma* and relatives, recently discovered cell wall-lacking archaea making a living in extremely acid, heavy metal-rich environments. Environ Microbiol 7: 1277-1288.

Grützner T. 2001. Auswirkungen von *Acidithiobacillus ferrooxidans* auf die Flotierbarkeit sulfidischer Minerale. Diploma thesis, TU Clausthal, Clausthal-Zellerfeld.

Hallmann R, Friedrich A, Koops H-P, Pommerening-Röser A, Rohde K, Zenneck C, Sand W. 1993. Physiological characteristics of *Thiobacillus ferrooxidans* and *Leptospirillum ferrooxidans* and pyhsiochemical factors influence microbial metal leaching. Geomicrobiol J 10: 193-206.

Hansford GS. 1997. Recent developments in modelling the kinetics of bioleaching sulphide minerals. In: Rawlings DE, ed. Biomining: Theory, Microbes and Industrial Processes. Springer-Verlag, Berlin, 153-175.

Harneit K, Göksel A, Kock D, Klock JH, Gehrke T, Sand W. 2005. Adhesion to metal sulfide surfaces by cells of *Acidithiobacillus ferrooxidans*, *Acidithiobacillus thiooxidans* and *Leptospirillum ferrooxidans*. In: Harrison STL, Rawlings DE, Petersen J, eds. Proceedings of the 16th International Biohydrometallurgy Symposium, September 25–29, Cape Town, South Africa. Produced by Compress www.compress. co.za, 635-645.

Ingledew WJ, Cobley JG. 1980. A potentiometric and kinetic study on the respiratory chain of ferrous-iron-grown *Thiobacillus ferrooxidans*. Biochim Biophys Acta 590: 141-158.

Kai M, Yano T, Fukumori Y, Yamanaka T. 1989. Cytochrome oxidase of an acidophilic iron-oxidizing bacterium, *Thiobacillus ferrooxidans*, functions at pH 3.5. Biochem Biophys Res Commun 160: 839-843.

Kinzler K, Gehrke T, Telegdi J, Sand W. 2003. Bioleaching – a result of interfacial processes caused by extracellular polymeric substances (EPS). Hydrometallurgy 71, 83-88.

Kletzin A, Urich T, Müller F, Bandeiras TM, Gomes CM. Dissimilatory oxidation and reduction of elemental sulfur in thermophilic Archaea. J Bioenerg Biomem 36: 77-91.

Kletzin A. 1989. Coupled enzymatic production of sulfite, thiosulfate, and hydrogen sulfide from sulfur: purification and properties of a sulfur oxygenase reductase from the facultatively anaerobic archaebacterium *Desulfurolobus ambivalens*. J Bacteriol 171: 1638-1643.

Kletzin A. 1992. Molecular characterization of the sor gene, which encodes the sulfur oxygenase/ reductase of the thermophilic archaeum *Desulfurolobus ambivalens*. J Bacteriol 174: 5854-5859.

Lazar P. 2004. Optimierung der Gewinnung von extrazellulären polymeren Substanzen von Laugungs-bakterien. Diploma thesis, University of Hamburg, Hamburg.

Lemesle-Meunier D, Brasseur G, Tron P, Bennaroch D, Nitschke W, Elbehti A. 2001. The membrane-bound C type cytochromes and the interaction between the downhill and uphill electron transfer pathways in the acidophilic chemolithotrophic ferrous ion-oxidizing bacterium *Thiobacillus ferrooxidans*. In: Ciminelli VST, Garcia Jr. O, eds. Biohydrometallurgy: Fundamentals, Technology and Sustainable Development. Elsevier Science BV, Amsterdam, Part A, 299-308.

Little B, Ray B, Pope R, Franklin M, White DC. 2000. Spatial and temporal relationships between localised corrosion and bacterial activity on iron-containing substrata. In: Sequeira CAC, ed. Microbial Corrosion, European Federation of Corrosion Publications No 29, The Institute of Materials, London, 21-35.

Luther GW III. 1987. Pyrite oxidation and reduction: molecular orbital theory considerations. Geochim Cosmochim Acta 51: 3193-3199.

Medvedev D, Stuchebrukhov AA. 2001. DNA repair mechanism by photolyase: Electron transfer path from the photolyase catalytic cofactor $FADH^-$ to DNA thymine dimer. J theor Biol 210: 237-248.

Meruane G, Salhe C, Wiertz J, Vargas T. 2002. Novel electrochemical-enzymatic model which quantifies the effect of the solution Eh on the kinetics of ferrous iron oxidation with *Acidithiobacillus ferrooxidans*. Biotechnol Bioeng 80: 280-288.

Meyer G, Schneider-Merck T, Böhme S, Sand W. 2002. A simple method for investigations on the chemotaxis of *A. ferrooxidans* and *D. vulgaris*. Acta Biotechnol 22: 391-399.

Moses CO, Nordstrom DK, Herman JS, Mills AL. 1987. Aqueous pyrite oxidation by dissolved oxygen and by ferric iron. Geochim Cosmochim Acta 51: 1561-1571.

Müller FH, Bandeiras TM, Urich T, Teixeira M, Gomes CM, Kletzin A. 2004. Coupling of the pathway of sulphur oxidation to dioxygen reduction: characterization of a novel membrane-bound thiosulphate:quinone oxidoreductase. Mol Microbiol 53: 1147-1160.

Mustin C, de Donato P, Berthelin J, Marion P. 1993. Surface sulphur as promoting agent of pyrite leaching by *Thiobacillus ferrooxidans*. FEMS Microbiol Rev 11: 71-78.

Ndlovu S, Monhemius AJ. 2005. The influence of crystal orientation on the bacterial dissolution of pyrite. Hydrometallurgy 78: 187-197.

NIST. 2003. Critical selected stability constants of metal complexes database. NIST standard reference database 46, version 7.0, National Institute of Standards and Technology, Gaithersburg, MD.

Norris PR, Barr DW, Hinson D. 1988. Iron and mineral oxidation by acidophilic bacteria: affinities for iron and attachment to pyrite. In: Norris PR, Kelly DP, eds. Biohydrometallurgy, Proceedings of the International Symposium, Warwick, Antony Rowe Ltd. Chippenham, Wiltshire, UK, 43-59.

Nouailler M, Bruscella P, Lojou E, Lebrun R, Bonnefoy V, Guerlesquin F. 2006. Structural analysis of HiPIP from the acidophilic bacteria: *Acidithiobacillus ferrooxidans*. Extremophiles 10: 191-198.

Ohmura N, Kitamura K, Saiki H. 1993. Selective adhesion of *Thiobacillus ferrooxidans* to pyrite. Appl Environ Microbiol 59: 4044-4050.

Olson GJ, Brierley JA, Brierley CL. 2003. Bioleaching review part B: Progress in bioleaching: applications of microbial processes by the minerals industries. Appl Microbiol Biotechnol 63:249-257.

Pace DL, Mielke RE, Southham G, Porter TL. 2005. Scanning force microscopy studies of the colonization and growth of *A. ferrooxidans* on the surface of pyrite minerals. Scanning 27: 136-140.

Petersen J, Dixon DG. 2005. Competitive bioleaching of pyrite and chalcopyrite. n: Harrison STL, Rawlings DE, Petersen J, eds. Proceedings of the 16th International Biohydrometallurgy Symposium, September 25–29, Cape Town, South Africa. Produced by Compress www.compress. co.za, 55-64.

Pronk JT, Meulenberg R, Hazeu W, Bos P, Kuenen JG. 1990. Oxidation of reduced inorganic sulphur compounds by acidophilic thiobacilli. FEMS Microbiol Rev 75: 293-306.

Ram RJ, VerBerkmoes NC, Thelen, MP, Tyson GW, Baker, BJ, Blake RC II, Shah M, Hettich RL, Banfield JF. 2005. Community proteomics of a natural microbial biofilm. Science 308: 1915-1920.

Ramírez P, Guiliani N, Valenzuela L, Beard S, Jerez CA. 2004. Differential protein expression during growth of *Acidithiobacillus ferrooxidans* on ferrous iron, sulfur compounds, or metal sulfides. Appl Environ Microbiol 70: 4491-4498.

Rawlings DE, Dew D, du Plessis C. 2003. Biomineralization of metal-containing ores and concentrates. Trends Biotechnol 21: 38-44.

Rawlings DE, Tributsch H, Hansford GS. 1999. Reasons why '*Leptospirillum*'-like species rather than *Thiobacillus ferrooxidans* are the dominant iron-oxidizing bacteria in many commercial processes for the biooxidation of pyrite and related ores. Microbiology 145: 5-13.

Rawlings DE. 1997. Mesophilic, autotrophic bioleaching bacteria: description, physiology and role. In: Rawlings DE, ed. Biomining: Theory, Microbes and Industrial Processes. Springer-Verlag, Berlin.

Rawlings DE. 2002. Heavy metal mining using microbes. Annu Rev Microbiol 56: 65-91.

Rimstidt JD, Vaughan DJ. 2003. Pyrite oxidation: A state-of-the-art assessment of the reaction mechanism. Geochim Cosmochim Acta 67: 873-880.

Rodriguez-Leiva M, Tributsch H. 1988. Morphology of bacterial leaching patterns by *Thiobacillus ferrooxidans* on synthetic pyrite. Arch Microbiol 149: 401-405.

Rohwerder T, Gehrke T, Kinzler K, Sand W. 2003. Bioleaching review part A: Progress in bioleaching: fundamentals and mechanisms of bacterial metal sulfide oxidation. Appl Microbiol Biotechnol 63: 239-248.

Rohwerder T, Jozsa P-G, Gehrke T, Sand W. 2002. Bioleaching. In: Bitton G, ed. Encyclopedia of environmental microbiology. John Wiley & Sons, New York, Vol. 2, 632-641.

Rohwerder T, Sand W. 2003. The sulfane sulfur of persulfides is the actual substrate of the sulfur-oxidizing enzymes from *Acidithiobacillus* and *Acidiphilium* spp. Microbiology 149: 1699-1709.

Rohwerder T, Schippers A, Sand W. 1998. Determination of reaction energy values for biological pyrite oxidation by calorimetry. Thermochim Acta 309: 79-85.

Rojas-Chapana JA, Tributsch H. 2004. Interfacial activity and leaching patterns of *Leptospirillum ferrooxidans* on pyrite. FEMS Microbiol Ecol 47: 19-29.

Sampson MI, Phillips CV, Blake RC II. 2000. Influence of the attachment of acidophilic bacteria during the oxidation of mineral sulfides. Miner Eng 13: 373-389.

Sand W, Gehrke T, Hallmann R, Schippers A. 1995. Sulfur chemistry, biofilm, and the (in)direct attack mechanism – a critical evaluation of bacterial leaching. Appl Microbiol Biotechnol 43: 961-966.

Sand W, Gehrke T, Hallmann R, Schippers A. 1998. Towards a novel bioleaching mechanism. Min Pro Ext Met Rev 19: 97-106.

Sand W, Gehrke T, Jozsa P-G, Schippers A. 2001. (Bio)chemistry of bacterial leaching – direct vs. indirect bioleaching. Hydrometallurgy 59: 159-175.

Sand W, Gehrke T. 2006. Extracellular polymeric substances mediate bioleaching/biocorrosion via interfacial processes involving iron(III) ions and acidophilic bacteria. Res Microbiol 157: 49-56.

Sand W, Rohde K, Sobotke B, Zenneck C. 1992. Evaluation of *Leptospirillum ferrooxidans* for leaching. Appl Environ Microbiol 58: 85-92.

Sanhueza A, Ferrer IJ, Vargas T, Amils R, Sánchez C. 1999. Attachment of *Thiobacillus ferrooxidans* on synthetic pyrite of varying structural and electronic properties. Hydrometallurgy 51: 115-129.

Schippers A, Jozsa P-G, Sand W. 1996. Sulfur chemistry in bacterial leaching of pyrite. Appl Environ Microbiol 62: 3424-3431.

Schippers A, Rohwerder T, Sand W. 1999. Intermediary sulfur compounds in pyrite oxidation: implications for bioleaching and biodepyritization of coal. Appl Microbiol Biotechnol 52: 104-110.

Schippers A, Sand W. 1999. Bacterial leaching of metal sulfide proceeds by two indirect mechanisms via thiosulfate or via polysulfides and sulfur. Appl Environ Microbiol 65: 319-321.

Schippers A. 2004. Biogeochmistry of metal sulfide oxidation in mining environments, sediments and soils. In: Amend JP, Edwards KJ, Lyons TW, eds. Sulfur biogeochemistry – Past and present. Special Paper 379, Geological Society of America, Boulder, Colorado, 49-62.

Schrenk MO, Edwards KJ, Goodman RM, Hamers RJ, Banfield JF. 1998. Distribution of *Thiobacillus ferrooxidans* and *Leptospirillum ferrooxidans*: implications for generation of acid mine drainage. Science 279: 1519-1522.

Shrihari RK, Modak JM, Kumar R, Gandhi KS. 1995. Dissolution of particles of pyrite mineral by direct attachment of *Thiobacillus ferrooxidans*. Hydrometallurgy 38: 175-187.

Singer PC, Stumm W. 1970. Acidic mine drainage: the rate-determining step. Science 167: 1121-1123.

Skourtis SS, Beratan DN. 1997. High and low resolution theories of protein electron transfer. J. Biol Inorg Chem 2: 378-386.

Solari JA, Huerta G, Escobar B, Vargas T, Badilla-Ohlbaum R, Rubio J. 1992. Interfacial phenomena affecting the adhesion of *Thiobacillus ferrooxidans* to sulphide mineral surfaces. Colloids Surfaces 69: 159-166.

Steudel R. 1996. Mechanism for the formation of elemental sulfur from aqueous sulfide in chemical and microbiological desulfurization processes. Ind Eng Chem Res 35: 1417-1423.

Suzuki I. 1999. Oxidation of inorganic sulphur compounds: chemical and enzymatic reactions. Can J Microbiol 45: 97-105.

Telegdi J, Keresztes Z, Pálinkás G, Kalman E, Sand W. 1998. Microbially influenced corrosion visualized by atomic force microscopy. Appl Phys A66: S639-S642.

Tributsch H, Bennett JC. 1981. Semiconductor-electrochemical aspects of bacterial leaching. I. Oxidation of metals. J Chem Technol Biotechnol 31:565-577.

Tributsch H. 2001. Direct versus indirect bioleaching. Hydrometallurgy 59: 177-185.

Tyson GW, Chapman J, Hugenholtz P, Allen EE, Ram RJ, Richardson PM, Solovyev VV, Rubin EM, Rokhsar DS, Banfield JF. 2004. Community structure and metabolism through reconstruction of microbial genomes from the environment. Nature 428: 37-43.

Urich T, Gomes CM, Kletzin A, Frazao C. 2006. X-ray structure of a self-compartmentalizing sulfur cycle metalloenzyme. Science 311: 996-1000.

Vandevivere P, Kirchman DL. 1993. Attachment stimulates exopolysaccharide synthesis by a bacterium. Appl Environ Microbiol 59: 3280-3286.

Wakai S, Kikumoto M, Kanao T, Kamimura K. 2004. Involvement of sulfide:quinone oxidoreductase in sulfur oxidation of an acidophilic iron-oxidizing bacterium, *Acidithiobacillus ferrooxidans* NASF-1. Biosci Biotechnol Biochem 68: 2519-2528.

Winkler JR, Gray HB. 1997. Electron tunneling in proteins: role of the intervening medium. J. Biol Inorg Chem 2: 399-404.

Yamanaka T, Fukumori Y. 1995. Molecular aspects of the electron transfer system which participates in the oxidation of ferrous ion by *Thiobacillus ferrooxidans*. FEMS Microbiol Rev 17: 401-413.

Yamanaka T, Yano T, Kai M, Tamegai H, Sato A, Fukumori Y. 1991. The electron transfer system in an acidophilic iron-oxidizing bacterium. In: Mukohata Y, ed. New era of bioenergetics. Academic Press, Tokyo, 223-246.

Yarzábal A, Appia-Ayme C, Ratouchniak J, Bonnefoy V. 2004. Regulation of the expression of the *Acidithiobacillus ferrooxidans* rus operon encoding two cytochromes c, a cytochrome oxidase and rusticyanin. Microbiology 150: 2113-2123.

Yarzábal A, Brasseur G, Ratouchniak J, Lund K, Lemesle-Meunier D, DeMoss JA, Bonnefoy V. 2002b. The high-molecular-weight cytochrome c Cyc2 of *Acidithiobacillus ferrooxidans* is an outer membrane protein. J Bacteriol 184: 313-317.

Zimmermann P, Laska S, Kletzin A. 1999. Two modes of sulfite oxidation in the extremely thermophilic and acidophilic archaeon *Acidianus ambivalens*. Arch Microbiol 172: 76-82.

CHAPTER 3

ELECTROCHEMICAL TECHNIQUES USED TO STUDY BACTERIAL-METAL SULFIDES INTERACTIONS IN ACIDIC ENVIRONMENTS

DENISE BEVILAQUA[1], HELOISA A. ACCIARI[1], ASSIS V. BENEDETTI[2] AND OSWALDO GARCIA JR.[1]

[1] *Universidade Estadual Paulista, Instituto de Química, Departamento de Bioquímica e Tecnologia Química – Araraquara-SP, Brazil – Caixa Postal 355, CEP 14801-970. E-mail: oswaldo@iq.unesp.br*

[2] *Universidade Estadual Paulista, Instituto de Química, Departamento de Fisico – Química*

1. INTRODUCTION

Bioleaching or bacterial leaching is a dissolution process of metal sulfides in which the oxidative activity of some bacterial species, mainly *Acidithiobacillus ferrooxidans*, formerly *Thiobacillus ferrooxidans* (Kelly & Wood, 2000), plays a central role. As a consequence of its capacity, this bacterium is one of the most important microorganisms utilized in industrial operations to recover metals, such as copper, uranium and gold (Rawlings, 2002).

This chemolithotrophic and acidophilic species utilizes ferrous iron or metallic sulfides as the sole energy source for CO_2 fixation and its growth, according to the following equations:

(1) $$4FeSO_4 + O_2 + 2H_2SO_4 \rightarrow 2Fe_2(SO_4)_3 + 2H_2O$$

(2) $$MeS + 2O_2 \rightarrow MeSO_4$$

where Me is the metal of interest.

This species can also oxidize other reduced forms of sulfur to produce energy according to the equations:

(3) $$2S^0 + 3O_2 + 2H_2O \rightarrow 2H_2SO_4$$

(4) $$Na_2S_2O_3 + 2O_2 + H_2O \rightarrow Na_2SO_4 + H_2SO_4$$

The biological mechanisms and the reactions that come into play during the oxidative dissolution of metal sulfides by bacterium are controversial and still poorly

E.R. Donati and W. Sand (eds.), Microbial Processing of Metal Sulfides, 59–76.

understood with different authors suggesting that its presence enhances leaching, has no effect or is detrimental to the metal leaching (Third *et al.*, 2000; Tributsch, 2001).

The knowledge of the mechanisms of bacterial dissolution of sulfides has been focused in several copper minerals, in order to improve the efficiency of the bioleaching operations.

The oxidation of sulfides by a chemical or biological oxidant can be look upon as an electrochemical reaction with the cathodic reduction of the oxidizing agent (bacterial cell, for example) and the anodic oxidation of the reducing one (metal sulfide), and it can consequently be studied by electrochemical techniques.

Electrochemical studies of mineral sulfide dissolution in acid oxidizing solutions (bacterially produced or not) are powerful tools to elucidate mechanisms and kinetics of these reactions. The mineral oxidative dissolution is a corrosion reaction, with anodic and cathodic half-reactions, which in oxidizing media such as acidic ferric chloride or sulfate solutions is very slow. Therefore, mechanistic studies were proposed in order to explain this slow rate, especially from the viewpoint of the anodic process (Biegler & Horne, 1985; Gómez *et al.*, 1997; Gómez *et al.*, 1996; Pesic & Kim, 1990). In these studies the following aspects were considered:

- Influence of temperature;
- Addition of ferrous or ferric ions in sulfate medium;
- Addition of catalytic cation;
- Origin of the mineral sample;
- Effect of applied potential;
- Influence of pH.

As a general result of these studies it was observed the presence of passive layer that limits the oxidation rate of the mineral sulfide in the medium. Although the composition of this layer was not well characterized yet, it has been described as:

- Metal-deficient sulfide;
- Polysulfide;
- Solid electrolyte interphase;
- Precipitated iron compounds.

Its effect on the oxidation kinetics has been explained in terms of physical blockage, solid-state diffusion of metal ions and passive film growth.

Voltammetric studies complemented with XPS analysis demonstrated that in acidic solutions the surfaces of bornite and chalcopyrite are covered with a CuS layer which is formed according to the following reactions (Kudaikulova, 1989):

$$(5) \qquad CuFeS_2 + 2H^+ \rightarrow CuS + Fe^{3+} + H_2S + e^-$$

$$(6) \qquad 2Cu_2S + 2H_2S \rightarrow 4CuS + 4H^+ + 4e^-$$

In solutions of pH>4 a film of copper oxide and iron hydroxide is formed on the bornite surface which protects the mineral against further oxidation.

In the late 1980s, the electrochemical oxidation of natural mineral sulfide electrodes in the presence or absence of bacteria in sulfuric acid medium was studied for several authors using both voltammetric and chronoamperometric techniques, at different experimental conditions, using mineral massive or carbon paste electrodes (Biegler & Horne, 1985; Kudaikulova, 1989).

In the early 1990s the works on carbon paste electrodes (CPE) modified with mineral sulfides emerged in the electrochemistry field. The physical state of the mineral in the CPE electrodes is similar to that in industrial leaching processes; for this reason, the use of CPE is closer to the real leaching conditions (Lázaro *et al.*, 1995; Rodríguez *et al.*, 2003; Lu, 2000; Elsherief, 2002).

Lázaro and co-workers (1995) showed the reproducibility of the voltammetric response of CPE-chalcopyrite electrodes, as well as the differences in the kinetics of the leaching process in different acidic media. The voltammetric response is associated with chalcopyrite reduction to bornite and chalcocite.

However, the CPE are not always sensitive enough because they involve small quantities of mineral embedded in the paste, which is not necessarily representative of the diversity of mineral grains. Another alternative is the use of Fixed Grains Electrode (FGE) that consists of ground mineral sulfide disperses on carbon paste previously prepared (Toniazzo *et al.*, 1999).

In recent years the voltammetric studies were utilized to characterize the products formed and their adhesion on mineral surfaces. An alternative for this study was to evaluate the anodic dissolution of chalcopyrite in acidic solutions and products formed using ring-disc electrode (RRDE) experiments. The oxidation-reduction reactions of metal sulfides are also susceptible to stirring that can be achieved by air bubbling (Holliday & Richmond, 1990).

The chalcopyrite was also studied in alkaline solutions in order to characterize corrosion products formed on mineral surface. Yin and co-workers (2000) investigated the surface oxidation of chalcopyrite in alkaline solutions using chronoamperometry. The current-time transient was recorded in a 0.1 M borax solution, stepping to various positive potentials from a rest potential, in the range of predicted $CuFeS_2$ thermodynamic stability. The integrated form of these data when plotted as Q against $t^{0.5}$ indicated that oxidation charge increased linearly with $t^{0.5}$ and the current decayed linearly with $t^{-0.5}$. Such behavior, independent of solution stirring suggests that the oxidation process was controlled by solid-state mass transport, i.e. Fe^{3+} ion transport from the bulk of the chalcopyrite through the passive film up to reach the solid/electrolyte interface.

Velásquez and co-workers (1998) evaluated the application of EIS to a chalcopyrite electrode in alkaline solution of borate (pH 9.2) to determine the influence of applied potential on the behavior of electrode/electrolyte interface and the equivalent circuit for modeling the electrode/solution interface. The authors concluded that the existence of a surface layer modified the double layer capacitance, which was associated with the mobile species in the surface layer. On the mineral surfaces several electrochemical processes can modify the surface to such an extension that it can be considered as a heterogeneous film of modified mineral, which is quite different from the substrate (Velásquez *et al.*, 2000, 2001).

In the last ten years a new concern has been raising up in the bacterial dissolution of sulfides, that is the importance of the exopolymeric substances (EPS) present in cell envelope and its role in the bacterial adhesion. However, this remains as a controversial topic, since the EPS can promote or inhibit the dissolution of mineral sulfides. Its formation and composition depend on the mineral characteristics, culture medium and bacterial species (Keresztes *et al.*, 2001; Beech *et al.*, 2002; Bevilaqua *et al.*, 2004).

Nowadays, the electrochemical impedance spectroscopy and electrochemical noise analysis are utilized for the same purpose, with emphasis in the proposition of models that explain the adhesion of the passive film or biofilm on mineral substrate.

Details of each technique and some applications to the bacterial oxidative dissolution of mineral sulfides are presented in the next sections of this chapter.

2. ELECTROCHEMICAL TECHNIQUES: PRINCIPLES, FUNDAMENTALS AND APPLICATIONS

Leaching or bioleaching of metal sulfides often involve a number of intermediate electrochemical reactions. The intermediate reactions control the rate of the overall mineral dissolution reactions, which can be assessed using cyclic voltammetry and chronoamperometry.

2.1. Cyclic Voltammetry (CV)

Cyclic voltammetry (CV) is one of the most commonly used electrochemical techniques, and is based on the application of a linear potential waveform, that is, the potential is changed as a linear function of time. The rate as the potential is changed with time is referred as the scan rate, ν.

Cyclic voltammetry is a simple and direct method for measuring the formal potential of a half reaction when both oxidized and reduced forms are stable during the time required to obtain the cyclic voltammogram (current-potential curve) (Evans *et al.*, 1983).

A triangular potential change is imposed on an electrode and the current response is measured. Analysis of the current response as a function of potential can give information about the thermodynamics and kinetics of electron transfer at the electrode-solution interface (Gosser, 1994).

A potentiostat system sets the control parameters of the experiment. Its purpose is to impose on an electrode (the working electrode) a cyclic linear potential sweep and to output the resulting current-potential curve. The term linear sweep voltammetry (LSV) is used for a half-cycle CV. Its initial (E_i), switching (E_s), final (E_f) potentials, and sweep (or scan) rate (ν, in V/s) describe the sweep. The potential as a function of time is (Gosser, 1994):

(7) $$E = E_i + \nu t \text{ (forward sweep)}$$

(8) $$E = E_i - \nu t \text{ (reverse sweep)}$$

During the anodic potential sweep, the generalized reduced species, R, is oxidized to O and electrons are released to the external circuit. The associated peak current is registered. The half peak current is then estimated and the corresponding half peak potential read from the potential scale. If the potential is measured against a reference electrode (SCE or Ag/AgCl), this potential value can be converted to a potential versus standard hydrogen electrode scale.

When using cyclic voltammetry to study a system by the first time, it is usual to start by carrying out qualitative experiments in order to get a feel from the system, before proceeding to semi-quantitative and finally quantitative ones from which kinetics parameters may be calculated. In a typical qualitative study it is usual to record voltammograms over a wide range of sweep rates and for various values of switching potentials. Commonly, there will be several peaks, and by observing how they appear and disappear as the potential limits and sweep rate are varied, and also by noting the differences between the first and subsequent cycles, it is possible to determine how the processes represented by the peaks are related. At the same time, from the sweep rate dependence of the peak amplitudes the role of adsorption, diffusion, and coupled homogeneous chemical reactions may be identified. The difference between the first and subsequent cyclic voltammograms frequently provides useful mechanistic information.

Voltammetry is widely applied for fundamental studies of oxidation and reduction processes, adsorption processes on surfaces and electron-transfer mechanisms, indicating the presence of intermediates in oxidation/reduction reactions (Skoog & Leary, 1992).

Its main advantage in electroanalysis is its ability to characterize an electrochemical system, but it is not usually a good technique for quantitative analysis.

Although cyclic voltammetry is widely used for the initial redox characterization of a molecule (i.e., the redox potentials and the stability of the different oxidations states) and for qualitative investigation of coupled chemical reactions with electron transfer, there is an apparent disadvantage inherent to this technique: the effects of slow heterogeneous electron transfer and fast chemical reactions are difficult to be separated. To distinguish both mechanisms more experimental work must be done as changing temperature and scan rate.

In spite of this limitation, cyclic voltammetry is very well suited to determine mechanisms and rates of oxidation/reduction surfaces.

Some examples of bacterial dissolution of sulfides using cyclic voltammetry can be mentioned. Choi and co-workers (1993) characterized some electrochemical aspects of the zinc sulfide leaching by *A. ferrooxidans* employing CV measurements and using carbon paste – ZnS working electrodes. Cyclic voltammograms showed that the oxidation of zinc sulfide-containing past to sulfate occurs through the oxidation of sulfur.

Bioleaching of pyrite by *A. ferrooxidans* was also studied employing cyclic voltammetry associated with Raman spectroscopy by Toniazzo and co-workers. In this study the authors analyzed the evolution of the voltammograms with bioleaching time and observed the presence of superficial ferrous sulfate, ferric and elemental sulfur (Toniazzo *et al.*, 1999).

Teixeira *et al.* (2002) studied the oxidation of covellite by *A. ferrooxidans* using open circuit potentials with the time and CV. The experiments were carried out in acid medium (pH 1.8), containing or not Fe^{2+} as additional energy source for the bacterium, and in different periods of incubation. The results showed that a sulfur layer was formed spontaneously due the acid attack, covering the sulfide surface in the initial phase of incubation, blocking the sulfide oxidation. However, the bacterium was capable to oxidize this sulfur layer. In the presence of Fe^{2+} as supplemental energy source, the corrosion process was facilitated, because occurred an indirect oxidation of covellite by Fe^{3+}, which was produced by *A. ferrooxidans* by the oxidation of Fe^{2+} ions added in the medium.

2.2. Chronoamperometry (CA)

Chronoamperometry is a type of electroanalytical transient technique in which an instantaneous potential perturbation is applied to the electrode and the flowing current is monitored as the system relaxes towards its new steady state. When this potential step is enough to cause an electrochemical reaction, a variation of the resultant current with time occurs. The analysis of CA data for reversible systems is based on the Cottrell equation, which defines the current-time dependence for a linear diffusion controlled process (Greef *et al.*, 1985):

$$i = nFACD^{0.5}\pi^{-0.5}t^{-0.5} \tag{9}$$

where:

i is the current density (A cm^{-2});
n the number of electrons transferred / molecule;
F the Faraday's constant (96,500 C mol^{-1});
A the electrode area (cm^2);
D the diffusion coefficient (cm^2s^{-1}) of electroactive specie;
C its concentration (mol cm^{-3}).

The equation indicates that, under these conditions, there is a linear relationship between the current and the 1/square root of time. A plot of i versus $t^{-0.5}$ is often referred as the Cottrell plot.

Using the Cottrell equation, the diffusion coefficient can be calculated by simple rearranging of the equation, providing the number of electrons obtained from the CV experiment and values for other variables used to determine the reaction mechanism and kinetic parameters (Greef *et al.*, 1985).

Harvey and Crundwell (1996, 1997) reported the bioleaching experiments utilizing *A. ferrooxidans*, designed to obtain kinetic data utilizing chronoamperometry. Fowler and Crundwell (1998) examined the decay of the current with time to evaluate the oxidation of ferrous ions during chemical and bacterial leaching of the zinc sulfide mineral. The parameters obtained from the batch culture and the redox-controlled batch culture were the same, indicating that the only role of the bacteria under the conditions used is regeneration of ferric ions in solution.

Palencia and co-workers (1991) evaluated the effect of the presence of bacteria on the anodic dissolution of pyrite, using chronoamperometry as the electrochemical technique. Chronoamperometric curves were obtained at three different potentials in the anodic region and the current density decaying was analyzed. The current densities were slightly higher for the sterile controls than for the inoculated when 800 and 600 mV were applied. At 400 mV, the current density reached a quasi-steady state value after 5 hours and slightly decreased with time. In this stage, most of the iron was present as ferrous ions and the current density experienced an initial decay due to the formation of some kind of passive film on the pyrite surface. After this initial decay, a steady state value was reached, indicating that the rate of pyrite dissolution was constant. At 600 mV, when the most of the iron was as ferric ion, sterile and inoculated tests showed a progressive decay of the current density with time due to the formation of a thick passive film. As a consequence, the rate of the pyrite dissolution decreased with time. At 800 mV, there was no current density decay with time, probably due to preferential formation of sulfate compared with the elemental sulfur.

2.3. Electrochemical Impedance Spectroscopy (EIS)

Electrochemical impedance spectroscopy (EIS) has been widely utilized in the last years as a tool, together with surface analysis, to investigate the electrical properties of metal surfaces, semiconductor electrodes, yielding information on the structure of the solid/electrolyte interface as well as on the mechanisms of electrochemical reactions (Bonora *et al.*, 1996; Jüttner, 1990; Gabrielli & Keddam, 1992).

The purpose of the EIS characterization of electrode is to obtain information about the properties of the system such as the presence of defects, reactivity of the interface, adhesion, barrier properties, etc. Knowledge of these parameters is very useful for predicting corrosive behavior (Bonora *et al.*, 1996).

The impedance $Z(\omega)$ of the electrochemical interface is a complex number which can be represented either in polar or in Cartesian coordinates:

$$Z(\omega) = |Z|\, e^{j\phi} \tag{10}$$

$$Z(\omega) = ReZ + jImZ \tag{11}$$

where *ReZ* and *ImZ* are the real and imaginary parts of the impedance. The relationships between these quantities are:

$$|Z|^2 = (ReZ)^2 + (ImZ)^2 \tag{12}$$

$$(13) \qquad \phi = arctan\frac{ImZ}{ReZ}$$

$$(14) \qquad Re(Z) = |Z|cos\phi$$

$$(15) \qquad Im(Z) = |Z|sin\phi$$

As a first approximation it was endeavored to fit the experimental results to a Randles equivalent circuit. Figure 1 shows the Randles circuit and the corresponding impedance plots. The impedance Z of the Randles circuit may be written as:

$$(16) \qquad Z(\omega) = R_{\Omega} + \frac{R_t}{1 + j\omega R_t C_d}$$

where R_{Ω} = electrolyte resistance
R_t = charge transfer resistance
C_{dl} = double-layer capacitance
$j = \sqrt{-1}$
$\omega = 2\pi f$ (f = frequency in Hz)

The usual approach is to curve-fit the semicircle that results in a capacitive response with a single time constant. This approach allows an estimation of the charge transfer resistance from the low frequency intersection of the semicircle response with the real axis (Silverman, 1989). However, this simple model is only applicable to single time constant responses, those modeled by a resistor and capacitor in parallel. Diffusion processes, porous electrodes require more complex models to explain such complex interfaces (Silverman & Carrico, 1988).

A more detailed analysis of the experimental data shows that in many systems appreciable deviations from the ideal semicircle exist, corresponding to the rotation of the centre of the capacitive semicircle by an angle ϕ below the real axis. Deviations of this kind have been attributed to heterogeneities of solid surfaces (Lemaitre *et al.*, 1989), more general to inhomogeneous current distribution at the electrode surface.

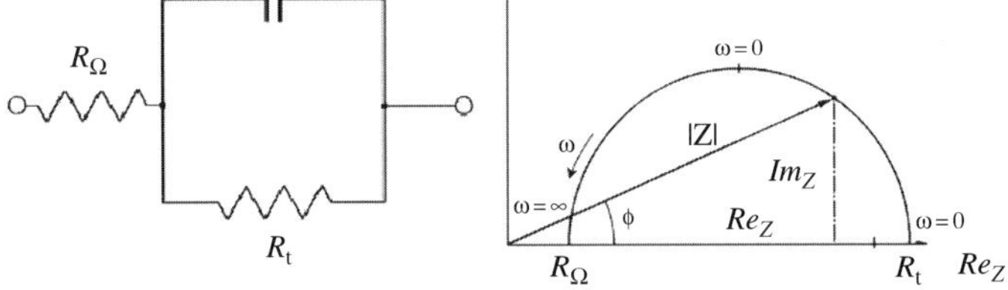

Figure 1. Randles equivalent circuit and corresponding complex plane plot (Appl. Note, 1985)

The new expression for the impedance Z is an adaptation of eq. 16, including the parameter $\alpha(<1)$:

$$(17) \qquad Z(\omega) = R_{\Omega} + \frac{Rt}{1 + (j\omega R_t C_d)^{\alpha}}$$

In this equation a constant phase element (CPE) was inserted in the nonfaradaic branch of the equivalent circuit, in series with the double-layer capacitance (Lemaitre *et al.*, 1989).

Analysis of EIS data by fitting to appropriate electrical equivalent circuits, which are designed based on the physical and chemical properties of the system under investigation, can lead to information concerning changes in the resistive properties of electrode/solution interface. In other words, the study of a physical phenomenon often leads to the elaboration of a model and can predict the behavior of a system under various conditions.

Development of appropriate models for the interpretation of impedance spectra in terms of physical properties requires, in addition to insight into the chemistry and physics of the systems, an understanding of the measurement error structure (Agarwal *et al.*, 1995).

Many of the concepts used in the analysis of electrical circuits can be applied in analyzing electrochemical impedance spectroscopy data. However, the analogy between electrical and electrochemical systems should not be taken too far, even if it is a common practice. Thus, there are several important differences between electrochemical and electrical systems: electrical systems are commonly composed of passive elements, whereas electrochemical interfaces frequently contain recognizable active elements; electrochemical systems are generally non-linear, with the relationship between current and voltage being exponential in nature (Butler-Volmer equation), rather than being linear (Ohm's Law) as for electrical systems (MacDonald *et al.*, 1998).

The models (equivalent electrical circuits) developed to explain the electrochemical impedance must obey two conditions: all the elements of the proposed circuit must have a clear physical meaning. They have to be associated with physical and/or chemical properties of the system that can generate this kind of electrical response; the equivalent electrical circuit must generate spectra which, with suitable values of the elements, are different from the experimental results for a small enough defined quantity only (the error is acceptable if it is not only small but also not periodical or regular as a function of the frequency). The equivalent electrical circuit has to be as simple as possible; in other words, if by eliminating one element the previously described condition is still valid and then the circuit must be simplified (Bonora *et al.*, 1996).

Heterogeneous surfaces can be identified by the application of EIS and many of the observed phenomena can be explained quantitatively, taking into account the specific influence of surface heterogeneities on potential and current distribution, charge transfer, adsorption, mass transport or crystallization steps in electrode processes (Jüttner, 1990; Mansfeld, 1990; Thompson, 1992).

The aim of the electrochemistry is generally to identify the processes occurring at the interface, either by elaborating a model for the interface behaviour or by trying to find the value of some parameters of the system of interest when the reaction mechanism involved is already known. However, in some practical cases, empirical relationships are sufficient.

Cabral and Ignatiadis (2000) studied the oxidative bacterial dissolution of pyrite using EIS. This work led to the elaboration of a model taking account the formation of an adherent compact layer at the pyrite surface, for the high frequency; a charge transfer reaction, for the medium frequency; and a process of material transport (diffusion-convection) at the pyrite/solution, for the low frequency. The layer formed in the region of high frequency appears from the first moments of bioleaching and tends to develop as the bioleaching progress to be constantly present at the end of the batch.

Recently the oxidative dissolution of a massive chalcopyrite electrode by *A. ferrooxidans* was evaluated by EIS associated with atomic force microscopy (AFM) (Bevilaqua *et al.*, 2002, 2004). In these works the differences in the impedance diagrams as a function of the immersion time (Figure 2) were correlated with the adhesion process of bacteria on the mineral surface. The presence of a film was evident in the atomically flat surface of the inoculated sample. Exopolymeric substances (EPS) could be responsible by the formation of this film. The adhesion of bacteria on the mineral surface was detected early of testing and it could be associated with the difference observed in the impedance diagrams during first hours of test. After about 71 h of incubation, the composition of adsorption layer

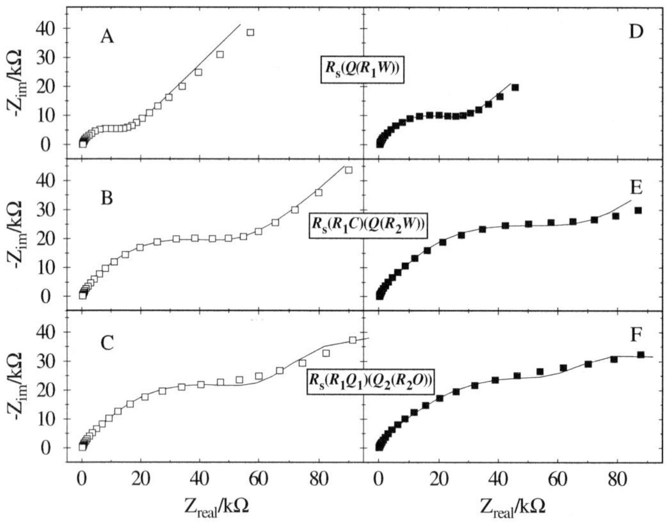

Figure 2. Experimental (symbol) and simulated (solid line) Nyquist plots (A and D) 7 h, (B and E) 71 h and (C and F) 288 h of immersion. Open symbols: impedance diagram without *A. ferrooxidans*-LR; closed symbols: impedance diagram with *A. ferrooxidans*-LR. (Bevilaqua *et al.*, 2004)

included cells, biomolecules and sulfur. Collectively, the layer was formed on the chalcopyrite surface and functions as a capacitor, which makes the diffusion of molecule and ions to and from the electrode surface difficult (Bevilaqua *et al.*, 2004).

EIS was also used to characterize bornite electrodes in the presence and absence of *A. ferrooxidans*. It was concluded that the bacterial action on bornite surface considerably accelerated the mineral dissolution rate due to their ability to maintain an oxidant environment rich in Fe^{3+} ion (Bevilaqua *et al.*, 2003).

2.4. Electrochemical Noise Analysis (ENA)

Electrochemical noise analysis is a relatively new technique, which finds growing application in corrosion monitoring, investigation of general and localized corrosion, microbiologically influenced corrosion (MIC) and evaluation of coatings on metal surfaces (Schauer *et al.*, 1998).

It is considered an alternative to the EIS and to other techniques applied in corrosion studies due to low cost of the equipment, which consists of a zero resistance ammeter (ZRA) and digital voltmeters (Mansfeld *et al.*, 2001). In addition, no external signal needs to be applied for the collection of experimental data, being a non-destructive technique.

ENA experiments require a different cell configuration. Shorting together two identical as possible working electrodes electrochemical noise measurements are performed (CorrWare for Windows, 1995).

In a noise experiment the potential and current are measured at a fixed rate for a chosen time interval. The mean potential E_{coup} is the potential of the coupled electrodes versus a stable reference electrode, and I_{coup} is the mean coupling current flowing between the two electrodes, measured by a ZRA (Lee & Mansfeld, 1998; Mansfeld *et al.*, 1997). Often, noise data are measured for days or weeks in an attempt to identify a change in the noise that corresponds to a change on the electrodes/solution interfaces, such as the onset of accelerated corrosion. If the data are continuously measured, the number of data points may soon become overwhelming. The noise data can be analyzed both in the time or frequency domain.

Prior to conduct any analysis of the electrochemical noise experiment trend must be removed from the raw data (Bertocci *et al.*, 2002).

In the time domain, the standard deviations of potential $\sigma[V(t)]$ and current $\sigma[I(t)]$ noise fluctuations can be followed as a function of time to determine the initiation of the corrosion process. In general, it has been observed that $\sigma[V(t)]$ decreases and $\sigma[I(t)]$ increases as the material is degraded. They are also used to compute R_n, the noise resistance, defined by Lee and Mansfeld (1998) and Mansfeld (1997) as:

$$R_n = \frac{\sigma[V(t)]}{\sigma[I(t)]} \tag{18}$$

A simplest method for data analysis is calculating the Room Mean Square (RMS) of the potential and current signals separately. The RMS value is effectively the average amplitude of the noise. First subtracting any linear background drift to the signal and then using the equation to calculate *RMS* values:

$$(19) \qquad RMS = \sqrt{\frac{\sum_{i=1}^{n} X_i^2}{n}}$$

Power spectral density (PSD) is one of the frequency domain representations of a time domain signal and PSD-plots can be obtained from the time domain signal by applying mathematical procedures such as fast Fourier transform (FFT) or maximum entropy method (MEM) (Bertocci *et al.*, 1998). The advantages of MEM over FFT are smoother plots and an easier determination of characteristic noise parameters such as spectral noise resistance, R_{sn} and slope S_{Rsn}.

Fast Fourier transforms of experimental electrochemical noise data converts the data collected in the time domain to the frequency domain, allowing observation of the frequency dependence of the noise phenomena under investigation (Mansfeld *et al.*, 1997, 1998; Mansfeld & Xiao, 1993). The number of points determines the limit of low frequencies and the sampling rate determines the limit of high frequencies of the resultant spectra. Spectral noise resistance, R^{o}_{sn}, is obtained from the spectral noise plots and can be determined as dc limit of these plots, defined as:

$$(20) \qquad R^{o}_{sn} = \lim_{f \to 0} R_{sn}(f)$$

When this dc limit is not observed, R^{o}_{sn} can be determined as the average of the last 10 data points of the PSD plot (Lee & Mansfeld, 1998).

The main question in the analysis of ENA is the relation between characteristic noise parameters such as the noise resistance R_n and $R_{sn}(f)$, defined as the spectral noise response or spectral noise impedance, and the impedance $|Z|_{f \to 0}$ or the polarization resistance R_p, which can be obtained from EIS measurements at a dc limit.

Potential and current PSD plots have the general frequency dependence:

$$(21) \qquad \log V_{PSD} = A_V + S_V \log f$$

$$(22) \qquad \log I_{PSD} = A_I + S_I \log f$$

where S_V and S_I are the slopes of the potential and current spectral noise plots, respectively; while for the spectral noise resistance plot the following equation is valid:

$$(23) \qquad \log R_{sn} = A_{Rsn} + S_{Rsn} \log f$$

Since

$$(24) \qquad R_{sn} = \left[\frac{V_{PSD}(f)}{I_{PSD}(f)}\right]^{0.5}$$

it follows that:

$$(25) \qquad S_{Rsn} = 0.5(S_V - S_I) = S_Z$$

where S_{Rsn} and S_Z are the slopes of the spectral noise and Bode plots (Lee & Mansfeld, 1998; Mansfeld *et al.*, 1997).

More recently ENA has been used in association with EIS for monitoring continuously the bacterial and chemical dissolution of bornite and chalcopyrite (Acciari *et al.*, 2005; Bevilaqua *et al.*, 2005). These data can provide information about physical, chemical and biochemical process occurring in the electrode/electrolyte (sulfide/culture medium) interface in the presence and absence of *A. ferrooxidans*. In addition, ENA showed to be adequate for evaluating the oxidative dissolution of bornite in the absence and presence of *A. ferrooxidans* and was sensitive enough to confirm that chalcopyrite is refractory to both chemical and bacterial dissolution. These results confirm the advantages of ENA as a simpler and non-destructive technique. The potential and current variations due to spontaneous oxidation were correlated with the particular corrosion processes, especially in the presence of bacteria for both mineral sulfides.

The typical noise plots are shown in Figure 3 and the corresponding PSD spectra are shown in Figure 4.

From analysis in the frequency domain it is possible to calculate the values of the PSD slopes S_V, S_I and S_{Rsn}. In general, there is a good agreement between PSD slopes with equation (2.4.8), as can be seen according to slopes of the plots

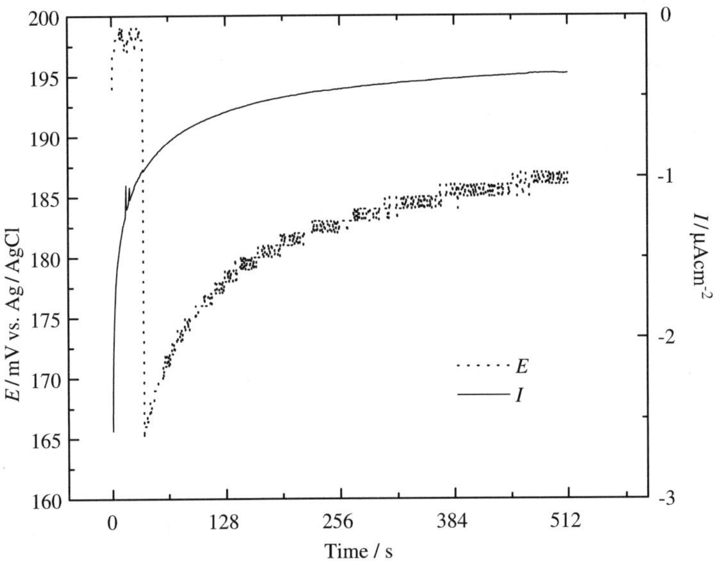

Figure 3. Potential and current noise recorded for bornite electrodes after 5 h immersion in control medium without bacteria (Bevilaqua *et al.*, 2006)

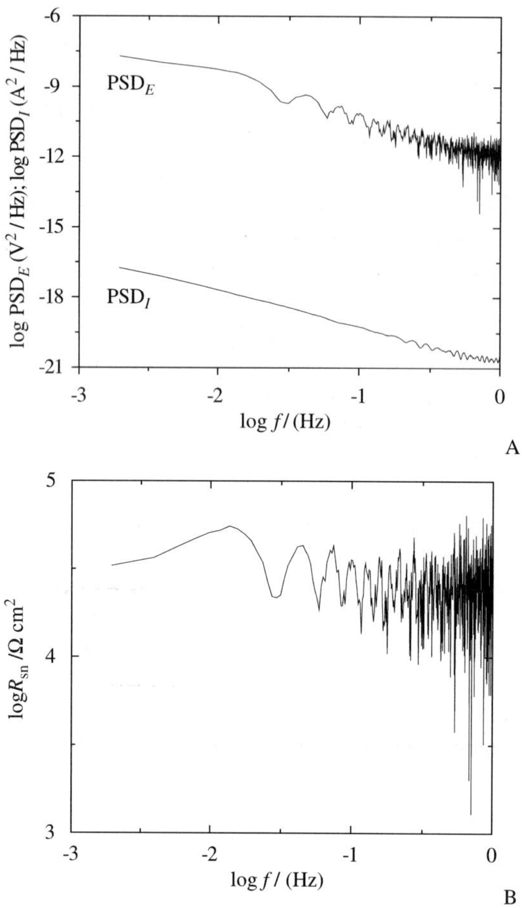

Figure 4. Potential and current PSD plots (a) and corresponding spectral noise plot (R_{sn}) (b) for bornite electrodes in control medium after 5 h of immersion (Bevilaqua *et al.*, 2006)

of Figures 4a and 4b. In addition, R^{o}_{sn} can be calculated from R_{sn} plot (determined as the average of the last 10 points).

3. CONCLUDING REMARKS

Bacterial leaching is a dissolution process of metal sulfides in which the oxidative activity of some bacterial species, mainly *A. ferrooxidans* play a central role. Since the finding of the microorganisms participation in the "natural" leaching of metals from low grade ores (1950s) several industrial applications has been set up and a huge number of works about this emerging (bio)technology has been published. The great majority of these studies deal essentially with microbiological/biochemical and technological aspects of bacterial leaching. In the last twenty years molecular biology of these microorganisms has made a great contribution to the knowledge of biological and biochemical aspects of the process. However the mechanisms and the reactions that come into play during the oxidative dissolution of metal sulfides by bacterium are controversial and still poorly understood.

As the oxidation of sulfides by a chemical or biological oxidant can be look upon as an electrochemical reaction it can, consequently, be studied by electrochemical techniques, which provide information about reaction mechanisms. In some early studies CV has been utilized to investigate qualitative aspects of electrochemical behavior of mineral sulfides in the presence or absence of bacteria, arising interesting results. However, there are some apparent disadvantages inherent to this technique: besides the fact that slow heterogeneous electron transfer and fast chemical reactions are difficult to be distinguished, it is a destructive technique.

EIS and ENA has been utilized in the last years, together with surface analysis techniques such as SEM and AFM, to investigate the electrical and electrochemical properties of metal surfaces, semiconductor electrodes, yielding information on the structure of the solid/electrolyte interface as well as on the mechanisms of electrochemical reactions. The purpose of the EIS electrode characterization is to obtain information about the properties of the system such as the presence of defects, reactivity of the interface, adhesion, barrier properties, etc. Knowledge of these parameters is very useful for predicting corrosive behavior of sulfides under bioleaching process. Similar information can be obtained using ENA with the advantage that no external electrical perturbation is applied.

It is possible to model impedance spectra during sulfide electrode oxidation in acid solution proposing a base equivalent electrical circuit that describes the response of a charge transfer process with diffusion into a biofilm. Derivations of this model can be also used for fitting the sulfide dissolution in the presence of microorganisms involved in the bioleaching. ENA data complement the impedance response analysis and are enough to investigate the behavior of sulfides to both chemical and bacterial dissolution.

The comparison between the impedance and noise spectra in some systems shows a good agreement among the results obtained by different methods. The combination of EIS and ENA can be a suitable approach for monitoring the corrosion behavior of sulfides in the absence and presence of leaching bacteria.

4. ACKNOWLEDGMENTS

The authors are very grateful to Vale do Rio Doce Company (CVRD), Brazil, which is funding a project to our Biohydrometallurgy group. Acknowledgements are also due to FAPESP for pos-doctoral fellowship to HAA and CNPq for researching fellowships to DB, OGJ and AVB.

5. REFERENCES

Acciari HA, Bevilaqua D, Benedetti AV, Fugivara CS, Tremiliosi-Filho G, Garcia Jr O. 2005. Electrochemical noise analysis of bioleaching of chalcopyrite ($CuFeS_2$) by *Acidithiobacillus ferrooxidans*. In: Proceedings of The European Corrosion Congress, EUROCORR 2005, Lisbon, Portugal, 1-6.

Agarwal P, Crisalle OD, Orazem ME, Garcia-Rubio LH. 1995. Application of measurement models to impedance spectroscopy. J Electrochem Soc 142: 4149-4158.

Application Note AC-1 1985. Basics of AC impedance measurements. EG&G Princepton Applied Research.

Beech IB, Smith JR, Steele AA, Penegar I, Campbell SA. 2002. The use of atomic force microscopy for studying interactions of bacterial biofilms with surfaces. Colloids and Surfaces B: Biointerfaces 23: 231-247.

Bevilaqua D, Acciari HA, Benedetti AV, Fugivara CS, Tremiliosi GF, Garcia Jr O. 2006. Electrochemical noise analysis of bioleaching of bornite (Cu_5FeS_4) by *Acidithiobacillus ferrooxidans*. Hydrometallurgy 83: 50–54.

Bevilaqua D, Diéz-Perez I, Fugivara CS, Sanz F, Benedetti AV, Garcia Jr O. 2004. Oxidative dissolution of chalcopyrite by *Acidithiobacillus ferrooxidans* analyzed by electrochemical impedance spectroscopy and atomic force microscopy. Bioelectrochemistry 64: 79-84.

Bevilaqua D, Diéz-Perez I, Fugivara CS, Sanz F, Garcia Jr O, Benedetti AV. 2003. Characterization of bornite (Cu_5FeS_4) electrodes in the presence of the bacterium *Acidithiobacillus ferrooxidans*. J Braz Chem Soc 14: 637-644

Bevilaqua D, Díez-Pérez I, Fugivara CS, Sanz F, Benedetti AV, Garcia Jr O. 2002. Analysis of chalcopyrite (CuFeS2) electrodes in the presence of *Acidithiobacillus ferrooxidans* by electrochemical impedance spectroscopy (EIS) and atomic force microscopy (AFM) In: Proceedings of 15th International Corrosion Congress, Granada – Spain, September, 22–27, Frontiers in Corrosion Science and Technology.

Biegler T, Horne MD. 1985. The electrochemistry of surface oxidation of chalcopyrite. J Electrochem Soc 132: 1363-1369.

Bertocci U, Huet F, Nogueira RP, Rousseau P. 2002. Drift removal procedures in the analysis of electrochemical noise. Corrosion 58: 337-347.

Bertocci U, Frydman J, Gabrielli, Huet F, Keddam M. 1998. Analysis of electrochemical noise by power spectral density applied to corrosion studies – maximum entropy or fast Fourier transform? J Electrochem Soc 145: 2780-2786.

Bonora PL, Deflorian F, Fedrizzi L. 1996. Electrochemical impedance spectroscopy as a tool for investigating underpaint corrosion. Electrochem Acta 41: 1073-1082.

Cabral T, Ignatiadis I. 2001. Mechanistic study of the pyrite-solution interface during the oxidative bacterial dissolution of pyrite (FeS_2) by using electrochemical techniques. Int J Miner Process 62: 41-64.

Choi WK, Torma AE, Ohline RW, Ghali E. 1993. Electrochemical aspects of zinc sulfide leaching by Thiobacillus ferrooxidans. Hydrometallurgy 33: 137-152.

CorrWare for Windows: electrochemistry/corrosion software. 1995. version 1.2. Operating manual. Scribner Associates, Charlottesville.

Elsherief AE. 2002. The influence of cathodic reduction, Fe^{2+} and Cu^{2+} ions on the electrochemical dissolution of chalcopyrite in acidic solution. Miner Eng 15: 215-223.

Evans DH, O'Connell MO, Petersen RA, Kelly MJ. 1983. Cyclic voltammetry. J Chem Edu 60: 290-292.

Fowler TA, Crundwell FK. 1998. Leaching of zinc sulfide by Thiobacillus ferrooxidans: experiments with a controlled redox potential indicate no direct bacterial mechanism. Appl Environ Microbiol 64: 3570-3575.

Gabrielli G, Keddam M. 1992. Review of applications of impedance and noise analysis to uniform and localized corrosion. Corrosion 48: 794-811.

Gómez C, Figueroa M, Muñoz J, Ballester A, Blázquez ML. 1997. A study of bioleached chalcopyrite surfaces in the presence of Ag(I) by voltammetric methods. Miner Eng. 10: 111-116.

Gómez C, Figueroa M, Muñoz J, Blázquez ML, Ballester A. 1996. Electrochemistry of chalcopyrite. Hydrometallurgy 43: 331-344.

Gosser-Jr DK. 1994. Cyclic voltammetry: simulation and analysis of reaction mechanisms. VCH Publishers, New York.

Greef R, Peat R, Peter LM, Pletcher D, Robinson J. 1985. Instrumental Methods in Electrochemistry. Ellis Horwood, Chichester.

Harvey PI, Crundwell FK. 1997. Growth of *Thiobacillus ferrooxidans*: a novel experimental design for batch growth and bacterial leaching studies. Appl Environ Microbiol 63: 2586-2592.

Harvey PI, Crundwell FK. 1996. The effect of As(III) on the growth of *Thiobacillus ferrooxidans* in an electrolytic cell under controlled redox potentials. Miner Eng 9: 1059-1068.

Holliday RI, Richmond WR 1990. An electrochemical study of the oxidation of chalcopyrite in acidic solution. J Electroanal Chem 288: 83-98.

Jüttner K. 1990. Electrochemical impedance spectroscopy (EIS) of corrosion processes on inhomogeneous surfaces. Electrochimica Acta 35: 1501-1508.

Kelly DP, Wood AP. 2000. Reclassification of some species of *Thiobacillus* the newly designated genera *Acidithiobacillus* gen. nov. *Halothiobacillus* gen. nov. and *Thermithiobacillus* gen. nov. Int J Syst Evol Micr 50: 511-515.

Keresztes Z, Felhosi I, Kálman E. 2001. Role of redox properties of biofilms in corrosion processes. Electrochimica Acta 46: 3841-3849.

Kudaikulova GA, Radyushkina KA, Tarasevich MR. 1989. Electrochemistry of sulfide minerals. Redox changes. Élektrokhimiya 25: 627-631.

Lázaro I, Medina NM, Rodríguez E, Arce I, González I. 1995. The use of carbon paste electrodes with non-conducting binder for the study of minerals: chalcopyrite. Hydrometallurgy 38: 277-287.

Lee CC, Mansfeld F. 1998. Analysis of electrochemical noise data for a passive system in the frequency domain. Corrosion Sci 40: 959-962.

Lemaitre L, Moors M, Van Peteghem AP. 1989. A model for the corrosion behaviour of dental amalgams. J Biomed Mater Res 23: 241-252.

Lu ZY, Jeffrey MI, Lawson F. 2000. An electrochemical study of the effect of chloride ions on the dissolution of chalcopyrite in acidic solutions. Hydrometallurgy 56: 145-155.

MacDonald D, Sikora E, Engelhardt G. 1998. Characterizing electrochemical systems in the frequency domain. Electrochim Acta 43: 87-107.

Mansfeld F. 1990. Electrochemical impedance spectroscopy (EIS) as a new tool for investigating methods of corrosion protection. Electrochim Acta 35: 1533-1544.

Mansfeld F, Han LT, Lee CC, Chen C, Zhang G, Xiao H. 1997. Analysis of electrochemical impedance and noise data for polymer coated metals. Corrosion Sci 39: 255-279.

Mansfeld F, Han LT, Lee CC, Zhang G. 1998. Evaluation of corrosion protection by polymer coatings using electrochemical impedance spectroscopy and noise analysis. Electrochim Acta 43: 2933-2945.

Mansfeld F, Sun Z, Hsu CH. 2001. Electrochemical noise analysis (ENA) for active systems in chloride media. Electrochim Acta 46: 3651-3664.

Mansfeld F, Xiao H. 1993. Electrochemical noise analysis of iron exposed to NaCl solutions of different corrosivity. J Electrochem Soc 140: 2205-2209.

Palencia I, Wan RY, Miller JD. 1991. The electrochemical behavior of a semiconducting natural pyrite in the presence of bacteria. Metallurgical Transactions 17: 765-774.

Pesic B, Kim I. 1990. Electrochemistry of *T. ferrooxidans* interactions with pyrite. In: EPD Congress'90: The minerals, Metals & Materials Society, 133-160.

Rawlings DE. 2002. Heavy metal mining using microbes. Annu Rev Microbiol 56: 65-91.

Rodríguez Y, Ballester A, Blásquez F, González F, Muñoz JA. 2003. New information on the pyrite bioleaching mechanism at low and high temperature. Hydrometallurgy 71: 37-46.

Schauer T, Greisiger H, Dulog L. 1998. Details on MEM analysis of electrochemical noise data and correlation with impedance measurements for organic coatings on metals. Electrochim Acta 43: 2423-2433.

Silverman DC, Carrico JE. 1988. Electrochemical impedance technique – a practical tool for corrosion prediction. Corrosion 44: 280-287.

Silverman DC. 1989. Corrosion rate estimation from pseudo-inductive electrochemical impedance response. Corrosion 45: 824-830.

Skoog DA, Leary JJ. 1992. Principles of instrumental analysis. Saunders College Publishing, USA.

Teixeira CM, Ramires I, Garcia Jr O, Guastaldi AC. 2002. Aplicação de técnicas eletroquímicas no estudo da dissolução oxidativa da covelita (CuS) Por *Thiobacillus ferrooxidans*. Química Nova 25: 20-26 (in portuguese).

Third KA, Cord-Ruwisch R, Watling HR. 2000. The role of iron-oxidizing bacteria in stimulation or inhibition of chalcopyrite bioleaching. Hydrometallurgy 55: 225-233.

Thompson NG, Syrett BC. 1992. Relationship between conventional pitting and protection potentials and a new, unique pitting potential. Corrosion 48: 649-659.

Toniazzo V, Lázaro I, Humbert B, Mustin C. 1999. Bioleaching of pyrite by Thiobacillus ferrooxidans fixed grains electrode to study superficial oxidized compounds. C R Acad Sci Paris/ Earth & Planetary Sciences 328: 535-540.

Tributsch H. 2001. Direct versus indirect bioleaching, Hydrometallurgy. 59: 177-185.

Velásquez P, Gómez H, Leinen D, Ramos-Barrado JR. 1998. Electrochemical impedance spectroscopy analysis of chalcopyrite $CuFeS_2$ electrodes. Colloids and Surfaces – A: Physicochemical and Engineering Aspects 140: 177-182.

Velásquez P, Leinen D, Pascual J, Ramos-Barrado JR, Cordova R, Gómez H, Schrebler R. 2000. SEM, EDX and EIS study of an electrochemically modified electrode surface of natural enargite (Cu_3AsS_4). J Electroanalytical Chem 494: 87-95.

Velásquez P, Leinen D, Pascual J, Ramos-Barrado JR, Cordova R, Gómez H, Schrebler R. 2001. XPS, SEM, EDX and EIS study of an electrochemically modified electrode surface of natural chalcocite (Cu_2S). J Electroanalytical Chem 510: 20-28.

Yin Q, Vaughan DJ, England KER, Kelsall GH, Brandon NP. 2000. Surface oxidation of chalcopyrite ($CuFeS_2$) in alkaline solutions. J Electrochem Soc 147: 2945-2951.

CHAPTER 4

CATALYTIC ROLE OF SILVER AND OTHER IONS ON THE MECHANISM OF CHEMICAL AND BIOLOGICAL LEACHING

ANTONIO BALLESTER, MARÍA LUISA BLÁZQUEZ, FELISA GONZÁLEZ AND JESÚS A. MUÑOZ

Departamento de Ciencia de Materiales e Ingeniería Metalúrgica, Facultad de Ciencias Químicas, Universidad Complutense de Madrid, 28040 Madrid, Spain

1. INTRODUCTION

Leaching, defined as the chemical attack of minerals for the recovery of metals, is the first step of any hydrometallurgical process. From a basic chemical point of view, this type of reactions can be: acid/alkaline, complexing or redox. However, most of the commercial processes are redox, 90% of which are oxidizing using reagents such as: oxygen, ferric iron, chlorine, etc. In these cases, the mechanism involved is electronic since for that the reaction takes place it is necessary an electronic transfer through the solid mineral and between this solid and the leaching reactant.

Taking into account bioleaching, it seems clear that the indirect mechanism is getting more support over time. According to this mechanism, the mineral leaches through chemical species in solution with an oxidizing character. Therefore, it is important for any biohydrometallurgical researcher to know, previously to the possibilities that bioleaching offers, the mechanism involved in this type of systems in which the electronic transfer is fundamental.

2. ELECTROCHEMICAL MECHANISM

During leaching, a specific reactant is chosen to selectively dissolve a certain mineral from several ones included in the ore, leaving a solid residue which is separated by sedimentation and filtration. As previously mentioned, the dissolution process can be chemical or electrochemical, playing an important role the crystalline nature of the solid, its division state, its structural defects and many other factors.

E.R. Donati and W. Sand (eds.), Microbial Processing of Metal Sulfides, 77–101.

In the leaching of several minerals, such as sulfides, the electrochemical mechanism is the most important to be considered. Habashi, 1983 has mentioned the following characteristics for this mechanism:

1) The mineral is conductor or semiconductor.
2) There is a transfer of electrons from the mineral solid phase to the reactant species. Thus, the process takes place through a redox couple in which the ions or molecules in dissolution are transported towards the solid surface where interact with electrons.
3) The minerals dissolve in certain localized sites, whereas the electronic transfer to some species present in solution takes place in other different separated sites.
4) Oxidation and reduction reactions take place simultaneously, but each one has its own characteristic with respect to the control kinetic mechanism (chemical or mass transport).
5) The dissolution rate is a complex function of reactants concentration.
6) The kinetics of these reactions is affected by the own mineral crystal lattice and by the presence of defects in it. In this sense, several aspects are important: impurities in solid solution (that modify mineral conductivity), the presence of different mineral phases (that provoke the formation of galvanic couples) and the presence of different ions in solution (that modify the electrochemical behavior when fixed onto the mineral).

Sulfides are, in most cases, semiconductor solids and can be leached in the presence of an oxidant like oxygen dissolved in water. The reaction can be schematically represented by Figure 1 in which several facts could be remarked:

1) Oxygen is adsorbed over some zones of the solid sulfide surface where takes place the electronic transfer according to the following cathodic reaction:

$$(1) \qquad 1/2O_2 + H_2O + 2e^- \rightarrow 2OH^-$$

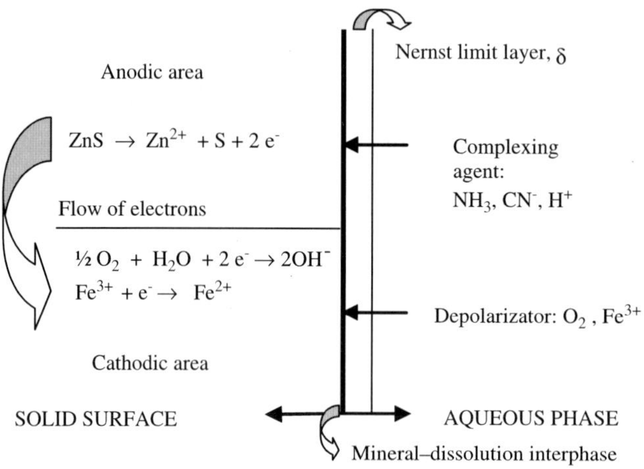

Figure 1. Scheme of the electrochemical mechanism of ZnS dissolution

2) Simultaneously, the solid is anodically oxidized in different zones, through the following semi-reaction:

(2) $MS \rightarrow M^{2+} + S^{o} + 2e^{-}$

Thus, the sulfide sulfur converts into elemental sulfur, though its oxidation can be complete until the sulfate form:

(3) $S^{o} + 3/2O_2 + H_2O \rightarrow 2H^{+} + SO_4^{2-}$

Therefore, the overall reaction will be:

(4) $MS + 2O_2 \rightarrow MSO_4$

3. KINETICS OF ELECTROCHEMICAL REACTIONS

An important fact to be considered in these processes is that kinetics depends on the electronic transfer between anodic and cathodic sites placed on the mineral. Thus, any action capable of modifying its conductivity can positively affect the kinetics of the process. In this case, the presence of impurities in solid solution, in metal sulfides, can modify appreciably its conductivity.

On the other hand, each sulfide has a fixed rest potential in a given medium. Then, when two or more mineral sulfides are in electrical contact, the one with the highest potential tends to act cathodically enhancing the anodic dissolution of the other sulfide.

Additionally, many oxidizing reactions of dissolution of sulfides can be considerably accelerated by incorporating to the leaching solution adequate catalyzing ions, as it was suggested many years ago by Björling (1954).

A last factor that can also affect the kinetics of the electrochemical leaching reactions is the presence of bacteria in the medium, both iron-oxidizing and those involved in the sulfur cycle. The former ones are able to maintain a high oxidizing level in the medium, due to its ability to oxidize ferrous to ferric iron, and the latter ones are capable to oxidize different reduced forms of sulfur to sulfate, with the subsequent recycling of the acid medium.

Following, these four mentioned factors will be considered in more detail.

3.1. Influence of Mineral Impurities

The most studied and known case is the acid and oxidizing leaching of the zinc sulfide, which can be used as an example. Its oxidation at high temperatures and pressures takes place according to the following reactions:

(5) $ZnS + 1/2O_2 + H_2SO_4 \rightarrow ZnSO_4 + S^{o} + H_2O$

(6) $ZnS + 2O_2 \rightarrow ZnSO_4$

Both reactions have a very slow kinetics when using chemically pure ZnS. An Australian patent (Sherrit Gordon Mines Ltd, 1963) remarks the catalytic role played by iron in this reaction. Fortunately, iron is present in almost every mineral concentrate either in the form of sulfide or by substituting zinc in the crystal lattice of sphalerite. This fact improves enormously the leaching of this sulfide and has contributed greatly to the development of commercial direct leaching processes for sphalerite and to the performance of many investigations on this topic. Among them, it is worthy to mention the works by Frenay (1984) who studied the influence of the content of iron in the leaching of sphalerite and found that the best zinc dissolution results corresponded to a mineral with a higher content of iron.

These results were explained as due to an improvement of the sphalerite conductivity by the appearance of this impurity in the composition of the raw material.

Vaughan (1984) gave a very accurate description of the electronic structure of ZnS and drew a diagram of energy levels for the molecular orbitals of this compound, describing the bond as a group of simple tetrahedric units of ZnS_4. A strong overlapping between 4s and 4p atomic orbitals of zinc and 3s and 3p atomic orbitals of sulfur leads to a set of bonding molecular orbitals, σ, full and stabilized with respect to contributing atomic orbitals, and to a set of antibonding molecular orbitals, σ^*, empty and unstabilized. The overlapping between molecular orbitals of close tetrahedric ZnS_4 units, in a crystal of sphalerite, produces a broadening of the bands energy levels: a valence band, σ, completely full and another of conduction, σ^*, empty.

All this leads to a simple and qualitative energy band diagram as that shown in Figure 2, corresponding to an electrical insulator material with an energy gap between bands of 3.65 eV, which also contains uncoupled electrons and, therefore, it is diamagnetic.

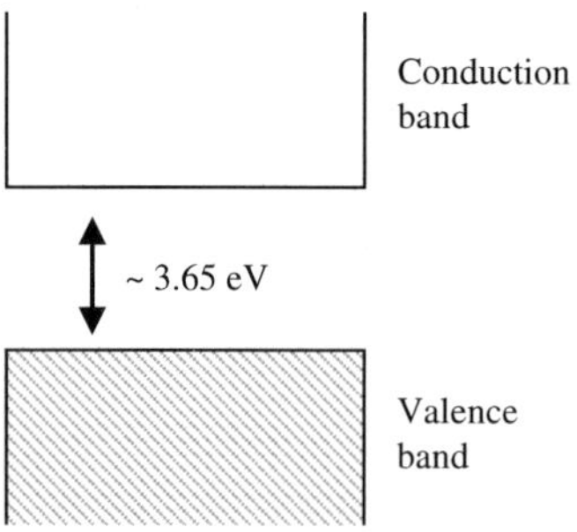

Figure 2. Energy gap between valence and conduction bands for pure ZnS

Figure 3 shows another type of diagram, called of energy levels, which has been calculated using mathematical functions, more or less complex, for a tetrahedrical lattice of one Zn^{2+} ion coordinated by four S^{2-} ($ZnS_4{}^{6-}$) ions. On the left hand of that figure there is a set of energy levels of molecular orbitals for $ZnS_4{}^{6-}$. The calculations also provide information on compositions of particular molecular orbitals in terms of both metal and sulfur atomic orbitals.

Thus, $4t_2$, lt and 2e molecular orbitals are related to 3p orbitals of sulfur and they are antibonding, while $3t_2$ and $2a_1$ orbitals are the main metal-sulfur bonding orbitals. 1e and $2t_2$ orbitals are known as metal orbitals type 3d.

When a great number of tetrahedrical $ZnS_4{}^{6-}$ units are grouped in a crystal of sphalerite, the overlapping between orbitals leads to the formation of bands shown in the simplified diagram of figure 2. The band with the highest energy corresponds to the metal-sulfur antibond, which is empty. Below that band there is a gap of energy until the full band of non-bonding of sulfur. Besides this last band, there is another one, even more stabilised, which corresponds to the main metal-sulfur bond, although the 3d levels of zinc shown in this model are localised on the cation without forming a band.

In natural sphalerite samples it is very unusual to find pure ZnS and often zinc is substituted in tetrahedral sites by Fe^{2+} ions. The effect of this ion on the electronic structure can be represented using the data related to the $FeS_4{}^{6-}$ association in which the 3d levels of crystalline field are not totally full. The 3d electrons of Fe^{2+} give rise to a division of the energetic levels of the valence orbitals in levels with spins of opposite sign.

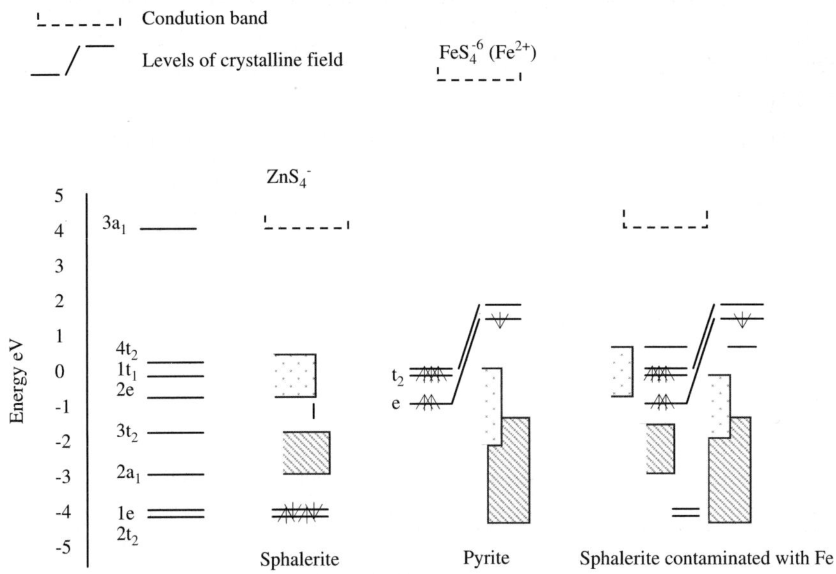

Figure 3. Model of bands for sphalerite, pyrite and sphalerite contaminated with Fe

The bands model for sphalerite with iron (figure 3) shows significant changes in the energies relative to the sulfur antibonding band, the metal-sulfur bonded band and particularly the 3d iron levels, when compared to those of pure ZnS.

For ZnS, the general trend is a decrease of its hardness and an increase of its lattice parameters when iron appears as impurity. This is related to the inherent unbalance due to the addition of t_2 electrons with downwards spin of Fe^{2+} and with the total unbalance of iron-sulfur bonding orbitals with respect to those of zinc-sulfur. This is also related to a decrease in the thermodynamic stability when iron is present. Finally, the energy gap between bands is smaller than 3,65 eV and therefore the conductivity in the ZnS impurified with Fe is better.

These models can be used to interpret and predict not only physical properties, but also the mineralogical behavior during the leaching of solids. For instance, reaction (5), which takes place through an electrochemical mechanism, is an indication that ZnS dissolves anodically according to:

(7) $$ZnS \rightarrow Zn^{2+} + S^{o} + 2e^{-}$$

UV radiation increases the dissolution rate appreciably. This is related to the model of the electronic structure, since UV rays favor the jump of electrons from the valence band to the conduction band, resulting in a higher conductivity and, in turn, in a faster leaching rate.

Something similar occurs when sphalerite is impurified with iron. According to Figure 3, the 3d electrons of iron produce a situation in the valence band, with non-occupied orbitals, that enhances conductivity and the attack rate as well. Additionally, the energy gap between valence and conduction bands decreases from 3.65 eV in pure ZnS to 2.3 eV in sphalerite impurified with iron, and this improves conductivity in the impure compound.

3.2. Influence of Galvanic Couples between Different Mineral Phases

The rest potential of a given mineral phase is the potential assumed by an electrode of such mineral in contact with an aqueous solution against a reference electrode, without having current flow through the circuit. In consequence, each mineral in contact with a specific aqueous medium has a rest potential value. These values have been calculated by Majima (1969) for several metal sulfides (Table 1).

Therefore, when two metal sulfides are in electrical contact in a conductor medium, that one with the highest rest potential acts cathodically, while the other with the lowest one will act anodically. The final result is that the second sulfide will dissolve faster than when it is alone. In this way, it has been demonstrated that pyrite enhances oxidation of minerals such as: galena, sphalerite or chalcopyrite, due to galvanic interactions in agreement with the mechanism shown in Figure 4.

If the mineral is a good electrical conductor, there are different electrochemical processes that can happen on the surface when it is in contact with an aqueous solution. Among the possible anodic processes, the oxidation of ions or molecules

Table 1. Rest potentials for several mineral sulfides at pH = 4, against the hydrogen electrode

Mineral	Potential(V)
Pyrite	0.66
Marcasite	0.63
Chalcopyrite	0.56
Sphalerite	0.46
Covellite	0.45
Bornite	0.42
Galena	0.40
Argentite	0.28
Stibine	0.12
Molybdenite	0.11

present in solution or the oxidation of the own mineral can be mentioned. In hydrometallurgical processes, the oxidation of the mineral generally predominates. Among the many cathodic processes that can occur are: the reduction of metal ions nobler than the constituent mineral metal and the reduction of dissolved oxygen. Being the last possibility the most common.

Thus, the difference between the rest potentials of both minerals is responsible for the formation of the galvanic couple and the driving force of which is a true galvanic corrosion. Among the many factors that limit the value of the galvanic current, the most important ones are:

1) The electrical resistance of both the solution and the minerals.
2) The type of contact between minerals.
3) The concentration overpotentials on both electrodes, which depend on the transport of reactants to and from the electrodes.
4) The activation overpotentials of processes of electrode on its surface.

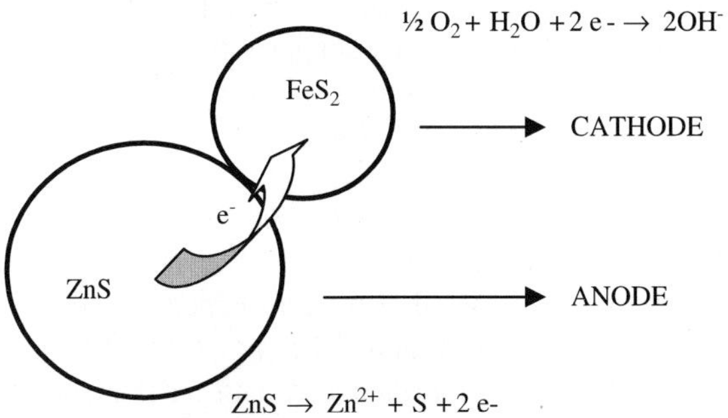

Figure 4. Mechanism of galvanic interactions between pyrite and sphalerite

When the concept "overpotential" is used, the reader must understand an additional potential, with regard to the theoretical one, which is necessary in order that a certain electrochemical reaction takes place. It is due, firstly, to transport phenomena close to the electrode surface provoke by the diffusion of either the reactants or the reaction products. On the other hand, overpotential is also related to the necessary activation energy in order that the electrode reaction occurs with a concrete rate

In conclusion, galvanic effects, that take place in aqueous suspensions of minerals able to conduct the electrical current, play an important role in hydrometallurgical processes of these same minerals. Also they explain why minerals of different locations behave in a so different way, though its chemical composition is similar, since its mineralogical composition can greatly affect the leaching process.

At the beginning of the 70's, there were advances in this field, though not many, with theoretical and practical contributions supported by experimental researches. In this sense, there has been pioneer works with metal sulfides, especially chalcopyrite, by Dutrizac and MacDonald (1973), Wadsworth (1976), Peters (1977), and Linge (1976/77); or those by Wan *et al.* (1984), studying galvanic interactions not in the presence of other sulfides, but with charcoal. In the last years, the interest has continued with works by Nicol and Lazaro (2002) on pyrite, arsenopyrite and chalcopyrite; Abraitis *et al.* (2004), on complex systems containing pyrite, chalcopyrite, galena and sphalerite; Aghamiriam and Yen (2005) studying galvanic interactions between different mineral sulfides and noble metals like gold. Finally, it should be mentioned the works by Cruz *et al.* (2005), who studied the galvanic effect between different mineralogical phases by voltammetric techniques in different mineral concentrates containing pyrite, sphalerite, galena and achantite.

Even the existence of these galvanic effects has been applied by several investigators in order to justify the different floatability of mineral sulfides. In this sense, Ekmekçi and Demirel (1997), justified that chalcopyrite is depressed in the presence of pyrite due to the anodic oxidation of the first one (active mineral) in the presence of the second (noble mineral). In a more recent work, Woods (2003) has indicated that the recovery indexes by flotation depend on the value of the potential in the mineral-solution interphase. This suggests the necessity of monitorization of the potential value in flotation plants in order to control the redox couplings between different mineral phases present, and also justifies the depression of anodic sulfides by surface oxidation processes.

Similarly, Chander (2003) reviewed different electrodes that can be used to measure potential in order to determine the conditions of this variable and to establish an adequate response of the mineral towards flotation, which, in turn, is conditioned by its surface state. Peng *et al.* (2003), and Güler *et al.* (2005) have deepened in the relationship between the mineral surface state and its floatability using surface analysis techniques (XPS, DRIFT) or electrochemical ones like cyclic voltammetry.

3.3. Influence of Catalytic Ions on the Dissolution

Mulak (1987) collected a brief but correct summary related to the addition of soluble salts of specific ions to a certain leaching medium. The results indicated that these ions catalyze, sometimes in several orders of magnitude, the dissolution of the solid. But the pioneer works in this field were performed by Scott and Dyson (1968) by investigating the effect of several metals on the pressure leaching of zinc sulfide. The effect of these additions increased following this order: Cu > Bi > Ru > Mo > Fe. From then, several works on the leaching of the most common metal sulfides have appeared in the literature, both in the presence of the previous ions and others like silver, with a great activity in the chalcopyrite leaching (Miller & Portillo, 1979). However, the specific mechanism of actuation of these ions is still unknown.

Nevertheless, at the moment the mechanisms propose by Scott and Dyson (1968) remain valid. For the case of ZnS, the results point out that the mentioned ions act through an electrochemical mechanism, with an anodic reaction of ZnS dissolution and a cathodic reaction of reduction of dissolved oxygen. Moreover, the ions exhibit a catalytic action when the following two conditions are fulfilled:

(1) The ions must activate the sulfide crystal by incorporation to the surface lattice and releasing simultaneously ions of the initial solid.

(2) The catalyst must be able to form a redox couple which participates in one or more of the chemical reactions between the sulfide and the oxidant.

Additionally, Peters (1986) introduced a new factor in the sense that the catalyst can increase the surface roughness and, thus, the cathodic areas, increasing the rate of the process.

The activation of the ZnS surface with copper ions (the first condition established by Scott and Dyson (1968), has been sufficiently demonstrated in the sphalerite flotation. There is no doubt that zinc ions are displaced from the sulfide surface by copper ions which form a monolayer on the solid. The authors suggest a model with some surface areas not covered with copper and others with an important penetration in depth through dislocations and cracks. This is also supported by the fact that, once reaction starts, the reaction is favored by subsequent additions of copper. This would be related to the necessity of catalyst during solid attack, since the grinding created by attack increases its effective surface area.

This mechanism for copper ions is also supported by works of Vaughan (1984) that established a relationship between the models of atomic and molecular structures of sulfides and their aptitudes to leaching. In this case, copper substitutes zinc in the sulfide surface layers and produces groups of $CuS_4{}^{7-}$. In this type of structures, 3d copper orbitals form a band of electrons partially full placed above the valence band and below the conduction band. Under these conditions, conductivity increases substantially with the corresponding beneficial effect on the reaction rate.

Scott and Dyson (1968) also used other catalytic ions: silver, mercury and lead, but they did not produce the desired effect. Vaughan (1984) indicates that the models of electronic structures for the sulfides of these metals suggest that the highest part of the valence band would be formed by sulfur antibonding orbitals producing a full band. Unlikely, in the case of copper the band of 3d orbitals is partially empty.

With respect to the second condition established by Scott and Dyson (1968), the scarcity of adequate data on electrode potentials related to sulfide-catalyst systems makes very difficult to explain why some metals catalyze while others not.

3.4. Influence of the Presence of Bioleaching Microorganisms

A last factor that can affect the kinetic of electrochemical reactions is the presence of bacteria, this situation would be close to what happens during bioleaching. In this sense, the action of microorganisms in the mineral dissolution process is related to both the location of galvanic couples and the catalytic effect exerted by several ions present in solution, since the attack mechanism is still electrochemical.

To gain further knowledge on the mechanisms of the process is very complex due to limitations of experimental techniques to investigate galvanic cells in minerals. These difficulties are much more important in an inoculated medium, due to the complexity introduced by the biological system.

Several authors have worked taking into consideration both the effect of galvanic couples and bacteria (Sand *et al.*, 1995; Rohwerder *et al.*, 2000; Crundwell, 2003; Fowler *et al.*, 1999). They are in agreement with works by Berry and Murr (1978), Berry *et al.* (1978) who suggested that bacteria accelerate, through an indirect mechanism, the galvanic conversion of sulfides with a lower rest potential: by oxidizing the ferrous to ferric ion –regenerating the leaching agent -, or by solubilizing the elemental sulfur formed on the mineral, improving the kinetics of the process. Two decades ago, Natarajan and Ywasaki, (1986) summarized the interactions between bacteria and mineral in the bioleaching of complex sulfides, referred to the mechanism previously mentioned and illustrated by the scheme of Figure 5.

Similarly, more recent studies such as those by Hansford and Vargas (2001) point out the important role played by the oxidation reaction of Fe^{2+} to Fe^{3+} in

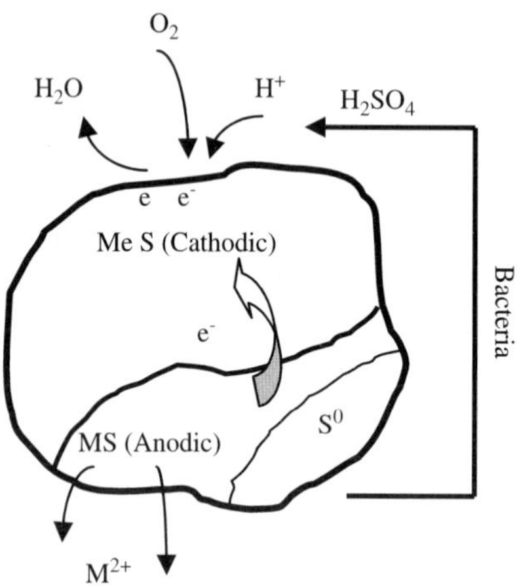

Figure 5. Role of bacteria in electrochemical mechanism of dissolution of a metal sulfide

the cell exopolymers (EPS), and propose its direct participation in the electronic transmission involved in the dissolution of pyrite. Silva *et al.* (2003) have described the contribution of bacteria in the galvanic dissolution of sphalerite, with a lower rest potential, in complex minerals of sphalerite-galena.

There are also several initiatives in the research field oriented to take advantage of the catalytic effect of some ions in solution, as silver, with the simultaneous bioleaching effect of some microorganisms (mesophiles, moderately and extremely thermophiles) on different metal sulfides, especially chalcopyrite. So the next paragraphs will be addressed to this topic.

4. CATALYSIS OF CHALCOPYRITE WITH SILVER IONS

The leaching with ferric ions in acidic media has been applied extensively to process base metal sulfides (Gupta & Mukherjee, 1990). However, an acidic ferric sulfate solution is not a very effective leaching agent for the dissolution of copper from chalcopyrite (Jones & Peters, 1976). The explanation to this behavior lies in the formation of a passivating layer, which inhibits further reaction. For reasons not completely understood, the copper dissolution from chalcopyrite stops after 20-30% copper extraction has been reached. Even for slow leaching chalcopyrite ores, a recovery of copper between 6 and 15% in 4-5 years is probably the best one that could be expected, with most of the copper dissolved according to reaction (Dutrizac, 1989):

$$(8) \qquad CuFeS_2 + 4Fe^{3+} \rightarrow Cu^{2+} + 5Fe^{2+} + 2S^o$$

As early as 1960s it was believed that bacteria were more effective than Fe^{3+} for leaching chalcopyrite (Razell & Trusell, 1963). However, the microbiological leaching of chalcopyrite is a relatively slow reaction by comparison with the biological leaching rates of secondary copper sulfide minerals (Habashi, 1978). The low copper extraction from the chalcopyrite is the result of changes on the mineral surface that renders difficult the attack by the bacteria or by the reactants generated by them.

According to different authors (Ballester *et al.*, 1990, 1992; Barriga *et al.*, 1987), the dissolution of metals from mineral sulfides can be accelerated by the addition of soluble foreign ions that modify the electrochemical behavior of the solid after they remain fixed onto its surface. These ions usually act through the formation of a metal sulfide dissolving the original sulfide material either by galvanic action or substitution in the crystal lattice.

Several authors have demonstrated the ability of silver to catalyse copper dissolution from chalcopyrite (Miller *et al.*, 1979, 1981; Price & Warren, 1986, Ahonen & Tuovinen, 1990a) in both chemical and biological leaching systems and, apparently, the reaction mechanisms involved are identical in both processes. The role of the microorganisms in the silver-catalyzed chalcopyrite leaching has been related to the regeneration of the oxidizing agent or to the oxidation of the elemental sulfur (Ahonen & Tuovinen, 1990a; Ballester, 1987).

Silver-catalyzed chalcopyrite experiments have shown that most of the silver remains in the solid phase during the process. But silver stays in solution during the early stages of bioleaching and this could delay the bacterial growth. Therefore, the main problems associated to the catalyzed process effectiveness that have been described are silver availability (for both chemical and biological processes) and silver toxicity (for the biological process).

Following, the most relevant aspects of both systems are reviewed.

4.1. Silver-Catalyzed Chemical Leaching of Chalcopyrite

The first studies in the absence of bacteria gave rise to two patents (Pawlek, 1974; Snell, 1975). Later, other authors published their results on the reaction mechanism (Miller, 1979, 1981; Price & Warren, 1986; Warren *et al.*, 1984) and a flowsheet for a large-scale operation was proposed (Snell & Sze, 1977). More recently, Hiroyoshi *et al.*, (2002) proposed a new model to explain the role of silver ions. Finally, Carranza and co-workers (1997, 2004) have developed an ingenious process in which the biological step is separated from the chemical step.

It was not until 1979 when Miller and Portillo proposed the first model to explain the catalytic effect of silver. According to this model, the formation of a Ag_2S film on the chalcopyrite surface, by an exchange chemical reaction, favors the copper release from the crystal lattice. This film neutralizes the effect of the passive layer (elemental sulfur) acting a channel for electrons generated during the anodic process and facilitating the cathodic reaction. In the absence of oxidant, the exchange reaction is described by:

$$(9) \qquad CuFeS_2 + 4Ag^+ \rightarrow 2Ag_2S_{(chalcopyrite\ surface)} + Cu^{2+} + Fe^{2+}$$

The silver film formed, most likely as Ag_2S, is oxidized to Ag^+ and S^o by excess Fe^{3+}, thus regenerating the silver catalyst:

$$(10) \qquad Ag_2S + 2Fe^{3+} \rightarrow 2Ag^+ + 2Fe^{2+} + S^o$$

and giving the global reaction [8]. While reaction [9] has been checked experimentally, reaction [10] has not. In fact, Miller *et al.* (1981) pointed out that the catalyzed chalcopyrite leaching is controlled by the intermediate electrochemical reaction of the silver sulfide film with the ferric ions (reaction [10]), since the first stage, the fixation of silver on chalcopyrite, is very rapid and occurs instantly. Additionally, Dutrizac (1994) showed that Ag_2S dissolution is negligible in sulfate media ($Fe_2(SO_4)_3$-H_2SO_4) at temperatures lower than 100°C in contrast to the relatively fast rates observed in chloride media ($FeCl_3$-HCl).

The existence of the Ag_2S layer can be explained in terms of Eh-pH diagrams (Warren *et al.*, 1984). In addition, it has been established (Miller & Portillo, 1979) that the elemental sulfur formed in reaction [10], unlike that of reaction [8], forms a non-protective porous layer and, therefore, favors the diffusion of the leaching reagents across it. However, Parker *et al.* (2003) were unable to detect elemental sulfur by using X-ray photoelectron spectroscopy (XPS). Moreover, in this mechanism, it is considered that Ag_2S has a cathodic behavior with respect to anodic chalcopyrite. So, galvanic couples will also favor the chalcopyrite dissolution.

Electrochemical studies (Price & Warren, 1986) have demonstrated that the formation of metallic silver in this process is also possible:

$$(11) \qquad CuFeS_2 + 4Ag^+ \rightarrow Cu^{2+} + Fe^{2+} + 4Ag^o + 2S^o$$

These studies also confirmed an increase of the film (mixture of Ag_2S and Ag^o) thickness with increasing exposure time and silver concentration.

Miller and Portillo (1979) and Price and Warren (1986) observed that the excess of silver does not improve the chalcopyrite dissolution rate but the opposite. With a large excess of silver, Price and Warren (1986) proposed the formation of a silver sulfate film on the sulfide surface to explain the decrease of the kinetics. Warren *et al.* (1984) also confirmed the presence of a passivating Ag_2SO_4 film by both SEM and X-ray diffraction studies. The formation of this film is the result of high oxidation potentials that modify the solution near the Ag_2S surface, which becomes saturated of Ag^+. This is undesirable since silver is consumed without catalyzing chalcopyrite dissolution.

On the other hand, silver catalysis increases with increasing ferric ion in solution while ferrous ion suppresses this effect (Miller *et al.*, 1981). However, with an increase of the ferric concentration in the medium, the ferric sulfate undergoes hydrolysis according to the following equilibrium:

$$(12) \qquad 3Fe_2(SO_4)_3 + 14H_2O \rightarrow 2(H_3O)Fe_3(SO_4)_2(OH)_6 + 5H_2SO_4$$

precipitating iron from the solution and releasing acid to the medium. Jarosite precipitation can also incorporate silver in its structure, as $AgFe_3(SO_4)(OH)_6$, with the consequent lack of continuing catalytic effect. Contrarily to chloride media, silver is preferentially incorporated in jarosite precipitated from sulfate media. Thus, if AgCl is formed in a sulfate-based hydrometallurgical process, it will not react further to form significant amounts of jarosite and the silver losses as jarosites formed from chloride media will be negligible (Dutrizac & Jambor , 1987). Bolorunduro *et al.* (2003) proposed that silver sequestering by jarosites is due to differences in the precipitation rate between silver sulfate and jarosite. The rapid precipitation of the former favors its occlusion by the later.

Additionally, the occurrence of other mineral sulfides together with chalcopyrite can affect the degree of its dissolution by trapping part of the added silver in their structure. Price and Warren (1986) explored the catalytic effect of silver ions on ZnS and PbS and concluded that, in both cases, elemental silver is formed on the surface. However, no other silver species, e.g. Ag_2S, were detected on the mineral surface. Then, several questions arise: How selective is the silver deposition in a mixture of different sulfide minerals? And, in that case, is the chemistry of the silver deposition the same?

More recently, Hiroyoshi *et al.* (2002) have proposed a different mechanism to interpret the catalytic effect of silver ions on chalcopyrite leaching. Their reaction model involves, besides the formation of Ag_2S, the reductive transformation of chalcopyrite in an intermediate Cu_2S, which is more easily leached by the oxidizing agent (ferric ion or dissolved oxygen). In order for this transformation to occur, the redox potential of the solution must be placed below the oxidation potential of Cu_2S and above a certain critical redox potential of Cu_2S formation. Under these conditions, the overall reaction is given by:

$$(13) \qquad 2CuFeS_2 + 4Ag^+ \rightarrow Cu_2S + 2Fe^{2+} + 2Ag_2S + S^o$$

Unlike Miller and Portillo's model, this new model does not postulate an interchange reaction between silver ions and chalcopyrite but the formation of Ag_2S by reaction of silver ions with the hydrogen sulfide formed during the previous chalcopyrite reduction in the presence of protons.

Additionally, the formation of metallic silver is explained through an electrochemical reaction between silver and ferrous ions in solution:

$$(14) \qquad Ag^+ + Fe^{2+} \rightarrow Ag^o + Fe^{3+}$$

This reaction is not essential for interpreting the catalytic effect of silver ions but may influence the leaching, because the formation of metallic silver causes a decrease in silver ion concentration and a shift in the critical potential for Cu_2S formation.

Finally, the University of Seville (Spain) has developed a process for treating metallic sulfide concentrates, known as IBES process (Indirect Bioleaching with Effect Separation), which consists of the ferric sulfate leaching of the concentrate followed by the biooxidation of the ferrous iron generated in the leaching to regenerate the ferric sulfate leaching agent (Carranza *et al.*, 1997; Palencia *et al.*, 1998; Romero *et al.*, 1998). The leaching is performed at 70 °C firstly, in the absence of silver, to recover zinc (primary leaching) and secondly, in the presence of silver (2 mg Ag/g concentrate), to recover copper from chalcopyrite (catalytic leaching). Then, the ferrous iron generated is biooxidized with mesophilic bacteria (*A. ferroxidans*) and the silver is recovered from the residue by acidic brine leaching.

These researchers developed a similar process for the treatment of secondary copper sulfides (chalcocite and covellite) extending the treatment to chalcopyrite (Palencia *et al.*, 2002) (Carranza *et al.*, 2004). The BRISA process (Fast Indirect Bioleaching with Actions Separation) basically consists in: a) a chemical stage performed in stirred tank reactors in which copper extraction can be enhanced by thermal activation (70-90 °C) and the use of catalysts (silver ion) in order to accelerate the chalcopyrite leaching and b) a biological stage of biooxidation of ferrous iron with bacteria (*A. ferroxidans*) attached on inert material particles forming a biofilm (31°C) in flooded packed-bed reactors.

Two reasons are given by these authors to separate the chemical from the biological stage: 1) the possibility to perform the chemical leaching at high temperature in order to increase kinetics; and 2) the inhibition of the bacterial growth by silver and the harmful abrasive effect that high pulp densities exert on the bacteria when using a single bioleaching reactor.

4.2. Silver-Catalyzed Biological Leaching of Chalcopyrite

Silver has also been used as catalyst in chalcopyrite bioleaching. Several patents (McElroy & Duncan, 1974; Bruynesteyn 1986; Muñoz *et al.*, 2000) and interesting research on this topic has been published (Bruynesteyn *et al.*, 1983; Lawrence *et al.*, 1993; Ahonen & Tuovinen, 1990a; Ballester *et al.*, 1992; Banerjee *et al.*, 1990).

Lawrence *et al.* (1993) studies of the silver-catalyzed biological leaching of chalcopyrite have shown that, in addition to enhance both rates and extractions of copper, the quantitative conversion of sulfide sulfur to elemental sulfur, instead of sulfate, can be achieved. This is possible at an oxidation potential below 680 mV SHE. The addition to the leach solution of sodium thiosulfate and a minimum initial concentration of cupric ions had two effects: to keep a low potential and to form complexes with silver ions, which remain in solution. Under these conditions, the cupric/cuprous couple was responsible for the dissolution of chalcopyrite.

Ballester and co-workers (Ballester, 1987; Ballester *et al.*, 1990) and Ahonen and Tuovinen (1990b) proposed a mechanism similar to the mechanism of Miller and Miller and Portillo (1979) and Price and Warren (1986). According to these authors, the dissolution of chalcopyrite takes place through reactions (9) and (10). The bacteria play an indirect role in the process facilitating the oxidation of Fe^{2+} to Fe^{3+} and depolarizing the cathodic half-reaction in accordance with:

$$(15) \qquad 2FeSO_4 + 0.5O_2 + H_2SO_4 \xrightarrow{\text{microorganisms}} Fe_2(SO_4)_3 + H_2O$$

Reaction (15) plays an important role in the process. When the bacterial activity decreases, the ferrous ion is not oxidized to ferric ion and the dissolution of chalcopyrite stops or takes place at a very slow rate.

The regeneration of Ag^+ give rise to a cyclic process that increases the rate of copper dissolution. The silver effect is amplified in the presence of iron and sulfur oxidizing microorganisms. On one hand, they maintain a favorable Fe^{3+}/Fe^{2+} ratio contributing to the oxidation and, on the other hand, they oxidize the elemental sulfur layer produced on the chalcopyrite surface in reaction [10] preventing chalcopyrite passivation:

$$(16) \qquad S^\circ + H_2O + 3/2O_2 \xrightarrow{\text{microorganism}} H_2SO_4$$

Studies on the bioleaching of chalcopyrite in the presence of silver have included the use of mesophiles (Coto *et al.*, 1996; Pooley & Sherstha 1996; Banerjee *et al.*, 1990; Sukla *et al.*, 1990), moderate thermophiles (Gómez *et al.*, 1999) and extreme thermophiles (Blázquez *et al.*, 1999). It is a proven fact that silver exerts its catalytic effect with mesophilic and moderately thermophilic microorganisms but not with extremely thermophilic microorganisms.

Sukla *et al.* (1990) studied the silver-catalyzed bioleaching of a chalcopyrite concentrate with mesophilic microorganisms and found an optimum silver concentration of 30 ppm Ag in the range studied (10 to 500 ppm). The negative effect of silver at concentrations higher than 30 ppm was attributed to the metal effect on the microorganism metabolism. These authors concluded that "the kinetics of copper dissolution change from diffusion control to chemical control in the presence of silver".

Similarly to the chemical process, the silver catalytic effect on the chalcopyrite bioleaching is negligible in the absence of an oxidizing agent (Ahonen and Tuovinen 1990b; Banerjee *et al.*, 1990). However, whereas in abiotic systems the redox potential is subjected to the chemical reactions between the solution and the ore, in biotic systems the biochemical reactions play an important role on Eh variations. As a consequence, the potential is practically not affected in the absence of bacteria, being dramatically affected in its presence.

The catalytic role of silver ions in chalcopyrite dissolution using static systems was investigated by Ahonen and Tuovinen (1990a, 1995) and Canfell *et al.* (1997). They found that the addition of silver (between 0.75 to 2.5 g Ag/kg of Cu) favors the copper leaching from chalcopyrite. Nevertheless, they observed a transient catalytic effect of silver on the rate of copper leaching which disappeared after 30 days. As a result, the final extractions of copper from a low-grade ore were as low as 30% after 2 years of bioleaching. The lack of a continuing catalytic effect observed was attributed to silver loss in jarosite precipitates.

The consumption of silver by other mineral sulfides different from chalcopyrite has also been considered in the bioleaching process. Ahonen and Tuovinen (1990b) described that the leaching of zinc from sphalerite and iron from pyrite from a complex sulfide ore was inhibited by silver addition. Similarly, Nakazawa *et al.* (1993) using a flotation concentrate containing chalcopyrite and pentlandite, observed catalysis of silver on the bioleaching of chalcopyrite but not of pentlandite, with increasing initial silver concentrations.

Gómez *et al.* (1997b) in a bioleaching study with a mixed culture of mesophilic microorganisms from mine waters, found that silver increased the extraction of copper from a bulk concentrate obtained by continuous flotation from a Spanish complex sulfide ore. However, silver did not improve sphalerite dissolution. Zinc extraction decreased about 25% in the experiment with silver (Figure 6).

Different authors (Muñoz *et al.*, 1998; Yuehua *et al.*, 2002; Wang *et al.*, 2004) have detected the formation of silver sulfide on the catalyzed chalcopyrite bioleaching. Gómez *et al.* (1997a, 1997c) using scanning electron microscopy (SEM), Auger electron spectroscopy (AES) and electrochemical techniques detected the presence of silver sulfide on the chalcopyrite surface and confirmed both its greater thickness for the microbial catalyzed leaching and its attack by microorganisms.

The ability of bacteria to regenerate ferric ion can create the adequate conditions for jarosite precipitation. Jarosites hinder bacterial attack and consume silver to form argentojarosites. Silver losses in the residues as argentojarosites have also been observed in the biological catalyzed process (Ahonen & Tuovinen, 1990; Gómez *et al.*, 1999). Ahonen and Tuovinen (1990a) estimated that about 5% of the total silver added is associated with the jarosite fraction.

Gómez *et al.* (1999) studied the effect of silver on the bioleaching of a chalcopyrite concentrate with mixed cultures of moderate thermophiles isolated from a drainage water of Rio Tinto mines. They found a three fold copper increase in bioleaching tests at 45°C with silver than in its absence. However, at 50°C, the effect was masked by the strong jarosite precipitation which increased with the amount of silver used; this produced a considerable decrease of the amount of ferric ion in solution (Figure 7). They suggested that the problem can be overcome by using more diluted culture media.

Logically, a negative aspect of the process is the toxic effect of silver on bacterial metabolism. In this sense, bacterial cultures of *Sulfolobus* are especially sensitive to small amounts of silver and this has been one of the main drawbacks in the use of this cation for the bioleaching of chalcopyrite at high temperature (Mier, 1996). (Figure 8) shows the results obtained during the bioleaching of a chalcopyrite concentrate with *Sulfolobus* BC in the presence of silver. Adapted cultures showed less inhibition and slightly more activity in comparison with non-adapted ones.

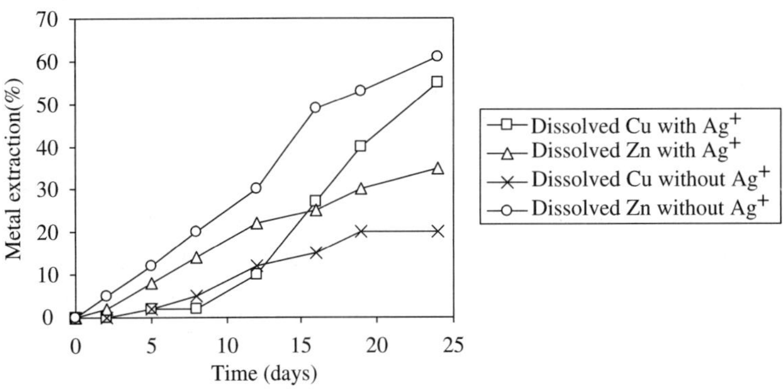

Figure 6. Influence of silver in the copper and zinc dissolution from a bulk concentrate of a complex sulfide ore with a mixed mesophilic culture (35° C, 5% w/v pulp density, 1 g Ag^+/kg concentrate)

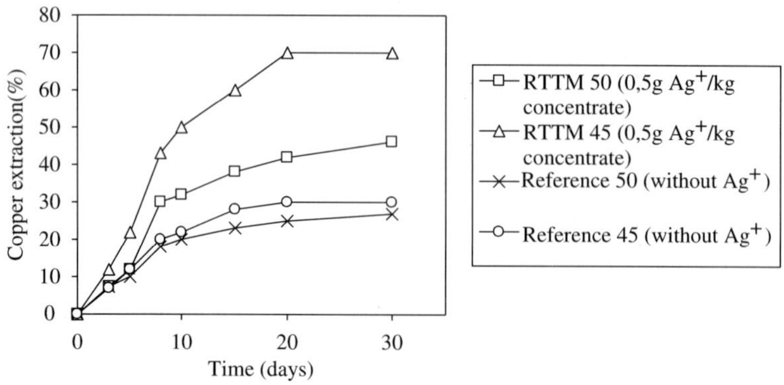

Figure 7. Bioleaching of a chalcopyrite concentrate with RTTM moderately thermophilic culture at 45°C and 50°C (5% w/v pulp density)

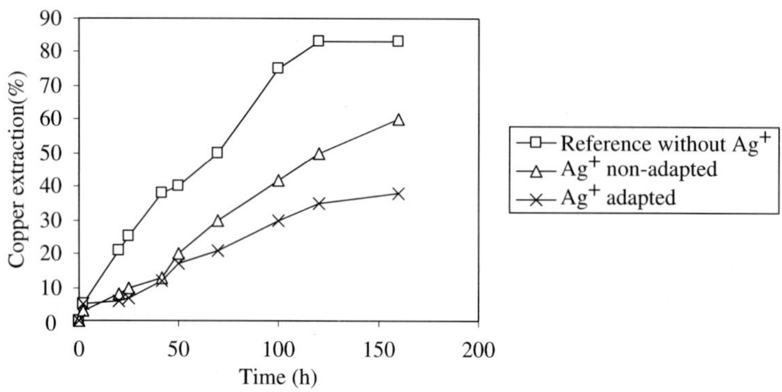

Figure 8. Bioleaching of a chalcopyrite concentrate with *Sulfolobus* BC cultures at 68°C in the presence of silver

Nevertheless, the dissolution rate was smaller than for reference test without silver. Silver inhibited the oxidation of ferrous ion to ferric and also the oxidation of elemental sulfur to sulfate. The use of cultures gradually adapted to the presence of silver increased the oxidation rate of both substrates (Figure 9).

Unlike the chemical process, silver can affect bacterial metabolism by suppressing iron- and sulfur-oxidizing ability of microorganisms. The toxic effect of silver on leaching microorganisms is collected in the literature (Norris & Kelly, 1978; Norris & Barr, 1985; Norris, 1989; Grogan, 1989; Hoffman & Hendrix, 1976; De *et al.*, 1996). These studies have shown that metal tolerance is greater for mesophilic than for thermophilic microorganisms. Norris and Barr (1985) established that silver concentrations of 2.5 ppm exert inhibition on bacterial growth. However, bacteria could maintain its normal activity in the presence of a small concentration of silver (< 1 ppm). Norris (1989) obtained a moderate inhibition of the bacterial growth

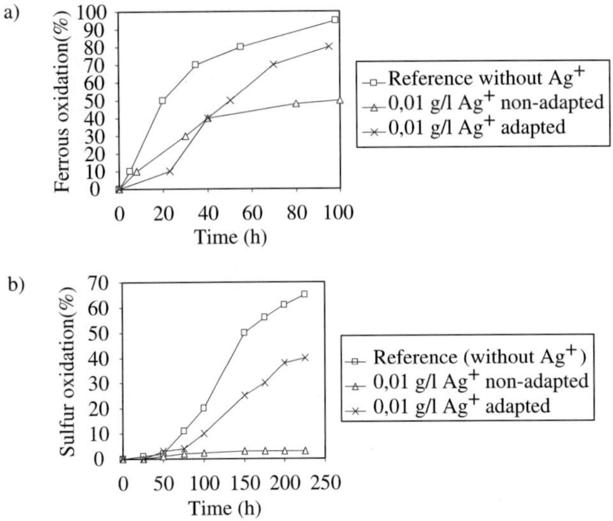

Figure 9. Ferrous ion (a) and elemental sulfur (b) oxidation by *Sulfolobus* BC cultures at 68°C in the presence of silver (1% w/v pulp density)

with a concentration of 0.5 μM of Ag for *Acithiobacillus ferrooxidans* and of 0.1 μM of Ag for *Sulfolobus*. Grogan (1998) reached a maximum silver concentration before inhibition of 8 μM of Ag for *Sulfolobus* and Hoffman and Hendrix (1976) of 0.1 ppm of Ag for *A. ferrooxidans*. De *et al.* (1996) found that *A. ferrooxidans* is able to accumulate different heavy metals on its cell wall and concluded that the adsorbed silver was responsible for the decrease of the iron-oxidizing ability of these microorganisms.

For these reasons, it is essential that the bacterial strains used in the catalyzed bioleaching process have a high resistance towards toxic elements, especially silver. In these sense, two actions can be adopted in order to obtain silver-resistant strains: 1) the progressive adaptation of the bacterial culture to increasing amounts of silver and 2) the mutation and selection of bacteria present in the culture. In both cases, it is expected that the length of the bacterial growth lag phase will be shorten.

Sato *et al.* (2000) have proposed the use of silver chloride instead of silver sulfate due to its lower toxicity to bacteria. In any case, fortunately, small amounts of silver are enough to catalyze the process and, for each mineral, there is an optimum silver concentration at which copper is preferentially extracted from chalcopyrite.

Yuehua *et al.* (2002) studied the bioleaching of chalcopyrite using different silver-bearing species: silver ions, silver sulfide and silver-bearing concentrates. This opens the possibility to use a variety of cheaper silver compounds as catalysts. These authors reached similar conclusions to Ahonen and Tuovinen (1990b) with respect to the necessity of having an oxidizing medium with a high Fe^{3+}/Fe^{2+} relationship.

Recently, Wang *et al.* (2004) have proposed a kinetic model based on the shrinking core model and conclude that the rate-controlling step during the catalyzed bioleaching is the interchange reaction between silver ions and chalcopyrite in clear contradiction with Miller and Portillo's idea. According to these authors, the transport of silver ions in the chalcopyrite particle would be the key step.

The lack of a continuing catalytic effect of silver due to its precipitation as elemental metal and as Ag_2S on surfaces other than chalcopyrite or its incorporation into Fe(III)-precipitates (jarosites-type) has led to the consideration of the addition of complexing agents. Bruynesteyn *et al.* (1986) and Lawrence *et al.* (1984) described a biological silver-catalyzed process for chalcopyrite which employs copper sulfate and thiosulfate as activating reagents and a low oxidation potential in the range 620-680 mV (SHE). The reaction mechanism for silver catalysis in this scheme has not been elucidated, but is presumed to involve copper reduction to form cuprous-thiosulfate complexes which subsequently decompose to copper sulfides that are quickly oxidized (Blancarte-Zurita *et al.*, 1988). Whatever the biological mechanism, the oxygen required to oxidize chalcopyrite in catalyzed leaching is 3.4 times less than the requirement in non-catalyzed leaching. Thus, copper leach rates could be expected to be higher in catalyzed leaching. The presence of thiosulfate favors the solubilization of insoluble silver chloride or phosphate. For viability of this process, a high recovery of silver added as catalyst is required. According to these authors, silver is more easily recovered in the middle of the leach than at the beginning or end since its recovery is inhibited by the formation of silver jarosite. Adding chloride to the bioleaching medium can prevent silver loss to jarosite.

4.3. Patents on Silver-Catalyzed Chalcopyrite Dissolution

In spite of the vast amount of knowledge available none commercial process has been developed using silver as a catalytic agent for the copper dissolution from its ores.

However, several inventions (McElroy & Duncan, 1974; Pawlek, 1974; Snell, 1975; Bruynesteyn *et al.*, 1983; Muńoz *et al.*, 2000) contemplate the step of adding silver to the leach pulp in a process for the chemical or bacteriological extraction of copper from particulate materials such as ores and concentrates containing chalcopyrite. The solid residue is removed from the pulping medium and the silver recovered by well-known recovery processes. The objective of these inventions is broadly accomplished by oxidative leaching of a copper ore in the presence of a catalytic amount of silver to increase the leaching rate Table 2.

The conditions for these inventions are referred to an acidic leaching medium containing an oxidizing agent or sulfide oxidizing bacteria, nutrient for said bacteria, ore or concentrate (from 2 to 60% pulp density or static leaching) and having a pH from about 1.5 to 3.0. Silver is added in finely divided or dissolved in the form of soluble or insoluble salts or as the metal to the leaching mixture. When silver is

Table 2. Data compiled from different patent sources on silver catalysis of copper ores

Reference	[Ag]	Silver compound	Copper Ore	Particle size (μm)	Pulp density (%)	Leaching process
(McElroy and Duncan)	0.2 to 7 g Ag/kg $CuFeS_2$	Soluble (chloride, nitrate, sulfate) or Insoluble (sulfide)	Chalcopyrite concentrate	<45	2 to 60	Bacterial leaching
(Snell)	25 to 10,000 ppm	Soluble	Complex mineral sulfides	<100	nd	Acid leaching with an oxidant (O_2 or $Fe_2(SO_4)_3$) from 65 to $113n^oC$
(Pawlek)	0.1% w	Nitrate or sulfate	Concentrates	<3	nd	Acid pressure leaching at 110^oC
(Bruynesteyn, Hackl, Lawrence and Vizsolyi)	0.1 to 4 g Ag/kg $CuFeS_2$	nd	Ore containing chalcopyrite	<75	2 to 60	Bacterial leaching at constant potential: 0.54 to 0.66 V
(Muñoz, Young and Dreisinger)	0.25 to 1.5 g Ag/kg Cu	Soluble silver	Low-grade chalcopyrite ores	<12,700	static leaching	Bacterial leaching at oxidation potentials > 0.7 V

nd: not described

added in the form of a highly insoluble salt, such as silver sulfide, the amount of added silver required to obtain a substantial increase in extraction might be higher. This addition may be made at the start of leaching of at any time thereafter, but is preferably made before the leach rate has slowed down or the pH of the leach slurry has become unfavorably low for continued leaching. Besides, the presence of small amounts of chloride in the leach solution is not detrimental to the process. The pulp is preferably agitated and simultaneously aerated with air preferably enriched with CO_2 in the case of the bacterial leaching processes.

The silver added to the leaching system remains primarily in the solid leach residue, and may be recovered for reuse by known methods, such as leaching with strong acids, chlorine or acid chloride solutions or leaching with cyanide salts or ammonia under alkaline conditions.

5. REFERENCES

Abraitis PK, Pattrick RAD, Kelsall GH, Vaughan DJ. 2004. Acid leaching and dissolution of major sulfide ore minerals: processes and galvanic effects in complex systems. Mineralogical Magazine 68: 343-51.

Aghamiriam MM, Yen WT. 2005. Mechanisms of galvanic interactions between gold and sulfide minerals in cyanide solution. Miner Eng 18: 393-407.

Ahonen L, Tuovinen OH. 1990a. Silver catalysis of the bacterial leaching of chalcopyrite containing ore material in column reactors. Miner Eng 3: 437-445.

Ahonen L, Tuovinen OH. 1990b. Catalytic effects of silver in the microbiological leaching of finely ground chalcopyrite-containing ore materials in shake flasks. Hydrometallurgy 24: 219-236.

Ahonen L, Tuovinen OH. 1995. Bacterial leaching of complex sulfide ore sample in bench-scale column reactors. Hydrometallurgy 37: 1-21.

Ballester A. 1987. A study of the mechanism of silver-catalyzed bioleaching of chalcopyrite. In: Davies GA, ed. Separation processes in hydrometallurgy. Ellis Horwood Publishers, Chichester, UK, 99-110.

Ballester A, González F, Blázquez ML, Mier JL 1990. The influence of several ions in the bioleaching of metal sulfides. Hydrometallurgy 23: 221-235.

Ballester A, González F, Blázquez ML, Gómez C, Mier JL. 1992. The use of catalytic ions in bioleaching. Hydrometallurgy 29: 145-160.

Banerjee PC, Chakrabarti BK, Bhattacharyya S, Das A. 1990. Silver-catalyzed hydrometallurgical extraction of copper from sulfide ores from Indian mines. Hydrometallurgy 25: 349-355.

Barriga F, Palencia I, Carranza F. 1987. The passivation of chacopyrite subjected to ferric sulfate leaching and its reactivation with metal sulfides. Hydrometallurgy 19: 159-167.

Berry V, Murr L. 1978. Direct observations of bacteria and quantitative studies of their catalytic role in the leaching of low-grade copper-bearing waste. In: Murr L, Torma A, Brierley J, eds. Metallurgical Applications of Bacterial Leaching and Related Microbiological Phenomena, Academic Press, New York.

Berry V, Murr L, Hiskey JB. 1978. Galvanic interaction between chalcopyrite and pyrite during bacterial leaching of low-grade ores. Hydrometallurgy 3: 309-326.

Björling G. 1954. Lixiviation of sulphidic minerals under oxygen pressure. Metall 8: 781-784.

Blancarte-Zurita MA, Branion RMR, Lawrence RW. 1988. Microbiological leaching of chacopyrite concentrates by *Thiobacillus ferrooxidans*. A comparative study of a conventional process and a catalyzed process. In: Norris PR, Kelly DP, eds. Biohydrometallurgy, Proceedings of the International Symposium, Warwick, Antony Rowe Ltd. Chippenham, Wiltshire, UK, 273-285.

Blázquez ML, Álvarez A, Ballester A, González F, Muñoz JA. 1999. Bioleaching behavior of chalcopyrite in the presence of silver at 35° and 68°C. In: Amils R, Ballester A, eds. Biohydrometallurgy and the environment toward the mining of the 21st century, Proceedings of the International Biohydrometallurgy Symposium IBS-99, El Escorial, Spain. Elsevier, Amsterdam, Part A, 137-147.

Bolorunduro SA, Dreisinger DB, Van Weert G. 2003. Fundamental study of silver deportment during the pressure oxidation of sulfide ores and concentrates. Miner Eng 16: 695-708.

Bruynesteyn A, Lawrence RW, Vizsolyi AI, Hackl RP. 1983. An elemental sulfur biohydrometallurgical process for treating sulfide concentrates. In: Proc. Conf. Progress in Biohydrometallurgy, Cagliari, Italy, 151-168.

Bruynesteyn A, Halck RP, Lawrence RW, Vizsolyi AI 1986. Biological-acid leach process. US Patent # 4, 571, 387.

Canfell AJ, Greenfield PF, Winbome DA. 1997. Silver catalyzed bioleaching of chalcopyrite ore in columns. In: Proceedings of the International Biohydrometalurgy Symposium and Biomine' 97; August 4–6; Australian Mineral Foundation Inc., Sydney. Glenside, Australia, M5.1.1-M5.1.10.

Carranza F, Palencia I, Romero R. 1997. Silver catalyzed IBES process: application to a Spanish copper-zinc sulfide concentrate. Hydrometallurgy 44: 29-42.

Carranza F, Iglesias N, Mazuelos A, Palencia I, Romero R. 2004. Treatment of copper concentrates containing chalcoyrite and non-ferrous sulfides the BRISA process. Hydrometallurgy 71: 413-420.

Chander S. 2003. A brief review of pulp potentials in sulfide flotation. Int J Miner Process 72: 141-150.

Coto O, Blázquez ML, Ballester A, González F. 1996. Semicontinuous bioleaching of a copper-indium concentrate. An Quim Int Ed 92: 31-36.

Cruz R, Luna-Sánchez RM, Lapidus GT, González I, Monroy M. 2005. An experimental stratrgy to determine galvanic interactions affecting the reactivity of sulfide mineral concentrates. Hydrometallurgy 78: 198-208.

Crundwell FK. 2003. How do bacteria interact with minerals? Hydrometallurgy 71: 75-81.

De GC, Oliver DJ, Pesic BM. 1996. Effect of silver on the ferrous iron oxidizing ability of *Thiobacillus ferrooxidans*. Hydrometallurgy 41; 211-229.

Dutrizac JE. 1989. Elemental sulfur formation during the ferric sulfate leaching of chacopyrite. Can Metall Quart 28: 337-344.

Dutrizac JE. 1994. The leaching of silver sulfide in ferric ion media. Hydrometallurgy 35: 275-292.

Dutrizac JE, Jambor JL. 1987. Behavior of silver during jarosite precipitation. Trans Inst Min Metall 96: C206-C218.

Dutrizac JE, MacDonald RJC. 1973. The effect of some impurities on the rate of chacopyrite dissolution. Can Metall Quart 12: 409-420.

Ekmekçi Z, Demirel H. 1997. Effects of galvanic interactions on collectorless flotation behavior of chalcopyrite and pyrite. Int J Miner Process 52: 31-48.

Fowler TA, Holmes PR, Crundwell FK. 1999. Mechanism of pyrite dissolution in the presence of *Thiobacillus ferrooxidans*. Appl Environ Microbiol 65: 2987-2993.

Frenay J. 1984. Leaching of sphalerite: Influence of its iron content. In: International Congress on Applied Mineralogy in the Mineral Industry, Los Angeles, USA.

Gómez C, Figueroa M, Muñoz JA, Ballester A, Blázquez ML. 1997a. A study of bioleached chalcopyrite surfaces in the presence of Ag (I) by voltammetric methods. Miner Eng 10: 111-116.

Gómez C, Blázquez ML, Ballester A. 1997b. Influence of various factors in the bioleaching of a bulk concentrate with mesophilic microorganisms in the presence of Ag(I). Hydrometallurgy 35: 271-287.

Gómez C, Román E, Blázquez ML, Ballester A. 1997c. SEM and AES studies of chalcopyrite bioleaching in the presence of catalytic ions. Miner Eng 10: 825-835.

Gómez E, Ballester A, Blázquez ML, González F. 1999. Silver-catalyzed bioleaching of a chalcopyrite concentrate with mixed cultures of moderately thermophilic microorganisms. Hydrometallurgy 51:37-46.

Grogan DW. 1989. Phenotypic characterization of the archaebacterial genus *Sulfolobus*: comparison of five wild-type strains. J Bacteriol 171: 6710-6719.

Güler T, Hiçyilmaz C, Gökagaç G, Ekmekçi Z. 2005. Electrochemical behavior of chalcopyrite in the absence and presence of dithiophosphate. Int J Miner Process 75: 217-228.

Gupta CK, Mukherjee TK. 1990. Hydrometallurgy in extraction processes. CRC Press, Boca Raton, USA.

Habashi F. 1978. Chalcopyrite: Its chemistry and metallurgy. McGraw-Hill, New York.

Habashi F. 1983. Dissolution of minerals and hydrometallurgical processes. Naturwissenschaften 70: 403-411.

Hansford GS, Vargas T. 2001. Chemical and electrochemical basis of bioleaching processes. Hydrometallurgy 59: 135-145.

Hiroyoshi N, Arai M, Miki H, Tsunekawa M, Hirajima T. 2002. A new reaction model for the catalytic effect of silver ions on chalcopyrite leaching in sulfuric acid solutions. Hydrometallurgy 63: 257-267.

Hoffmann LE, Hendrix JL. 1976. Inhibition of *Thiobacillus ferrooxidans* by soluble silver. Biotechnol Bioeng 18: 1161-1165.

Jones DL, Peters E. 1976. The leaching of chalcopyrite with ferric sulfate and ferric chloride. In: Yannopoulos JC, Agarwal JC, eds. Extractive metallurgy of copper, Vol. II, Chapter 34, The Metallurgical Society of AIME, New York, 633-653.

Lawrence RW, Vizsolyi A, Vos RJ, Rule C, Chua G. 1984. The biology and chemistry involved in a silver-catalyzed bioleaching process. BC Research Council, Report 1-42-487, Vancouver, Canada.

Lawrence RW, Vizsolyi A, Vos RJ. 1993. The silver catalyzed bioleach process for copper concentrate. In: Hiskey JB,Warren GW, eds. Hydrometallurgy: fundamentals, technology and innovations. AIME, Littletown, Colorado, 65-82.

Linge HG. 1977. Reactivity comparison of Australian chalcopyrite concentrates in acidified ferric solution. Hydrometallurgy 2: 219-233.

Majima H. 1969. How oxidation affects selective flotation of complex sulfide. Can Metal Quar 8: 269-273.

McElroy RO, Duncan DW. 1974. Copper extraction by rapid bacteriological process. US Patent # 3, 856, 913.

Mier JL, Ballester A, Gónzález F, Blázquez ML, Gómez E. 1996. The influence of metallic ions on the activity of *Sulfolobus BC*. J Chem Tech Biotechnol 65: 272-280.

Miller JD, Portillo HQ. 1979. Silver catalysis in ferric sulfate leaching of chalcopyrite. In: Laskowski J, ed. XIII International Min Proc Conf. Elsevier, Amsterdam, 851-901.

Miller JD, McDonough PJ, Portillo HQ. 1981. Electrochemistry in silver catalyzed ferric sulfate leaching of chalcopyrite. In: Kuhn MC, ed. Process and fundamental considerations of selected hydrometallurgical systems. SME-AIME, New York, 327-338.

Mulak W. 1987. The catalytic action of cupric and ferric ions in nitric acid leaching of Ni(III)S(II). Hydrometallurgy 17: 201-214.

Muñoz JA, Gomez C, Ballester A, Blázquez ML, González F. 1998. Electrochemical behavior of chalcopyrite in the presence of silver and *Sulfolobus* bacteria. J Appl Electrochem 28: 49-56.

Muñoz JA, Young S, Dreisinger DB. 2000. Silver-catalyzed bioleaching process for copper extraction from chalcopyrite heap. US Patent # WO0037690.

Nakazawa H, Hashizume T, Sato H. 1993. Effect of silver ions on bacterial leaching of flotation concentrate of copper-nickel sulfide ores. Metallurgical Review 10: 87-97.

Natarajan K, Ywasaki I. 1986. Microbe-Mineral Interactions in the Leaching of Complex Sulfides. In: Clum J, Haas L, eds. Microbiological Effects on Metallurgical Processes, TMS, Warrendale, USA.

Nicol MJ, Lázaro I. 2002. The role of Eh measurements in the interpretation of the kinetics and mechanisms of the oxidation and leaching of sulfide minerals. Hydrometallurgy 63: 15-22.

Norris PR. 1989. Mineral-oxidizing bacteria: metal-organism interactions. In: Poole RK, Gadd GM, eds. .Metal-Microbe interactions. Oxford University Press, Oxford, 99-117.

Norris PR, Barr DW. 1985. Growth and iron oxidation by acidophilic moderate thermophiles. FEMS Microbiol Lett 28: 221-224.

Norris PR, Kelly DP. 1978. Toxic metals in leaching sistems. In: Murr L, Torma A, Brierley J, eds. Metallurgical applications of bacterial leaching and related microbiological phenomena. Academic Press, New York, 83-102.

Palencia I, Romero R, Carranza F. 1998. Silver catalyzed IBES process: application to a Spanish copper-zinc sulfide concentrate. Part 2. Biooxidation of the ferrous iron and catalyst recovery. Hydrometallurgy 48: 101-112.

Palencia I, Romero R, Mazuelos A, Carranza F. 2002. Treatment of secondary copper sulfides (Chalcocite and Covellite) by the BRISA process. Hydrometallurgy 66: 85-93.

Parker A, Klauber C, Kougianos A, Watling HR, van Bronswijk W. 2003. An X-ray photoelectron spectroscopy study of the mechanism of oxidative dissolution of chalcopyrite. Hydrometallurgy 71: 265-276.

Pawlek F. 1974. Verfharen zur hudrometallurgischen gewinnung von kupfer aus kupferkiesbzw buntkupferkieskonzentraten. Austrian Patent # 319,616.

Peng Y, Grano S, Fornasiero D, Ralston J. 2003. Control of grinding conditions in the flotation of chalcopyrite and its separation from pyrite. Int J Miner Process 69: 87-100.

Peters E. 1977. The Electrochemistry of Sulfide Minerals. In: Bockis J, Rand D, Welch B, eds. Trends in Electrochemistry. Plenum Press, New York.

Peters E. 1986. Leaching of sulfides. In: Somansudaran P, ed. Advances in Mineral Processing. SME/AIME, New York, 445-462.

Pooley FD, Sherstha GN. 1996. The distribution and influence of silver in pyrite bacterial leaching systems. Miner Eng 9: 825-836.

Price DW, Warren GW. 1986. The influence of silver ion on the electrochemical response of chacopyrite and other mineral sulfide electrodes in sulfuric acid. Hydrometallurgy 15: 303-324.

Razell WE, Trusell PC. 1963. Microbiological leaching of metallic sulfides. Appl Microbiol 11: 105-110.

Rohwerder T, Gherke T, Kinzler K, Sand W. 2003. Bioleaching review part A: Progress in bioleaching: fundamentals and mechanisms of bacterial metal sulfide oxidation. Appl Microbiol Biotech 63: 239-248.

Romero R, Palencia I, Carranza F. 1998. Silver catalyzed IBES process: application to a Spanish copper-zinc sulfide concentrate. Part 3. Selection of the operational parameters for a continuous pilot plant. Hydrometallurgy 49: 75-86.

Sand W, Gehrke T, Hallmann R, Schippers A. 1995. Sulfur chemistry, biofilm and the (in)direct attack mechanism: a critical evaluation of bacterial leaching. Appl Microbiol Biotech 43: 961-966.

Sato H, Nakazawa H, Kudo Y. 2000. Effect of silver chloride on the bioleaching of chalcopyrite concentrate. Int J Miner Process 59: 17-24.

Scott TR, Dyson NF. 1968. The catalyzed oxidation of zinc sulfide under acid pressure leaching conditions. Transactions of the Metallurgical Society of AIME 242: 1815-1821.

Sherrit Gordon Mines. 1963. Ltd. Australian Patent 274,816.

Silva G, Lastra MR, Budden JR. 2003. Electrochemical passivation of sphalerite during bacterial oxidation in the presence of galena. Miner Eng 16: 199-203.

Snell GJ. 1975. Sulfate leaching of copper ores using silver catalyst. US Patent # 3,886,257.

Snell GJ, Sze MC. 1977. New oxidative leaching process uses silver to enhance copper recovery. Eng Mining J. 178: 100-105.

Sukla LB, Chaudhury GR, Das RP. 1990. Effect of silver ion on kinetics of biochemical leaching of chalcopyrite concentrate. Trans Inst Min Metall 99: C43-C46.

Vaughan DJ. 1984. Electronic Structures of Sulfides and Leaching Behavior. In: Bautista RG, ed. Hydrometallurgical Process Fundamental. Plenum Press, New York.

Wadsworth M. 1976. Physical chemistry of hydrometallurgy. Electrochemical Processes in Leaching. In: Physical Chemistry in Metallurgye. US Steel Research Lab., Monroeville.

Wan RY, Miller JD, Simkovich G. 1984. Enhanced ferric sulfate leaching of copper fron CFeS/C particulate aggregates. In: International Conference on recent advances I Mineral Science and technology. Council for Mineral Technology, Johannesburg, South Africa.

Wang M, Zhang Y, Deng T, Wang K. 2004. Kinetic modeling for the bacterial leaching of chalcopyrite catalyzed by silver ions. Hydrometallurgy 17: 943-947.

Warren GW, Drouven B, Price DW. 1984. Relationships between the Pourbaix diagram for Ag-S-H_2O and electrochemical oxidation and reduction of Ag_2S. Metallurgical Transactions B 15B: 235-242.

Woods R. 2003. Electrochemical potential controlling flotation. Int J Miner Process 72: 151-162.

Yuehua H, Guanzhou Q, Jun W, Dianzuo W. 2002. The effect of silver-bearing catalysts on bioleaching of chalcopyrite. Hydrometallurgy 64: 81-88.

CHAPTER 5

RECOVERY OF ZINC, NICKEL, COBALT AND OTHER METALS BY BIOLEACHING

MARISA VIERA, CRISTINA POGLIANI, EDGARDO DONATI
Centro de Investigación y Desarrollo en Fermentaciones Industriales (CINDEFI, UNLP-CONICET), Facultad de Ciencias Exactas, 47 y 115 (1900) La Plata, Argentina. E-mail: mviera@biotec.org.ar

1. GENERAL CONSIDERATIONS

Bioleaching is currently an economical alternative for treating specific sulfidic ores. Commercial scale bioleaching processes are in operation for the recovery of copper from sulfide minerals, especially in developing countries. These operations treat copper ores by using heap-leaching technology; in addition, this technology has been used for the recovery of uranium. Bioleaching is also successfully applied in the pre-treatment (biooxidation) of refractory gold minerals. Since 1986, large-scale biooxidation plants have been operating at several locations in the world. These plants use tank-leaching technologies, although there are some commercial applications in which refractory gold ores are treated in heaps. Bioleaching may be used as an alternative process for the extraction and recovery of other metals like nickel, zinc, cobalt, and molybdenum among others. Such metals can occur in significant quantities in sulfidic ores or can be associated with them. Although bioleaching could have the same advantages over other technologies for the recovery of these metals, as occurs in the cases of copper and gold, there are practically no commercial applications of this technique. The exceptions include a plant where cobalt is recovered from sulfidic tailings and some reports concerning the technical feasibility of applying the technology to zinc, nickel and cobalt concentrates (Brierley & Brierley, 2001; Acevedo, 2002; Harvey *et al.*, 2002; Rawlings *et al.*, 2003; Olson *et al.*, 2003; da Silva, 2004a; Clark *et al.*, 2005; Morin *et al.*, 2005).

The extraction of base metals from sulfide ores is mainly based on the activity of acidophilic microorganisms. The most important are the iron- and/or sulfur-oxidizing bacteria and archaea, which grow autotrophically by fixing CO_2 from the atmosphere. A complete description of these microorganisms can be found in chapter 1. The susceptibility of sulfide minerals to bioleaching depends on several different characteristics, including crystal structure, solubility and electrochemical

E.R. Donati and W. Sand (eds.), Microbial Processing of Metal Sulfides, 103–119.

potential. According to the widely accepted bioleaching mechanisms (Schippers & Sand, 1997; Sand *et al.*, 1999; Rohwerder *et al.*, 2003; chapter 2 in this book), sulfides can be classified into two different groups. Acid-insoluble sulfides, like pyrite (FeS_2), molybdenite (MoS_2) and tungstenite (WS_2), are attacked through the thiosulfate mechanism. Acid-soluble sulfides, like ZnS, $CuFeS_2$, NiS, CoS and PbS, are degraded through the polysulfide mechanism. More in-depth information on the bioleaching mechanisms is presented in chapter 2.

The conditions for leaching through the use of autotrophic microorganisms are very restrictive and extremely favorable for the dissolution of metal (especially the low pH) but the presence of large amounts of oxides and carbonates limits the use of these microorganisms. In addition, the extraction of metals from nonsulfidic minerals (like lateritic ores) can be catalyzed by fungus, yeasts or heterotrophic bacteria (Ehrlich, 2001) that can often tolerate high pH values. These microorganisms produce large amounts of organic acids such as citric, lactic and tartaric acids among others and these can complex and mobilize metals from nonsulfidic solids. Among the bacteria, members of the genus *Bacillus* are the most effective, while *Aspergillus* and *Penicillium* are the most important species among fungi (Viera & Donati, 2005). Several studies on the use of heterotrophic microorganisms in the leaching of different materials – including metallic oxides, metallic ores, quartz sands and silicates, as well as solid residues – have been published (Willscher & Bosecker, 2003).

An excellent review on studies involving the bioleaching and biobeneficiation of ores to recover metals like cobalt, nickel, zinc, molybdenum, manganese, silver, gallium, platinum group metals and uranium was written by Ehrlich (1997). That review also includes some discussion about leaching with heterotrophic bacteria. In this chapter, we present an update of that review on the use of autotrophic and heterotrophic microorganisms for the bioleaching of base metals other than iron and copper. The bioleaching of zinc will be covered in more detail and this discussion will include our own results on the bioleaching of zinc ores using pure and mixed cultures of iron-oxidizing and sulfur-oxidizing microorganisms.

1.1. Molybdenum

Though molybdenum is found in such minerals as wulfenite ($PbMoO_4$) and powellite ($CaMoO_4$), the main commercial source of molybdenum is molybdenite (MoS_2). Molybdenum is mined directly and is also recovered as a byproduct of copper mining. Although levels of sensitivity to molybdate vary with the tested strain (Rawlings, 1997), concentrations higher than 1–2 mM of molybdate are usually highly toxic for *A. ferrooxidans*. Experiments on molybdenum bioleaching from a low-grade copper ore (with 27% of molybdenum) were carried out using *A. ferrooxidans* adapted to grow in the presence of molybdate. Optimization of the culture conditions (pH, temperature and culture media composition) led to a solid residue with 7% of molybdenum (Nasernejad *et al.*, 1999). However, experiments on the chemical leaching and bioleaching of a molybdenite concentrate (with 46.7%

of molybdenum) made by Romano and co-workers did not show any dissolution of the molybdenite phase. These results show that bioleaching with mesophilic and moderate thermophilic bacteria partially attacked the chalcopyrite but the extreme thermophiles were able to selectively dissolve the chalcopyrite rendering practically pure molybdenite (Romano *et al.*, 2001).

Some molybdenum ores contain 0.002% to 0.2% rhenium, which can also be recovered by bioleaching. In order to recover the rhenium associated with molybdenite, bioleaching experiments on a molybdenite concentrate were conducted using a native strain of *A. ferrooxidans* (Askari Zamani *et al.*, 2005). The recovery of both molybdenum and rhenium was evaluated under different conditions. When the mineral concentrate was used as the only energy source, the extraction of both elements was very low. The addition of ferrous sulfate, sulfur or pyrite enhanced the recoveries. The use of pyrite as an energy source enabled *A. ferrooxidans* to grow in the presence of up to 250 $mg.L^{-1}$ of molybdenum, a concentration much higher than that reported as toxic.

1.2. Uranium

In the last decade there have been very few studies on the bioleaching of uranium. The effects of several factors, such as the presence of different concentrations of iron(II) and changes in the nutrient medium, on uranium bioleaching from black shale using *A. ferrooxidans* were recently published (Lee *et al.* 2005). The extent of uranium leaching was low and was not enhanced by the addition of iron(II).

1.3. Lithium

Lithium is used in many industrial processes such as the production of ceramics and glass, in primary aluminum production and in the manufacture of lubricants, greases and batteries. Spodumene, a lithium aluminosilicate, is an important source of lithium but it is very expensive to mine due to the large energy requirements. The bioleaching of an aluminosilicate (containing 95% spodumene) by two moulds (*Aspergillus niger* and *Penicillium purpurogenum*) and a yeast (*Rhodotorula rubra*) isolated from the mineral was investigated. The bioleaching by fungus was related to low molecular weight metabolites, while bioleaching by yeast was related to its metabolic activity and to structural macromolecules of the capsule and cell wall (Rezza *et al.*, 2000).

1.4. Lead

The main lead mineral is galena (lead sulfide), which is commonly found together with sphalerite. The microbial bioleaching of the insoluble lead sulfides produces lead sulfate (anglesite), which also has a very low solubility. Although lead is toxic for most microorganisms used in bioleaching, this effect is reduced due to the low solubility of the product. In complex ores and concentrates, the selective solubilization of copper, nickel and zinc, which often occurs with galena, could be the basis of the beneficiation of lead ores where lead remains in the residue. A diffusion-controlled indirect mechanism was proposed by da Silva (2004b) for the bacterial oxidation of galena by a mixed culture of iron- and sulfur-oxidizing bacteria. The product layer consisted of lead sulfate and elemental sulfur. During bioleaching processes galena was selectively oxidized to anglesite with partial passivation of sphalerite (da Silva *et al.*, 2003). Some studies have been carried out to recover lead – using chemical leaching – from the residues obtained in the bioleaching of complex sulfides containing sphalerite and galena (Liao & Deng, 2004; Frías *et al.*, 2002). In the latter case, silver and gold were also recovered from the residues. A consortium consisting of autotrophic and heterotrophic microorganisms was used in the bioleaching of metals from a copper-lead-zinc sulfide concentrate (Tipre & Dave, 2004). Results showed copper and zinc extraction levels above 80%; in addition, bioleaching produced 83% galena oxidation from the concentrate.

1.5. Manganese

Manganese occurs principally as pyrolusite (MnO_2) and, to a lesser extent, as rhodochrosite ($MnCO_3$). Bioleaching of manganese from mine tailings on a laboratory and pilot plant scale using *A. ferrooxidans* has been performed. Results showed that the use of a drum reactor overcame the initial bacterial inhibition observed when operating at over 15% pulp density in baffled flask and stirred reactors. The drum reactor allowed the authors to work with 50% pulp density and they achieved dissolution of up to 90% of manganese (Medrano-Roldan *et al.*, 2005). The bioleaching of a low grade manganese ore, containing around 25.7% of manganese (mainly as pyrolusite) and 25% of iron, was studied using a native fungal strain of *Penicillium citrinum* (a manganese-reducing fungus) (Acharya *et al.*, 2003). Around 68.3% of the manganese present in the ore was dissolved in 45 days. This fungus produced organic acids (mainly citric acid and oxalic acid) in the leaching medium and these reduced the manganese ore. The manganese ions (released as manganese oxalate and manganese citrate) formed a product layer that was precipitated on the ore. The reaction was controlled by diffusion of the reactants through the permeable product layer in a similar way to the bioleaching of ores by acidophilic bacteria. Veglio *et al.* (1997) studied the bioleaching of manganiferous ores (17–20% of manganese) by heterotrophic mixed cultures. They performed experiments on laboratory and pilot plant scales. Lab-scale studies resulted in 95–100% manganese extraction after 36–48 hours of treatment with a pulp density of 20%. A combined approach using two packed bed reactors was employed in the bioleaching with *A. ferrooxidans* of a low-grade sulfide ore from Capillitas (Catamarca, Argentina) (Curutchet *et al.*, 2001). The mineral contained 9.7% zinc, 1.89% copper, 11.6% iron and 11.4% manganese. Due to the high content of carbonate minerals, bacterial leaching was only feasible after neutralization. Therefore, the experimental set-up consisted of two packed bed reactors and an air-lift percolator containing the ore. Sulfuric acid was produced in one reactor by a biofilm of *A. ferrooxidans* formed on sulfur particles. In the other reactor, iron(III) was produced by a biofilm of *A. ferrooxidans* formed on glass beads. The effluents generated by these reactors were used for ore leaching. After 75 days, 99.5% of the manganese was recovered. The simultaneous bioleaching of zinc sulfide and manganese dioxide was studied in the presence of *A. ferrooxidans*; in these studies it was shown that the dissolution rate of zinc sulfide in particular was improved (Kai *et al.*, 2000).

1.6. Nickel

Nickel is widely used as an alloying element in steels and super alloys and, amongst other applications, in batteries, catalysts, coinage and plating. The nickel mineral most commonly mined is pentlandite ($(Fe,Ni)_9S_8$). Results obtained in bioleaching experiments of different nickel sulfides (millerite (NiS), nickeline (NiS), heazelwoodite (Ni_3S_2), pentlandite $(Fe,Ni)_9S_8$) show that the nickel recovery is highly dependent on the sulfide crystalline structure. Hence, the dissolution of heazelwoodite was higher than that obtained for NiS but lower than that reported for pentlandite (Giaveno & Donati, 2001). Experiments with *A. ferrooxidans* showed that bacteria attached to the mineral surface play an important role. In the case of bioleaching using the thermophilic bacteria, results showed that there were no bacteria attached to the mineral; the dissolution of the mineral was due to oxidation by Fe(III) and by the action of sulfuric acid (Zhang & Fang, 2005). Mason and Rice (2002) performed leaching experiments with *A. ferrooxidans* adapted over 18 weeks to a copper-nickel-iron sulfide concentrate. In this case nickel was completely leached in 5 weeks. The extraction curve correlated with the bacterial growth curve. Pilot plant tests have demonstrated the efficacy of mixed cultures of *A. ferrooxidans*, *A. thiooxidans* and *L. ferrooxidans* in the bioleaching of nickel from pentlandite in a complex sulfide concentrate (Brierley & Brierley, 2001). Billiton developed a process named BioNIC™ to treat low-grade nickel ores based on the gold bioleaching process (Clark *et al.*, 2005). This process has been tested but it has not been possible to identify an ore with a suitable concentration and size to allow economic recovery of nickel (Rawlings *et al.*, 2003). Nickel could be recovered from lateritic minerals by bioleaching using heterotrophic microorganisms (Valix *et al.*, 2001). Comparisons between the efficacy of bacteria from the genera *Bacillus* and *Pseudomonas* and fungi from the genera *Aspergillus* and *Penicillium* in the leaching process on garnierite ($Ni_3MgSi_6O_{15}(OH)_2.6H_2O$) were carried out. *A. Niger* was the most efficient organism in the bioleaching of nickel. Leaching experiments with citric acid concentrations similar to those obtained in the cultures showed a much lower dissolution compared to the process in which the biogenic acid was present (Castro *et al.*, 2000). Studies on the bioleaching of copper, zinc and nickel from a mining ore using organic acids produced by *A. niger* have been published (Mulligan *et al.*, 2004). The effectiveness of the organic acid was enhanced when sulfuric acid was added. The maximum metal dissolution obtained was 68% for copper, 46% for zinc and 34% for nickel. Ultrasonic pretreatment of

A. niger spores substantially improved the bioleaching of lateritic nickel ore due to a higher biomass formation. On using this technique, 95% nickel was obtained in 14 days (cf. only 24.9% nickel obtained on using conventional in situ bioleaching for the same time period) (Swamy *et al.*, 2005).

1.7. Cobalt

Cobalt is extracted from a number of minerals, the main ones being smaltite, $(CoNi)As_3$; linnaeite, Co_3S_4; cobaltite, CoAsS; and glaucodot, (CoFe)AsS. All commercial cobalt is obtained as a by-product from the production of other metals, usually copper. BRGM (Bureau de Recherches Géologiques et Minières) developed a bioleaching process that uses stirred-tank reactors for the recovery of cobalt at the Kasese plant in Uganda. The bacterial community was composed essentially by mesophilic sulfur- and iron-oxidizers. The process addressed environmental and economic objectives. Firstly, it aimed at extracting the valuable cobalt, trapped in sulfide minerals, from dredged mine tailings left behind by closed mining activity; secondly, the approach was aimed at reducing the pollution caused by the natural leaching of mine tailings. The plant treated 10.2t per hour of a cobaltiferous sulfide concentrate containing on average 80% pyrite and 1.4% cobalt. The reactors were designed to work at 42 °C, but during operation the temperature could only be kept between 46 and 50 °C. However, the increase in temperature did not significantly affect the bioleaching efficiency of the system due to the flexibility of bacterial communities in adapting themselves to new situations. The plant has been operating (with interruptions) since 1999 (Morin *et al.*, 2003; Brochot *et al.*, 2004). Bioleaching of cobalt from lateritic ores was assessed using a two-stage batch system. In the first stage, the leaching acids were produced by the fermentation of sugar cane by *A. niger*. The chemical leaching of the ore took place in the second stage on using a pulp density of 10% and a temperature of 60 °C. The recovery of cobalt increased from 40% (one stage system) to 68%. In this way, the interference produced in the fungal growth by the presence of the mineral was avoided (Coto *et al.*, 2005).

1.8. Zinc

According to the International Zinc Association, the mining, smelting and refining of zinc contributes US$ 18.5 billion to the world economy each year (International Zinc Association, 2005). Over the last 25–30 years the zinc industry has moved away from traditional pyrometallurgy to hydrometallurgy. Nowadays, about 80% of the world's total zinc is produced through conventional hydrometallurgical methods (roast-leach-electrowinning) (Harvey *et al.*, 2003).

The potential benefits of commercial bioleaching of zinc minerals are significant in the treatment of concentrate, which is difficult to process using conventional technologies. In addition, zinc concentrate bioleaching, compared to hydrometallurgy, has the advantage of not requiring roasting, sulfuric acid plants and washing of the gaseous effluents (Rodriguez *et al.*, 2003).

The main zinc mineral is sphalerite, which is degraded by acidophilic autotrophic bacteria through the polysulfide mechanism.

(1) $ZnS + H^+ + Fe^{3+} \rightarrow Zn^{2+} + 0.5\,H_2S_n + Fe^{2+} (n \geq 2)$

(2) $0.5\,H_2S_n + Fe^{3+} \rightarrow 0.125\,S_8 + Fe^{2+} + H^+$

(3) $0.125\,S_8 + 1.5\,O_2 + H_2O \rightarrow SO_4^{2-} + 2H^+$

(4) $2\,Fe^{2+} + 0.5\,O_2 + 2\,H^+ \rightarrow 2\,Fe^{3+} + H_2O$

The ferrous iron produced in reactions 1 and 2 can be reoxidized to ferric iron by iron-oxidizing microorganisms such as *A. ferrooxidans*, *Leptospirillum* or *Sulfobacillus*. According to reaction (2), when zinc sulfide is oxidized by ferric ions, a product layer of elemental sulfur is formed on the surface of the mineral. The diffusion of ferrous ions across this sulfur layer becomes the rate-limiting step (Lochmann & Pedlik, 1995). Elemental sulfur is relatively stable but it may be oxidized to sulfate by sulfur-oxidizing microbes. In such a case, the surface reaction with ferric ions becomes the rate-limiting step. Significant slowing of the leaching process confirmed the passivation of the sphalerite surface by a layer of elemental sulfur when there was no sulfur-oxidizing bacteria present (Fowler & Crundwell, 1999). Even in the presence of sulfur-oxidizing bacteria, if the rate of elemental sulfur oxidation (reaction 3) is inadequate, the diffusion through this layer may be the rate-limiting step (da Silva, 2004b). To sum up, the role of the microorganisms in the solubilization of zinc sulfides is to provide sulfuric acid (reaction 3) for a proton attack and to keep the iron in the oxidized ferric state (reaction 4) for an oxidative attack on the mineral (Rawlings, 2005).

Lizama *et al.* studied the bioleaching of a zinc sulfide ore in stirred-tank reactors operating in batch mode. The aim of their experiments was to study the events that took place at the start of the bioleaching process. Kinetic analyzes of these tests revealed that the leaching process can be divided into two stages: (1) a period of colonization of the ore by bacteria, and (2) a steady rate period where the reaction could be described by a classical shrinking-core model (Lizama *et al.*, 2003).

A direct relationship was found between the initial microbial attachment on the mineral and the zinc dissolution rate (Rodríguez *et al.*, 2003). In that study, a mixed culture of mesophilic microorganisms (*Acidithiobacillus* and *Leptospirillum*) in batch experiments was used to leach a sphalerite concentrate. The results showed that the presence of soluble iron and its oxidation state seem to be determinant factors in the bioleaching of zinc sulfide. Shi and co-workers (2006) performed bioleaching experiments on marmatite, sphalerite and synthetic ZnS using *A. ferrooxidans* and moderately thermoacidophilic iron-oxidizing bacteria. The bioleaching rate of zinc sulfide increased as the iron in the sample increased.

Thus, marmatite (the zinc sulfide with the highest iron content) had a higher dissolution rate than sphalerite and synthetic ZnS. Iron in both concentrates existed as pyrite. Iron(II) from the minerals was released into the solution and oxidized to iron(III) by bacteria during the bioleaching process. The dissolution of zinc sulfide was accelerated due to galvanic effects.

Pina and co-workers (2005) found that a zinc extraction level over 80% could be achieved in less than 100 hours in a shaken flask at 1% pulp density with *Acidithiobacillus* sp. The presence of ferric iron in the initial leaching step had a great impact on the leaching rate.

In our laboratory, bioleaching experiments on a polymetallic ore containing 17% zinc from La Resbalosa (province of Neuquén, Argentina) were performed in media without iron(II) and with the addition of 1 $g.L^{-1}$ of iron(II). Pure cultures of iron-oxidizing bacteria or sulfur-oxidizing bacteria and mixed cultures of both types of bacteria were employed. The major constituents of the ore were: sphalerite (12%), pyrite (8%), galena (2%) and chalcopyrite (1%). During the first 15 days of experiments, sulfuric acid solution (1.8 M) was added to maintain the pH value near the initial value of 4.0 in order to avoid the deleterious effect of higher pH values on the bacterial growth (Hallman *et al.*, 1992). Results obtained in shaken flasks with 5% pulp density showed that the addition of iron(II) improved the extraction efficiency (Figure 1).

Similar observations were made by Shi *et al.* 2005. The bioleaching of marmatite ((Zn, Fe)S) flotation concentrate by mixed cultures of *A. ferrooxidans* and *Leptospirillum ferrooxidans* was carried out in a 10 L air-lift reactor. Experimental results showed that the addition of iron(II) was favorable for improving the growth and activity of leaching bacteria and increasing the dissolution rate of marmatite. The addition of iron(II) led to a high weight loss of the concentrate (over 70%); the leaching ratio of zinc was higher than 80% and the concentration of zinc in the leach liquor was about 50 $g.L^{-1}$ after 10 days of bioleaching.

When iron was present in the mineral samples, the addition of iron(II) to the culture medium did not improve zinc extraction. Frizan and co-workers (2003) performed the bioleaching of a complex sulfide ore containing 9% of zinc from La Silvita (province of Neuquén, Argentina). Bioleaching experiments were carried out in glass columns with the medium inoculated with *A. ferrooxidans* percolated through the mineral. A total of 75% of zinc was leached in 100 days. Although the culture medium was iron-free 9K (pH = 1.8), within a few hours 0.5 $g.L^{-1}$ of iron(III) appeared in the solution. The presence of iron(III) in the initial leaching step has a beneficial effect on the leaching rate, despite the fact that the first step in the dissolution of sphalerite is chemical (Pina *et al.*, 2005).

Although the presence of soluble iron seems to be important, an excess of this species causes a reduction in the level of zinc extraction due to the formation of jarosite on the mineral surface, a situation that causes a negative effect on both bacterial growth and zinc sulfide leaching (Konishi *et al.*, 1992). The addition of 0.3 and 1.4 $Kg.m^{-3}$ of ferric ion to the *A. brierleyi* culture resulted in a significant decrease in the leaching rates in a batch stirred reactor, probably because of the

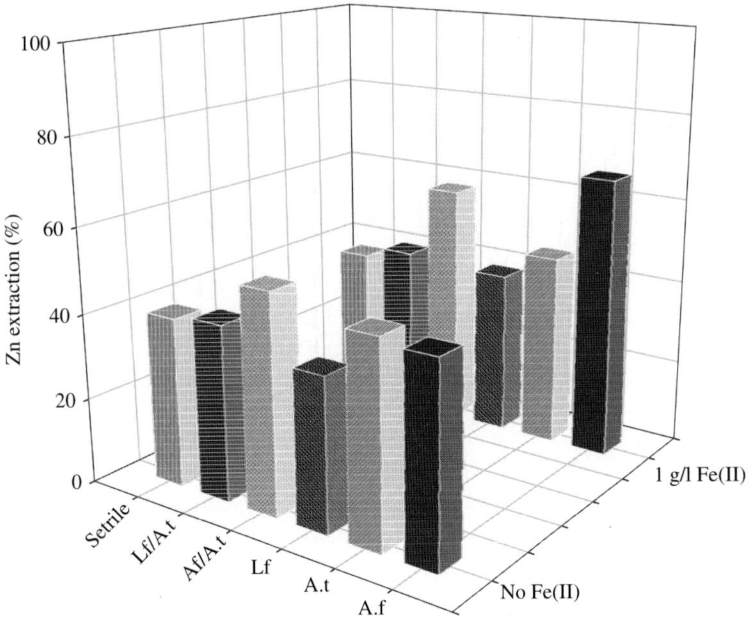

Figure 1. Maximum percentage of zinc extraction reached from La Resbalosa in flasks inoculated with different iron- and sulfur-oxidizing bacteria in the presence or the absence of iron(II). *A.f.*: inoculated with *A. ferrooxidans*; *A.t.*: inoculated with *A. thiooxidans*; *L.f.*: inoculated with *L. ferrooxidans*

formation of an iron precipitate such as jarosite (Konishi and co-workers *et al.*, 1998).

Optimal pH values for bioleaching are between 1.75 and 2.00. At pH values below 1.75, the microorganism metabolism is slower or retarded and at pH values above 2.00 the precipitation of iron as jarosite is very significant (Pogliani & Donati, 2000). The presence of jarosite could be observed in X-ray diffraction patterns of the mineral La Resbalosa. X-ray diffraction patterns of the ore without treatment and the residues from the system inoculated with *A. ferrooxidans* in the presence of iron(II) can be seen in Figure 2 along with that for the sterile control. The main mineralogical species' in the untreated mineral are quartz, pyrite, sphalerite, galena and iron sulfide. In the sterile system, peaks due to jarosite and elemental sulfur are detected. X-ray diffraction patterns of the residue in the flask inoculated with *A. ferrooxidans* in the presence of 1 $g.L^{-1}$ of iron(II) show a decrease in the intensity of the peaks corresponding to pyrite, sphalerite and pyrrothite and an increase in the peaks corresponding to quartz. In addition, peaks due to insoluble products such as anglesite ($PbSO_4$) can be detected. The peaks corresponding to sulfur are not detected in this system, indicating that *A. ferrooxidans* used this element as an energy source. Similar results were found by Shi and Fang (2004).

Bioleaching of La Resbalosa ore was also carried out in columns. Zinc extraction by *A. ferrooxidans* reached only 17.5%. In column experiments the pH was not

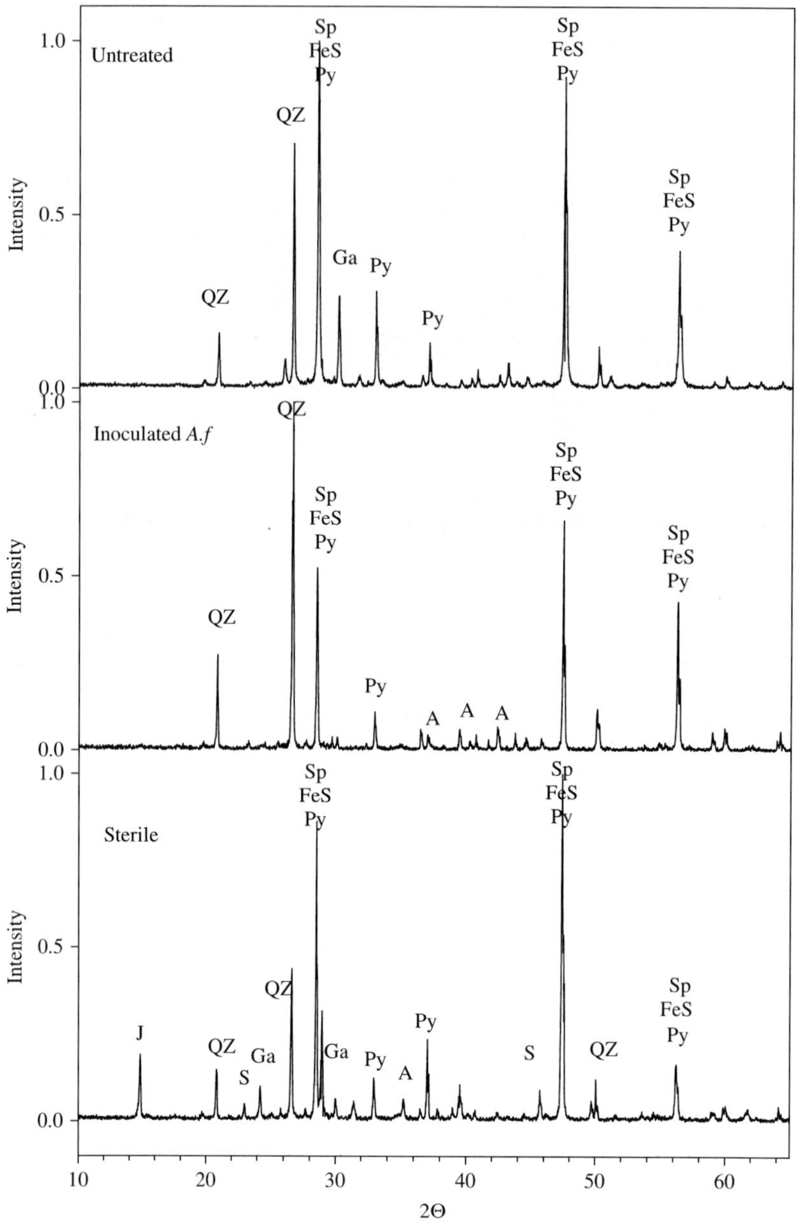

Figure 2. X-ray diffraction patterns for La Resbalosa. (A) untreated mineral, (B) inoculated with *A. ferrooxidans* in the presence of 1 $g.L^{-1}$ iron(II), (C) sterile control

controlled, thus a significant amount of oxide precipitated on the mineral and created a diffusion barrier, which could explain the low recovery obtained (Frizan *et al.*, 2003).

Several experiments on the bioleaching of sphalerite have been carried out using pure cultures or mixed cultures of iron-oxidizing and sulfur-oxidizing bacteria. Comparisons between the different published results are difficult because the experiments were carried out using different culture conditions and temperatures. Bioleaching of sphalerite can be improved by using mixed cultures of bacteria. In experiments using 20% pulp density, zinc extraction as high as 84% was reached using a consortium of acidophilic chemolithotrophic iron and sulfur-oxidizers (*A. ferrooxidans*, *L. ferrooxidans* and *A. thiooxidans*) and heterotrophic organisms (Tipre & Dave, 2004).

In experiments carried out in our laboratory, zinc extraction in the culture with *A. ferrooxidans* (DMS 11477) was scarcely higher than that obtained using a mixed culture (*A. ferrooxidans* and *A. thiooxidans*) (Figure 3). Similar results were obtained in bioleaching experiments on La Resbalosa ore in an air-lift reactor, where the highest zinc extraction reached 60% in 32 days (Giaveno *et al.*, 2005) when inoculated only with *A. ferrooxidans*.

It can be seen in Figure 3 that when the iron-oxidizing bacteria were *L. ferrooxidans*, the extraction was similar to the sterile control. For this reason additional experiments were performed to assess the effect of dissolved zinc and pulp density on the oxidation of iron(II) by *L. ferrooxidans*. It was found that the oxidation rate of iron(II) by *L. ferrooxidans* was not significantly modified by the presence

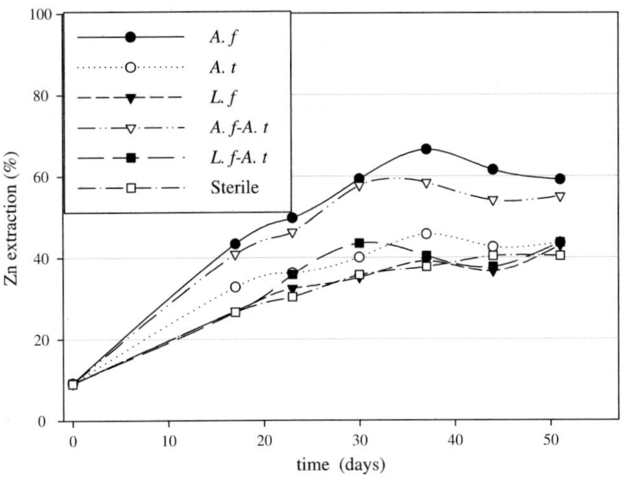

Figure 3. Percentage of zinc extraction vs. time from La Resbalosa in the presence of 1 g.L^{-1} of iron(II). *A.f.*: inoculated with *A. ferrooxidans*; *A.t.*: inoculated with *A. thiooxidans*, *L.f.*: inoculated with *L. ferrooxidans*; *A.f.-A.t.*: inoculated with *A. ferrooxidans/A. thiooxidans*; *L.f.-A.t.*: inoculated with *L. ferrooxidans/A.thiooxidans*

of different concentrations of zinc. Similar results were found by Sampson *et al.* 2005, who indicated that concentrations of 90 $g.L^{-1}$ of zinc in the column effluent did not appear to inhibit the microbial oxidation of iron-oxidizing bacteria. On the other hand, the presence of solids slowed down the process in batch cultures and this phenomenon could have been the cause of the low performance of *L. ferrooxidans*. Nevertheless, bioleaching experiments on a complex sulfide ore containing 9% of zinc from La Silvita (province of Neuquén, Argentina) in an air-lift reactor inoculated with an indigenous strain of *L. ferrooxidans* gave a zinc extraction level of 98% in 65 days (Giaveno *et al.*, 2005).

Rodriguez and co-workers (2003) studied the sphalerite bioleaching mechanism at low and high temperatures. A very pure zinc concentrate was bioleached at two different temperatures; at 35 °C in the presence of mesophilic microorganisms (mainly *Acidithiobacillus* and *Leptospirillum* spp) and at 68 °C in the presence of thermophilic microorganisms (mainly *Sulfolobus* spp). After 30 days, the zinc extraction was 27% at 35 °C and 76% at 68 °C. The differences found between mesophilic and thermophilic bioleaching efficiencies could be due to the higher sulfur-oxidizing activity of the thermophilic microorganisms, which would eliminate the passive layer of sulfur. In experiments in batch culture in the absence of iron, dissolution of zinc sulfide was enhanced by the action of *Acidithiobacillus caldus* at 40 °C. The bioleaching mechanism was similar to that observed for mesophilic species of *Acidithiobacillus* at 30 °C, although the metal recovery was lower (Semenza *et al.*, 2002).

A kinetic study of sphalerite leaching by thermophilic *A. brierleyi* in a batch stirred reactor was carried out by Konishi and co-workers *et al.* (1998). Stoichiometric parameters have demonstrated that the bioleaching rate with the thermophilic *A. brierleyi* is about seven times higher than that obtained with *A. ferrooxidans*. Moderate thermophiles can tolerate higher acidity and temperature, rendering the process more effective when compared with *A. ferrooxidans* (Shi *et al.*, 2005). Deveci and co-workers (2004) investigated the bioleaching of the complex Pb/Zn ore concentrate in shaken flasks using mesophilic (at 30 °C), moderately thermophilic (at 50 °C) and extremely thermophilic (at 70 °C) strains of acidophilic bacteria. Moderate thermophilic bacteria (*S. yellowstonensis*) showed a better bioleaching performance than the extremely thermophilic (*A. brierleyi*) bacteria. This finding could be related to a lower tolerance to dissolved metals and more fragile cell structure (Sampson *et al.*, 2005).

The use of native bioleaching microbes or adapted strains improved the extraction of zinc in bioleaching processes. Experiments performed by da Silva (2004b) showed a significant initial lag period of about 20 days, which represents a time of bacterial adaptation. This lag period can be overcome by using cells adapted to the mineral (Giaveno *et al.*, 2005). The extraction of zinc from a marmatite flotation concentrate improved from 35% to 90% on using adapted strains of *A. ferroxidans* (Shi & Fang, 2004). Bioleaching studies on sphalerite concentrate have demonstrated that zinc recovery is five times higher at 30 °C with N39-30-03 (native bioleaching microbes) compared with the reference strain *A. ferrooxidans* (Keeling *et al.*, 2005).

As shown by Pani and co-workers (2003), the bioleaching of zinc can be enhanced by the optimization of operational variables. Many factors affect the bioleaching of zinc sulfide and these include source and concentration of the microorganisms, leaching conditions (temperature, pH), pulp density, sphalerite content and particle size. The last two factors are different in ores and concentrates, thus the mechanism of ore leaching could be quite different to that of the concentrate (Lizama *et al.*, 2003). Due to the number of factors affecting bioleaching, the universal application of a kinetic model is restricted. Liu and co-workers (2004) proposed a simple three-parameter model for bacterial growth and bioleaching by *Acidithiobacillus* spp and this fitted quite well the experimental data obtained under different conditions by several authors.

GeoBiotics LLC has developed the GEOCOAT® process for the treatment of sulfide base metals and gold concentrates. In this process, thickened flotation concentrate is contacted during heap stacking with gravel-sized support rock, forming a thin adherent concentrate coating on the support rock particles. The stacked heap has an open structure and is highly permeable to flows of solution and air. Acid solution, which contains acidophilic bacteria, is introduced using a sprayer system and circulated through the heap to bioleach the contained metals. Leaching periods in this process are usually longer than those for tank leaching, but they are considerably shorter than those for typical heap-leaching systems.

GeoBiotics LLC and Kumba Resources (Pty) Ltd are investigating the feasibility of applying the GEOCOAT® process to the leaching and recovery of zinc from a low-grade sphalerite concentrate produced from the accumulation of flotation tailings by Kumba's Rosh Pinah zinc mine in Nambia. The heat produced by exothermic oxidation reactions is transferred to the percolating solution and the air blown up through the heap. The rates of solution application and aeration can be varied to control the heap temperature and maintain it within the optimum range for bacterial activity (autothermality). The inoculum used was a mixed mesophilic culture containing *A. ferrooxidans*, *A. thiooxidans* and *L. ferrooxidans*. Final zinc dissolution after 90 days was 91% with a corresponding sulfide oxidation of 89% (Sampson *et al.*, 2005). GEOCOAT® heap biooxidation technology was also applied in heaps inoculated with the moderate thermophiles *A. caldus* and *Sulfobacillus thermosulfidooxidans*, and the extreme thermophiles *Acidianus brierleyi*, *Acidianus infernus*, *Metallosphaera sedula*, *Sulfolobus acidocadarius*, *Sulfolobus shibatae* and *Sulfolobus metallicus*. Tests in six-meter columns indicated that a zinc extraction higher than 95% could be achieved (Harvey *et al.*, 2002). According to the company reports, the bioleaching of sphalerite is a relatively straightforward exercise but winning the zinc from the pregnant leach solution is much more complicated due to the presence of a variety of impurities. Elements such as fluorine, chlorine, manganese, cobalt, copper and iron need to be closely monitored, particularly in circulating solutions. Computer simulation for the sphalerite bioleaching and the downstream processing has been employed to aid in the development of the new process (Harvey *et al.*, 2003).

2. CONCLUSION

Bioleaching using iron- and sulfur-oxidizing microorganisms has been successfully applied on a commercial scale for the recovery of copper from sulfide minerals and in the pre-treatment (biooxidation) of refractory gold minerals since 1986. More recently, the use of this technology has been extended to the recovery of cobalt concentrates. In this chapter we have presented several reports concerning the technical feasibility of this technology for the recovery of several other metals. The presence of large amounts of oxides and carbonates in the minerals limits the use of iron- and sulfur-oxidizing microorganisms. The extraction of metals from non-sulphidic minerals (like lateritic ores) can be catalyzed by fungus, yeasts or heterotrophic bacteria, which can often tolerate high pH values. There are few laboratory studies that address the use of heterotrophic microorganisns but the applicability of this process remains a fertile area in terms of both research and commercial development.

3. REFERENCES

Acevedo F. 2002. Present and future of bioleaching in developing countries. EJB Electronic J Biotechnol 5: 18-19.

Acharya C, Kar RN, Sukla LB. 2003. Studies on reaction mechanism of bioleaching of manganese ore. Minerals Engineering 16: 1027-1030.

Askari Zamani MA, Hiroyoshi N, Tsunekawa M, Vaghar R, Oliazadeh M. 2005. Bioleaching of Sarcheshmeh molybdenite concentrate for extraction of rhenium. Hydrometallurgy 80: 23-31.

Brierley JA, Brierley CL. 2001. Present and future commercial applications of biohydrometallurgy. Hydrometallurgy 59: 233-239.

Brochot S, Durante MV, Villeneuve J, d'Hugues P, Mugabi M. 2004. Modelling of the bioleaching of sulfide ores: application for the simulation of the bioleaching/gravity section of the Kasese Cobalt Company Ltd process plant. Miner Eng 17: 253-260.

Castro IM, Fietto JLR, Vieira RX, Trópia MJM, Campos LMM, Paniago EB, Brandão RL. 2000. Bioleaching of zinc and nickel from silicates using *Aspergillus niger* cultures. Hydrometallurgy 57: 39-49.

Clark ME, Batty J, van Buuren C, Dew D, Eamon M. 2005. Biotechnology in minerals processing: Technological Breakthroughs creating value. In: In: Harrison STL, Rawlings DE, Petersen J, eds. Proceedings of the 16th International Biohydrometallurgy Symposium, September 25–29, Cape Town, South Africa. Produced by Compress www.compress.co.za, 17-29.

Coto O, Peguero M, Abín L, Bruguera N, Marrero J, Bosecker K. 2005. Bioleaching of laterite by *Aspergillus niger* strain O-5: an acidophilic, and nickel and cobalt resistant fungus. In: Harrison STL, Rawlings DE, Petersen J, eds. Proceedings of the 16th International Biohydrometallurgy Symposium, September 25–29, Cape Town, South Africa. Produced by Compress www.compress.co.za, 357-364.

Curutchet G, Donati E, Oliver C, Pogliani C, Viera M. 2001. Development of *Thiobacillus* biofilms for metal recovery. In: Doyle RJ, ed. Microbial Growth in Biofilms, Methods in Enzymology, Academic Press, San Diego, 337, 171-186.

da Silva G. 2004a. Relative importance of diffusion and reaction control during the bacteria and ferric sulfate leaching of zinc sulfide. Hydrometallurgy 73: 313-324.

da Silva GL. 2004b. Kinetics and mechanism of the bacterial and ferric sulfate oxidation of galena. Hydrometallurgy 75: 99-110.

da Silva G, Lastra MR, Budden JR. 2003. Electrochemical passivation of sphalerite during bacterial oxidation in the presence of galena. Miner Eng 16: 199-203.

Deveci H, Akcil A, Alp I. 2004. Bioleaching of complex zinc sulfides using mesophilic and thermophilic bacteria: comparative importance of pH and iron. Hydrometallurgy 73: 293-303.

Ehrlich HL. 1997. Technical potential for bioleaching and biobeneficiation of ores to recover base metals (other than iron and copper), platinum-group metals and silver. In: Rawlings DE, ed. Biomining: Theory, Microbes and Industrial Processes. Springer-Verlag, Berlin, 129-150.

Ehrlich HL. 2001. Past, present and future of biohydrometallurgy. Hydrometallurgy 59: 127-134.

Fowler TA, Crundwell FK. 1999. Leaching of zinc sulfide by *Thiobacillus ferrooxidans*: bacterial oxidation of the sulfur product layer increases the rate of the sulfide dissolution at high concentrations of ferrous ions. Appl Environ Microbiol 65: 5285-5292.

Frías C, Díaz G, Ocaña N, Lozano JI. 2002. Silver, gold and leach recovery from bioleaching residues using the PLINT process. Miner Eng 15: 877-878.

Frizán V, Giaveno A, Chiacchiarini P, Donati R. 2003. Bioleaching of Argentinean sulfide ores using pure and mixed cultures. In: Tzesos M, Hatzikioseyan A, Remoundaki E, eds. Biohydrometallurgy – A Sustainable Technology in Evolution, Proceedings of the International Biohydrometallurgy Symposium IBS 2003, September 14–19, Athens, Greece, National Technical University of Athens, 125-135.

Giaveno A, Donati R. 2001. Bioleaching of heazelwoodite by *Thiobacillus* spp. Process Biochem 36: 955-962.

Giaveno A, Chiacchiarini P, Lavalle L, Donati R. 2005. Reversed flow airlift characterization for bioleaching applications. In: Harrison STL, Rawlings DE, Petersen J, eds. Proceedings of the 16th International Biohydrometallurgy Symposium, September 25–29, Cape Town, South Africa. Produced by Compress www.compress.co.za, 125-135.

Hallman R, Friedrich A, Koops HP, Pommerening-Röser A, Rohde K, Zenneck C, Sand W. 1992. Physiological characteristics of *Thiobacillus ferrooxidans* and *Leptospirillum ferrooxidans* and physicochemical factors influence microbial metal leaching. Geomicrobiol J 10: 193-206.

Harvey TJ, Van Der Merwe W, Afewu K. 2002. The application of the GeoBiotics GEOCOAT® biooxidation technology for the treatment of sphalerite at Kumba resources' Rosh Pinah mine. Miner Eng 15: 823-829.

Harvey TJ, Joubert J, Van Der Merwe W. 2003. Use of Computer Simulation for the Design of a Bioheap Leach for Sphalerite. In: SME Annual General Meeting, February 2003, Cincinnati, Ohio. Available from www.geobiotics.com/pages/publications.cfm.

International Zinc Association: Zinc production and uses. 2005. Available from www.iza.com.

Kai T, Suenaga Y-i, Migita A, Takahashi T. 2000. Kinetic model for simultaneous leaching of zinc sulfide and manganese dioxide in the presence of iron-oxidizing bacteria. Chem Eng Scie 55: 3429-3436.

Keeling, SE, Palmer, ML, Caracatsanis, FC, Johnson, JA, Watling, HR. 2005. Leaching of chalcopyrite and sphalerite using bacteria enriched from a spent chalcocite heap. Miner Eng 18: 1289-1296.

Konishi Y, Kubo H, Asai S. 1992. Bioleaching of zinc sulfide concentrate by *Thiobacillus ferrooxidans.* Biotechnol Bioeng 39: 66-74.

Konishi Y, Nishimura H, Asai S. 1998. Bioleaching of sphalerite by the acidophilic thermophilic *Acidianus brierleyi*. Hydrometallurgy 47: 339-352.

Lee J-U, Kim S-M, Kim K-W, Kim IS. 2005. Microbial removal of uranium in uranium-bearing black shale. Chemosphere 59: 147-154.

Liao MX, Deng TL. 2004. Zinc and lead extraction from complex raw sulfides by sequential bioleaching and acidic brine leach. Miner Eng 17: 17-22.

Liu, H-L, Teng C-H, Cheng Y-C. 2004. Semiempirical model for bacterial growth and bioleaching of *Acidithiobacillus* spp. Chem Eng J 99: 77-87.

Lizama HM, Fairweather MJ, Dai Z, Allegretto TD. 2003. How does bioleaching start?. Hydrometallurgy 69: 109-116.

Lochmann J, Pedlik M. 1995. Kinetics anomalies of dissolution of sphalerite in ferric sulfate solution. Hydrometallurgy 37: 89-96.

Mason LJ, Rice NM. 2002. The adaptation of *Thiobacillus ferrooxidans* for the treatment of nickel–iron sulfide concentrates. Miner Eng 15: 795-808.

Medrano-Roldán H, Salinas-Martínez A, Hernández-Carbajal GR, Solís-Soto A, Morales-Castro J, Delgado E, Soto-Cruz NO, Pereyra-Alférez B, Galán-Wong JL, Dávila-Flores RT. 2005. Manganese bioleaching from mine tailings by a native *Acidithiobacillus ferrooxidans* strain. II. Data from laboratory and pilot plant non-conventional rotating Drum reactors. In: Harrison STL, Rawlings DE, Petersen J, eds. Proceedings of the 16th International Biohydrometallurgy Symposium, September 25–29, Cape Town, South Africa. Produced by Compress www.compress.co.za, 351-355.

Morin D, d'Hugues P, Mugabi M, 2003. Bioleaching of metallic sulfide concentrate in continuous stirred reactors at industrial scale experience and lessons. In: Tzesos M, Hatzikioseyan A, Remoundaki E, eds. Biohydrometallurgy – A Sustainable Technology in Evolution, Proceedings of the International Biohydrometallurgy Symposium IBS 2003, September 14–19, Athens, Greece, National Technical University of Athens.

Morin D, Lips A, Pinches A, Huisman J, Frias C, Norberg A, Forssberg E. 2005. BioMinE Integrated project for the development of biotechnology for metal-bearing materials in Europe. In: Harrison STL, Rawlings DE, Petersen J, eds. Proceedings of the 16th International Biohydrometallurgy Symposium, September 25–29, Cape Town, South Africa. Produced by Compress www.compress.co.za, 33-40.

Mulligan CN, Kamali M, Gibbs BF. 2004. Bioleaching of heavy metals from low-grade mining ore using *Aspergillus niger*. J Haz Mat 110: 77-84.

Nasernejad B, Kaghazchi T, Edrisi M, Sohrabi M. 1999. Bioleaching of molybdenum from low-grade copper ore. Process Biochem 35: 437-440.

Olson GJ, Brierley JA, Brierley CL. 2003. Bioleaching review part B: Progress in bioleaching: applications of microbial processes by the mineral industries. Appl Microbiol Biotechnol 63: 249-257.

Pani CK, Swain S, Kar RN, Chaudhury GR, Sukla LB, Misra VN. 2003. Biodissolution of zinc sulfide concentrate in 160 l 4-stage continuous bioreactor. Miner Eng 16: 1019-1021.

Pina PS, Leão VA, Silva CA, Daman D, Frenay J. 2005. The effect of ferrous and ferric iron on sphalerite bioleaching with *Acidithiobacillus* sp. Miner Eng 18: 549-551.

Pogliani C, Donati E. 2000. Immobilisation of *Thiobacillus ferrooxidans*: importance of jarosite precipitation. Process Biochem 35: 997-1004.

Rawlings DE. 1995. Restriction enzyme analysis of 16S rRNA genes for the rapid identification of *Thiobacillus ferrooxidans*, *Thiobacillus thiooxidans* and *Leptospirillum ferrooxidans* strains in bioleaching environments. In: Jerez JCA, Vargas T, Toledo H, Wiertz JV, eds. Biohydrometallurgical processing, Proceedings of the International Biohydrometallurgy Symposium IBS-95, Vina del Mar, Chile. University of Chile, Santiago de Chile, Vol. 2, 9-17.

Rawlings DE. 1997. Mesophilic, Autotrophic bioleaching bacteria: description, physiology and role. In: Rawlings DE, ed. Biomining: Theory, Microbes and Industrial Processes, Springer-Verlag, Berlin, 229-245.

Rawlings DE, Dew D, du Plessis C. 2003. Biomineralization of metal-containing ores and concentrates. Trends Biotechnol. 21: 38-44.

Rawlings DE. 2005. Characteristics and adaptability of iron- and sulfur oxidizing microorganisms used for the recovery of metals from minerals and their concentrates. Microbial Cell Factories 4: 13-28.

Rezza I, Salinas E, Elorza M, Sanz de Tosetti M, Donati E. 2000. Mechanisms involved in bioleaching of an aluminosilicate by heterotrophic microorganisms. Process Biochem 36: 495-500.

Rodríguez Y, Ballester A, Blázquez ML, González F, Muñoz JA. 2003. New information on the sphalerite bioleaching mechanisms at low and high temperature. Hydrometallurgy 71: 57-66.

Romano P, Blázquez ML, Ballester A, González F, Alguacil FJ. 2001. Reactivity of a molybdenite concentrate against chemical or bacterial attack. Miner Eng 14: 987-996.

Rohwerder T, Gehrke T, Kinzler K, Sand W. 2003. Bioleaching review part A: Progress in bioleaching: fundaments and mechanism of bacterial metal sulfide oxidation. Appl Microbiol Biotechnol 63: 239-248.

Sampson MI, Van der Merwe W, Harvey TJ, Bath MD. 2005. Testing the ability of a low grade sphalerite concentrate to achieve autothermality during biooxidation heap leaching. Miner Eng 18: 427-437.

Sand W, Gehrke T, Jozsa P-G, Schippers A. 1999. Direct versus indirect bioleaching. In: Amils R, Ballester A, eds. Biohydrometallurgy and the environment toward the mining of the 21st century,

Proceedings of the International Biohydrometallurgy Symposium IBS-99, El Escorial, Spain. Elsevier, Amsterdam, Part A, 27-49.

Schippers A, Sand W. 1997. Bacterial leaching of metal sulfides proceeds by two indirect mechanisms via thiosulfate or via polysulfides and sulfur. Appl Environ Microbiol 65: 319-321.

Semenza M, Viera M, Curutchet G, Donati E. 2002. The role of *Acidithiobacillus caldus* in the bioleaching of metal sulfides. Latin American Appl Res 32: 303-306.

Shi S-Y, Fang Z-H. 2004. Bioleaching of marmatite flotation concentrate by *Acidithiobacillus ferrooxidans*. Hydrometallurgy 75: 1-10.

Shi S-Y, Fang Z-H. 2005. Bioleaching of marmatite flotation concentrate by adapted mixed mesoacidophilic cultures in an air-lift reactor. Int J Miner Process 76: 3-12.

Shi S-Y, Fang Z-F, Ni J-R. 2005. Bioleaching of marmatite flotation concentrate with a moderately thermoacidophilic iron-oxidizing bacterial strain. Hydrometallurgy 18: 1127-1129.

Shi S-Y, Fang Z-F, Ni J-R. 2006. Comparative study on the bioleaching of zinc sulfides. Process Biochem 41: 438-446.

Swamy KM, Narayana KL, Misra VN. 2005. Bioleaching with ultrasound. Ultrasonics Sonochemistry 12: 301-306.

Tipre DR, Dave SR. 2004. Bioleaching process for Cu-Pb-Zn bulk concentrate at high pulp density. Hydrometallurgy 74: 37-43.

Valix M, Tang JY, Cheung WH. 2001. The effects of mineralogy on the biological leaching of nickel laterite ores. Miner Eng 14: 1629-1635.

Veglio F, Beolchini F, Gasbarro A, Toro L, Ubaldini S, Abbruzzese C. 1997. Batch and semi-continuous tests in the bioleaching of manganiferous minerals by heterotrophic mixed microorganisms. Int J Miner Process 50: 255-273.

Viera M, Donati E. 2004. Microbial processes to metal recovery from waste products. Curr Top Biotechnol 1: 117-127.

Willscher S, Bosecker K. 2003. Studies on the leaching behaviour of heterotrophic microorganisms isolated from an alkaline slag dump. Hydrometallurgy 71: 257-264.

Zhang G, Fang ZH. 2005. The contribution of direct and indirect actions in bioleaching of pentlandite. Hydrometallurgy 80: 59-66.

CHAPTER 6

BIOLEACHING OF METALS IN NEUTRAL AND SLIGHTLY ALKALINE ENVIRONMENT

ALEKSANDRA SKLODOWSKA AND RENATA MATLAKOWSKA
Warsaw University, Faculty of Biology, Warsaw, Poland

1. INTRODUCTION

Copper deposits are found in all around the world in a variety of geologic environments. Depending on the origin and composition they are grouped in five classes: porphyry and related copper deposit, sediment-hosted copper deposits, volcanic-hosted massive sulfide deposits, veins and replacement bodies associated with metamorphic rocks and deposits associated with ultramafic, mafic, ultrabasic and carbonatite rocks (U.S. EPA, 1994).

Porphyry copper and associated deposits are the predominant class located mainly in the western continental edges of North and South America and constitute up 59% of total world mining activity. Stratiform sedimentary and metasedimentary deposits are an important source of copper too making up 24% of the copper mining activity. Volcanic massive sulfide deposits located in the southern Australia made up 7% of the worldwide copper mining activity (U.S. EPA, 1994). Veins and replacement deposits made up 7% of worldwide copper mining activity and occur for example within the uplands in Wales (Ixer, 2001). Copper bearing alkaline ultrabasis and carbonatite deposits are very rare and made up 4% of worldwide copper mining activity. However, almost all copper deposits contain different concentrations of calcium compounds mainly as dolomite and calcium carbonates. For example, in Fairview (South Africa) carbonate contents reach 4.5%, in Ashanti Sansu (Ghana) 7%, Harbour Lights (Australia) 8% and in Sao Bento (Brazil) 9.4% (Karas & Sadowski, 2002). However, carbonatite copper deposits contain up to 62% of dolomite. This type of ore is located in the middle of Europe in a great area from Middle Germany (the eastern edge of the Harz Mountains and the Saale river) to Fore-Sudetic Monocline and North Sudetic Trough and originated of the lower Permian Zechstein formation.

Copper deposits of Fore-Sudetic Monocline and North Sudetic Trough are ranked as one of the largest and the richest of Cu and Ag deposits in the world and

E.R. Donati and W. Sand (eds.), Microbial Processing of Metal Sulfides, 121–129.

they are still under exploitation opposite to Mansfeld region (Bechtel *et al.*, 2002; Piestrzynski *et al.*, 2002). Ores formation is related with basal Zechstein rocks, formed during the marine transgression of southwestern Zechstein Sea. Copper sulfides occur in minerals merged with sandstone (24–85%) and dolomite (11–62%). In this situation all known methods of acid metals bioleaching widely use throughout the world are not useful in Poland. Acidification of ores containing copper up to 3% by sulfuric acid require 50 kg H_2SO_4 per ton of sandstone ore and more for dolomite and flotation wastes with maximum gain 30 kg of copper. The process is difficult from the ecological and technological points of view, however possible.

2. CHARACTERISATION OF ZECHSTEIN ORES INCLUDING KUPFERSCHIEFER POLYMETALLIC MINERALISATION

Copper ores are of the sandstone and limestone type laminated by shale, characterized by relatively high pH value reaching even 8.5. Typical lithostratigraphy of the deposit is presented in Figure 1.

The copper ores show high diversity concerning the structure of ore, ore mineralization and metals content. Generally, depending on the localization sandstone ore represents 24–85%, carbonaceous ore 11–62% and shale ore 5–14% of the total ore volume. The ore series consist of white sandstones (Weissliegendes), Basal Limestone, metal bearing shale (Kupferschiefer) and Zechstein Limestone. Kupfeschiefer are divided into two basic zones oxidized and reduced with transition layer. The deposition of mineralization in the deposit is very changeable and

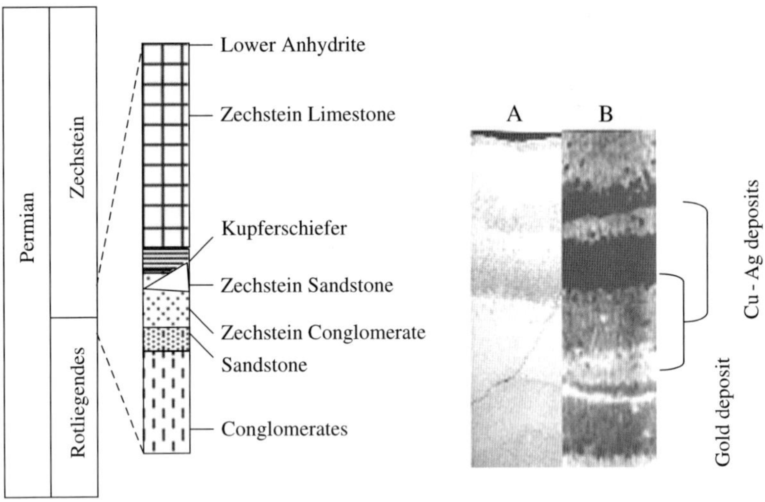

Figure 1. Dominating geological profile of Fore-Sudetic Monocline deposits. A. Drill core sample; B. Drill core sample at X-ray image (Ostrowski & Sklodowska, 1996; Oszczepalski, 1999; Piestrzynski *et al.*, 2002)

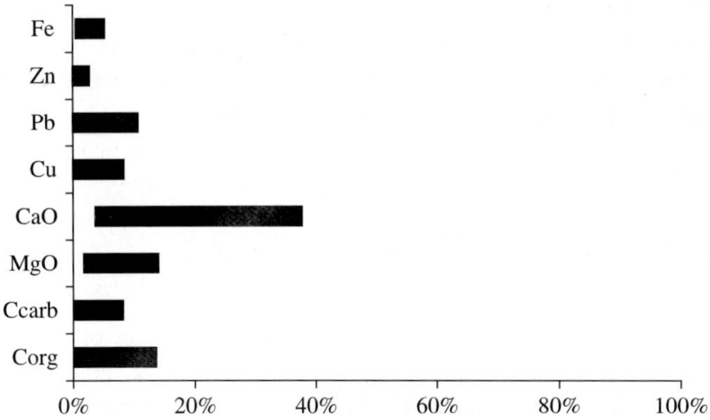

Figure 2. Organic carbon, carbonate carbon and base metal concentration in Fore-Sudetic Monocline copper deposits (Bechtel *et al.*, 2001; Bechtel *et al.*, 2002)

moves from sandstones and shales to shales and dolomites (Bechtel *et al.*, 2002; Piestrzynski *et al.*, 2002).

The most important copper sulfide minerals represented in ores are chalcocite, digenite and covellite with bornite and chalcopyrite. In the deposit commonly occur also silver, zinc, lead, cobalt, arsenic, iron, nickel and noble metals (Figure 2, Table 1).

The place of particularly high concentration of metal sulfide is the level of copper bearing shale (Kupferschiefer) and the contact zones in its bottom and top.

The organic carbon concentration in the Kupferchiefer profile decreases from bottom to top. Between organic compounds such substances as long chain saturated and unsaturated hydrocarbons (C-14–C-32), polycyclic aromatic hydrocarbons (e.g. phenantrene, chrysene, benzo(c)phenantrene), 1,2-benzendicarboxylic acid esters, farnesane, norpristane, pristane and phytane are present. S- and O-PAHs are observed too. The organic geochemical data shows clearly the participation of

Table 1. Selected trace elements concentration in Fore-Sudetic Monocline copper deposits (Bechtel *et al.*, 2001; Bechtel *et al.*, 2002)

Chemical element	Concentration (ppm)
Mo	1–290
Co	8–485
U	4–72
	(ppb)
Au	<2–1750
Pt	<5–883
Pd	<4–564

organic matter as reducing agent in metal sulfide precipitation especially within the Kupferschiefer. Enhanced noble metal concentrations are observed in this type of organic matter and this matter with metal organic complexes seems to be responsible for mobilization and transport of gold.

3. MICROORGANISMS POTENTIALLY USEFUL IN BIOLEACHING FROM NEUTRAL OR SLIGHTLY ALKALINE ORES

A variety of microorganisms are known to be able to catalyze leaching of metal from ore deposits and wastes containing copper sulfides (Kelly & Harrison, 1989; Bosecker, 1997). The best known are acidophilic iron- and sulfur-oxidizing chemolithoautotrophic bacteria. However, they are not useful in bioleaching of metals from neutral or slightly alkaline ores. In this case microorganisms able to oxidize reduced sulfur compounds or elemental sulfur at pH 6-8 are required. These microorganisms can be divided into four physiological groups according to the energy and carbon source:

1. Obligate chemolithoautotrophs – obtain energy by the oxidation of inorganic reduced sulfur compounds and elemental sulfur and assimilate carbon from CO_2 (*Halothiobacillus neapolitanus*, *Thiomicrospira* sp.).
2. Facultative chemolithoautotrophs – use organic as well as inorganic reduced sulfur compounds as energy source and assimilate carbon from CO_2 or organic compounds (*Starkeya novella*, *Thiomonas* sp.).
3. Chemolithoheterotrophs (mixotrophs) – use sulfur compounds as energy source and organic compounds as the source of carbon (*Beggiatoa* sp., *Thiotrix* sp.) (Figure 3).
4. Phototrophs – are able to oxidize reduced sulfur compounds under anaerobic conditions with the formation of sulfate and assimilate carbon from CO_2 (*Thiocystis*, *Chlorobiaceae*, *Chromatiaceae*, *Rhodospirillaceae*).

The most important in metal bioleaching in neutral or slightly alkaline environment are sulfur-oxidizing chemolithotrophic bacteria belonging to five genera: *Thiobacillus*, *Halothiobacillus*, *Thermithiobacillus*, *Thiomonas* and *Starkeya* (Beijerinck, 1904; Kelly & Harrison, 1989; Kelly & Wood, 2000; Kelly *et al.*, 2000; Moreira & Amils, 1997).

These gram-negative rods are members of α, β, and γ classes of *Proteobacteria* and exhibit a wide range of physical growth conditions, such as pH and temperature, but all microorganisms belonging to these classes obtain energy by the oxidation of reduced sulfur compounds, including sulfides, sulfur, thiosulfate, polythionates and thiocyanate. Some species can derive energy from organosulfur compounds and all of them fix carbon dioxide. All species with the exception of *Thiomonas* occur at pH about 6–8. Facultative chemolithoautotrophs *Thiomonas* are moderately acidophilic bacteria living at pH 3–6 and may be useful in bioleaching in neutral environment too although they role was not confirmed (Moreira & Amils, 1997) Table 2 presents a list of some neutrophilic sulfur-oxidizing bacteria.

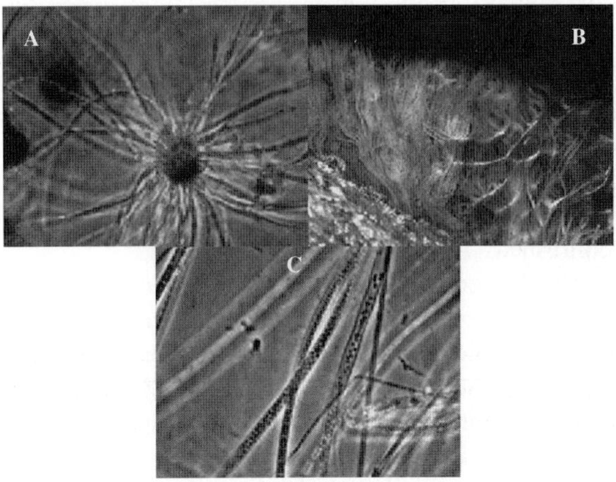

Figure 3. A. *Thiothrix* sp; B. Floating mat of sulfur oxidizing filamentous bacteria; C. *Beggiatoa alba* (Phot. Chlebicki A, Institute of Botany PAS, Krakow, Poland). A, C – magn. 1000x

Other groups of microorganisms useful in bioleaching at neutral pH are heterotrophs, which leach heavy metal by enzymatic reduction of oxidized metal compounds or by the formation of organic and inorganic acids as well as by the excretion of complexic agents (Burgestaller & Schinner, 1993). The most effective in metal solubilisation heterotrophs are *Bacillus* genus and fungi: *Aspergillus* and *Penicillium* (Bosecker, 1997). Such species as *B. circulans* and *B. mucilaginosus* were described as able to leach manganese from oxide ores using different organic compounds as reducing agents (Groudev, 1987).

Table 2. Neutrophilic microorganism potentially useful in metal bioleaching (Kelly & Wood, 2000; Kelly *et al.*, 2000)

Microorganism	Range of pH for growth
α-Proteobacteria	
Paracoccus versutus	8.0–9.0
Starkeya novella	5.0–9.0
β-Proteobacteria	
Thiobacillus thioparus	6.0–8.0
Thiobacillus denitrificans	6.0–8.0
Thiobacillus plumbophilus	3.0–7.0
γ-Proteobacteria	
Halothiobacillus neapolitanus	4.5–8.5
Halothiobacillus halophilus	6.5–8.4
Halothiobacillus hydrothermalis	5.5–9.0
Halothiobacillus kellyi	3.5–8.5
Thermithiobacillus tepidarius	5.5–8.0

Eleven heterotrophic strains were isolated from alkaline slag dump by Willscher and Bosecker (2003). The isolates were divided on the basis of their morphological characteristics into three large groups: slime-forming bacteria, coryneform bacteria and fungi. Between slime forming bacteria *Arthrobacter oxydans*, *Microbacterium* sp. and *Dietzia natronolimnaea* were identified and between coryneform bacteria *Promicromonospora* sp., *Pseudonocardia autotrophica* and *Nocardiopsis metallicus* were found. The role of all isolates in bioleaching of Fe, Zn, Ni and Co from alkaline slag dump has been confirmed.

Additionally, some iron-oxidizing bacteria may play important role in bioleaching at neutral and alkaline environment. *Galionella feruginea, Sphaerotilus natans* and *Leptotrix ochracea* are iron bacteria, which live at interfaces between anoxic and oxic conditions where ferrous iron is present at neutral pH (Kasama & Murakami, 2001). Besides iron, a few bacteria (*Leptotrix discophora*) oxidize manganese at neutral pH and play role in mobilization of this element in the environment.

Indigenous bacteria occurring in alkaline and neutral ores and wastes are still not well recognized, although a few groups of neutrophilic facultative autotrophic microorganisms able to oxidize copper sulfide and sodium thiosulfate were isolated from tailing ponds and mines (Sklodowska *et al.*, 1996).

A number of fungus as well as the occurrence of chemolithotrophic bacteria *Pseudomonas* sp. and the sulfide oxidizing aerobic bacteria *Beggiatoa alba* and *Thiothrix* sp. (Figure 3) were reported in ancient gold mine Zloty Stok (Chlebicki *et al.*, 2005).

4. BACTERIAL AND CHEMICAL LEACHING OF COPPER FROM NEUTRAL AND SLIGHTLY ALKALINE ORE AND ITS POTENTIAL ENVIRONMENTAL IMPACT

It has been reported that bacterial cells exhibit specific affinity for sites containing copper sulfide within neutral or slightly alkaline ores (Ostrowski & Sklodowska, 1993; Ostrowski & Sklodowska, 1996). Interestingly, the leaching pattern obtained with the use of wild type bacteria (freshly isolated from flotation wastes) strongly depends on the duration of the leaching process. In the first weeks, cells are rare on the mineral surface and the pattern of leaching is similar to chemical leaching (using different but known reagents like H_2SO_4 or EDTA).

But after two months of leaching by wild type bacterial strains, the corrosion is deeper, with sponge-like bases to the pits (Figure 4C) and it is possible to distinguish many cells on the mineral surface (Ostrowski & Sklodowska, 1993). These results indicate that the colonization of this type of ores requires a long time. Even after some weeks, it is possible to distinguish there are some sites containing sulfides which are not destroyed by bacteria and which bear no traces of cell adhesion (Figure 4). A similar phenomenon was described for bioleaching of pyrite crystals (Bennet & Tributsch, 1978; Rodriquez-Leiva & Tributsch, 1988).

Extracellular polymeric substances (EPS) seem to play a main role in bioleaching by acidophiles (Kinzler *et al.*, 2003). For neutral or slightly alkaline ores, the

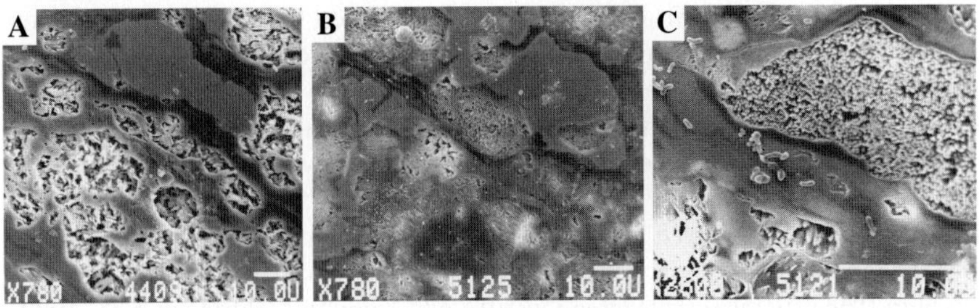

Figure 4. Bacterial and chemical leaching patterns. A. Chemical leaching with EDTA; B. Bacterial leaching with selected most active strains; C. Pits after two months' leaching with *Halothiobacillus neapolitanus*

mucosal film on the leaching surface and in the corrosion pits is also clearly visible (Figure 5), indicating these pits may arise as a result of the action of exopolymers which would be the responsible for initial leaching.

Those exopolymers can modify the mineral surface, solubilize metal sulfides among other effects (Ostrowski & Sklodowska, 1993; Sklodowska & Matlakowska, 1998). Due to this exopolymers, ecological microsites containing organic acids and chelates are created around bacteria (Ostrowski & Sklodowska, 1993; Ostrowski & Sklodowska, 1996; Sklodowska *et al.*, 2005). These microsites may enhance the metal mobilization. A similar role of extracellular substances was described in connection of bacterial corrosion of mild steels surfaces (Beech & Gaylarde, 1989).

A mixed culture of bacterial strains isolated from the flotation wastes sedimentation pond was shown to be able to leach copper from copper sulfides at initial alkaline pH (8.5). The experiments carried out in the medium with synthetic covellite (CuS) showed up to 90% efficiency of copper leaching after 170 days. The maximal leaching rate was 3–4 mg Cu per liter and hour (Lejczak *et al.*, 1980). However,

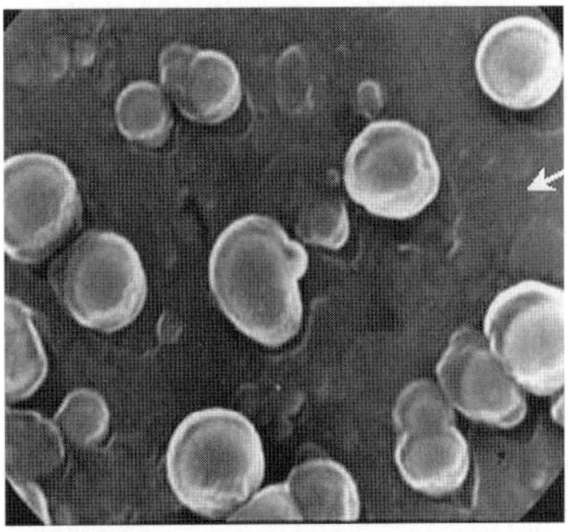

Figure 5. Pits formed under coccal bacterial cell during bioleaching of limestone type of copper ore (white arrow) – magnification 20 000x

metal bioleaching process from flotation tailings (which should be treated as low charge ore) in neutral or slightly alkaline pH indicated a very low efficiency. It was confirmed that the process may be enhanced by chelating agents such as EDTA or brown coal with high concentration of humic acids (Ostrowski *et al.*, 1990). The efficiency of copper bioleaching from flotation tailings was up to 77% after 120 days using a mixed culture of isolates and humic acids (5% v/w) (Ostrowski *et al.*, 1990). These tailings showed pH values around 8–9. The best result was obtained after 136 days of cultivation with addition of EDTA to a final concentration of 0.02% w/v. The efficiency of leaching was 93.3% (Kunicki-Goldfinger *et al.*, 1980).

The copper bioleaching from alkaline tailings has not any significant application in industrial scale due to it is a very time consuming process. However, the bioleaching connected with naturally occurring microorganisms should be taken into account in long-term environmental risk assessment deal with waste management.

All indigenous microorganisms occurring in tailing ponds as well as in ores are active in biogeochemical cycles and their activity may strongly affect the environmental conditions. In the region of copper exploitation from Fore-Sudetic Monocline deposits, some anomalies in amounts of metals (As, Fe, Pb, V) and sulfur have been observed. Higher concentration of metals can be detected in ground waters (during last twenty years increased threefold to geochemical background). Additionally, the contamination of soil by polycyclic aromatic hydrocarbons occurs in the same area. This type of pollution was observed not only in the neighborhoods of smelters but around tailing ponds too. The role of microorganisms in dissemination of this type of contaminant is significant and out of discussion, especially in the case of mine's water and tailing ponds. The incorporation of shale's organic compounds into the bacterial exopolymers and the role of exopolymers in metal bioleaching seem to confirm the important role of microorganisms in pollution dissemination (Sklodowska *et al.*, 2005). Such microbial activity should be taken into account in long-term environmental risk assessment of waste deposits, especially when they contain organic matter, such as black shale.

REFERENCES

Bechtel A, Elliott WC, Oszczepalski S. 2001. The occurrences of the rare earth elements and the platinum group elements in relation to base metal zoning in the vicinity of Rote Faule in the Kupferschiefer of Poland. Appl Geochem 16: 375-386.

Bechtel A, Gratzer R, Puttmann W, Oszczepalski S. 2002. Geochemical characteristics across the oxic-anoxic interface within the Kupferschiefer of the Lubin-Sieroszowice mining district. Chem Geol 185: 9–31.

Beech IB, Gaylarde CC. 1989. Adhesion of *Desulfovibrio desulfuricans* and *Pseudomonas fluorescens* to metal steels surfaces. J Appl Bacteriol 67: 201-207.

Beijerinck MW. 1904. Ueber die bakterien welche sich im Dunkels mit Kohlensaure als Kohlenstoffquelle ernahren konnen. Centralb Bacteriol Parasitenkb Infektionskr Hyg Abt II 11: 593-599.

Bennet IB, Tributsch H. 1978. Bacterial leaching patterns on pyrite crystals surfaces. J Bacteriol 134: 310-317.

Bosecker K. 1997. Bioleaching: metal solubilization by microorganisms. FEMS Microbiol Rev 20: 591-604.

Burgestaller W, Schinner F. 1993. Leaching of metals with fungi, J Biotechnol 27: 91-116.

Chlebicki A, Godzik B, Lorenc MW, Sklodowska A. 2005. Fungi and arsenic-tolerant bacteria in the hypogean environment of an ancient gold mine in lower Silesia, SW Poland. Polish Botanical Studies 19: 81-95.

Groudev SN. 1987. Use of heterotrophic microorganisms in mineral biotechnology. Acta Biotechnol 7: 299-306.

Ixer RA. 2001. An assessment of copper mineralisation from the Grat Orme Mine, Llandudno, North Wales, as ore in the Bronze Age. Proceedings of the Yorkshire Geological Society 53: 213-219.

Karas H, Sadowski Z. 2002. Biohydrometalurgia na swiecie. In: Charewicz W, ed. Biometalurgia metali nieżelaznych podstawy i zastosowanie, Wroclaw, CBPM Cuprum.

Kasama T, Murakami T. 2001. The effect of microorganisms on Fe precipitation rates at neutral pH. Chem Geol 180: 117-128.

Kelly DP, Harrison AP. 1989. Genus *Thiobacillus*. In: Staley JT, Bryant MP, Pfennig N, Holt JG, eds. Bergey's Manual of Systematic Bacteriology, Williams and Wilkins, Baltimore.

Kelly DP, Wood A. 2000. Reclassification of some species of *Thiobacillus* to the newly designated genera *Acidithiobacillus* gen. nov., *Halothiobacillus* gen. nov. and *Thermithiobacillus* gen. nov. Int J Syst Evolut Microbiol 50: 511-516.

Kelly DP, McDonald IR, Wood A. 2000. Proposal for the reclassification of *Thiobacillus novellas* as *Starkeya novella* gen. nov., comb. nov., in the α-subclass of the *Proteobacteria*. Int J Syst Evolut Microbiol 50: 1797-1802.

Kinzler K, Gehrke T, Telegdi J, Sand W. 2003. Bioleaching – a result of interfacial processes caused by extracellular polymeric substances (EPS). Hydrometallurgy 71: 83-88.

Kunicki-Goldfinger W, Lejczak-Sklodowska A, Ostrowski M. 1980. Leaching of copper. in alkaline environment. In: Proceedings of the International Conference on use of the microorganism in hydrometallurgy, Pecs, Hungary.

Lejczak A, Ostrowski M, Kunicki-Goldfineger W. 1980. Assay of bacterial copper leaching from covellin at alkaline initial pH. Acta Microbiol Polon 29: 69-74.

Moreira D, Amils R. 1997. Phylogeny of *Thiobacillus cuprinus* and other mixotrophic *Thiobacilli*: proposal for *Thiomonas* gen. nov. Int J Syst Bacteriol 47: 522-528.

Ostrowski M, Sklodowska A, Kunicki-Goldfinger W. 1990. Bacterial leaching of copper from alkaline and neutral postflotation wastes with the use of brown coal. Acta Microbiol Polon 39: 71-78.

Ostrowski M, Sklodowska A. 1993. Bacterial and chemical leaching pattern on copper ores of sandstone and limestone type. World J Microbiol Biotechnol 9: 328-333.

Ostrowski M, Sklodowska A. 1996. Acid leaching in alkaline environment. Bulletin of the Polish Academy of Sciences, Biol Sci 44: 279-283.

Oszczepalski S. 1999. Origin of the Kupferschiefer polymetallic mineralization in Poland. Mineralium Deposita 34: 599-613.

Piestrzynski A, Pieczonka J, Gżuszek A. 2002. Red-bed type gold mineralization, Kupfershiefer, southwest Poland. Mineralium Deposita 37: 512-528.

Rodriquez-Leiva M, Tributsch H. 1988. Morphology of bacterial leaching patterns by *Thiobacillus ferrooxidans* on synthetic pyrite. Arch Microbiol 149: 401-405.

Sklodowska A, Matlakowska R, Ludwig W. 1996. Unusual bacteria strain closely related to *Bacillus insolitus*. Acta Microbiol Polon 45: 131-141.

Sklodowska A., Matlakowska R. 1998. Relative surface charge, hydrophobicity of bacterial cells and their affinity to substrate during copper bioleaching from post-flotation wastes. Biotechnol Lett 20: 229-233.

Sklodowska A, Matlakowska R, Bal K. 2005. Extracellular polymer produced in the presence of copper minerals during bioleaching. Geomicrobiol J 22: 1-9.

U.S.EPA. 1994. Technical resource document. Extraction and beneficiation of ores and minerals. Copper, Vol 4, EPA 530-R-94-031, Washington.

Willscher S, Bosecker K. 2003. Studies on the leaching behavior of heterotrophic microorganisms isolated from alkaline lag dump. Hydrometallurgy 71: 257-264.

SECTION II

BIOREACTORS AND BIOHEAPS

CHAPTER 7

BIOLEACHING OF SULFIDE MINERALS IN CONTINUOUS STIRRED TANKS

DOMINIQUE HENRI ROGER MORIN

BRGM, 3 Avenue Claude Guillemin, 45060 Orléans cedex 2, France

1. INTRODUCTION

Bioleaching in stirred tanks for the selective recovery of precious and base metals from sulfide concentrates is a fascinating subject for at least two reasons.

The first reason is the high potential of industrial-scale applications of this technology that has acted as a driving force for countless investigations required to make the economics of this process really attractive. Around forty years ago, this was just a concept imagined by microbiologists and metallurgists, but already at the end of the 1970s studies at pilot-scale in Russia, Canada and South Africa showed that the concept was technically sound. At the middle of the next decade, the experiment in industrial context was undertaken by Gencor at Fairview mine after the works led by Eric Livesey Goldblatt on an arsenical sulfide concentrate containing gold refractory to direct cyanidation. Since that time, a steady growth of projects using this technology has persisted as illustrated by the Figure 1.

The second reason is the growth selectivity and the steadiness of the microbial ecosystems in the bioreactors compared to the diversity of the natural environment that have given the opportunity to study the interactions between micro-organisms and minerals in privileged conditions.

Bioleaching in agitated tanks at industrial scale is only relevant to the treatment of sulfidic concentrates and consist in using the catalytical enhancement exerted by some micro-organisms to oxidise sulfides to release valuable metals they contain. The sulfide compounds to (bio)-oxidize in the existing industrial plants are essentially pyrite and arsenopyrite in various proportions. Many successful investigations and demonstration operations have been run on other metal-bearing sulfides, like sphalerite (ZnS), pentlandite ($(FeNi)_9S_8$), covellite (CuS), chalcocite (Cu_2S), and chalcopyrite ($CuFeS_2$).

From the chemical engineering point of view, bioleaching in agitated tanks is a continuous-flow steady-state process. The circuit of bioreactors configured in

E.R. Donati and W. Sand (eds.), Microbial Processing of Metal Sulfides, 133–150.

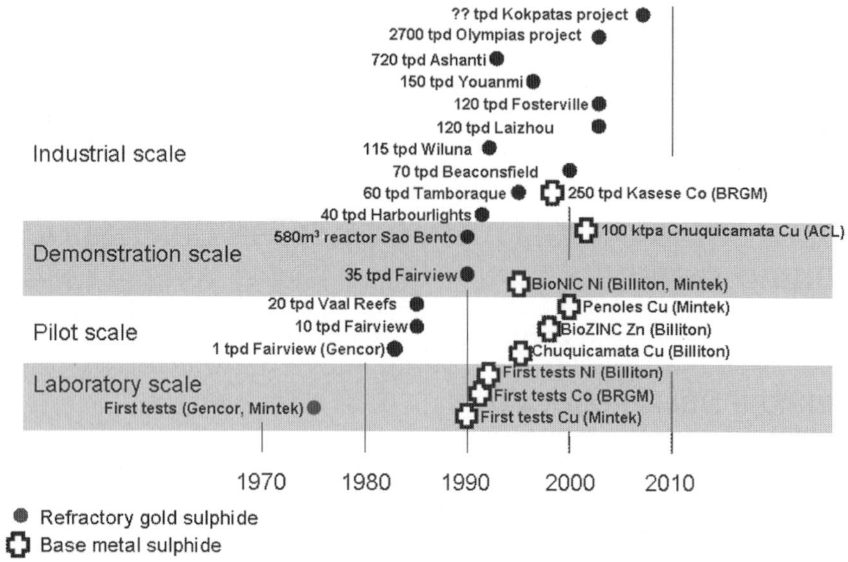

Figure 1. Main known industrial projects using bioleaching in stirred tanks during the last thirty years (modified of Clark *et al.*, 2005)

series or in parallel or combination of the two is fed with the slurry of the sulfidic concentrate, nutrients and air to bring the oxygen required to the sulfide oxidation. Micro-organisms are injected once at the beginning of the start-up of an operation and a batch culture is maintained until a certain point as close as possible to the middle of the exponential phase of the bacterial growth when the feed in fresh substrate can begin. A continuous flow of substrate and nutrients through the tanks is then ensured to keep optimum growth of micro-organisms required for the fastest degradation of the sulfidic minerals. The situation of equilibrium between substrate feed rate and stable microbial growth is called chemostat.

This chapter provides an overview of the operating conditions of this technology and therefore to emphasize the field of applications for the present and the future.

2. SPECIFIC CRITERIA OF THE BIOLEACHING IN CONTINUOUS-FLOW STEADY-STATE REACTORS

2.1. Bacterial Population Characteristics and Monitoring

Bioleaching ecosystems consist of several species living together even if the growth conditions are not optimal for all of them for a given set of operating conditions. For example it is known that *Acidithiobacillus ferrooxidans* and *Leptospirillum ferrooxidans* frequently isolated in the same ecosystems have different optimal pH and temperature of growth. However, it is also demonstrated that every species and especially *L. ferrooxidans* have a high capacity of adaptation to extreme conditions (Gomez *et al.*, 1999).

It appears that inoculation remains at industrial scale and with mesophilic bacteria subject to uncertainties regarding the identity of the micro-organisms. All natural sulfide minerals are the substrates of specific and adapted species, which may prevail over any other micro-organism inoculated. Of course, the studies at lab-scale for the evaluation of the amenability of an ore to bioleaching are carried out without any particular precautions in terms of contamination and it is justified to think that there is no reason that the medium at industrial scale will contain the same biological composition as in the best operating conditions at the lab. Moreover, the best operating conditions are generally the result of months of continuous growth at lab-scale which are not totally reproducible at industrial scale. Is there any influence of the change in operating conditions on the bacterial composition during transfer for inoculation and if any at the beginning how does it evolve with time?

Extreme conditions generated in bioleaching installations are highly selective and for example, heterotrophic micro-organisms that are generally encountered in acid mine drainage would not occur in biooxidation reactors (Johnson & Francisco, 1997). It may be a question of time to find heterotrophic micro-organims able to use mineral sources of energy in a competing way compared to autotrophic ones. Already, bioreactors are appropriately confined vessels where new micro-organisms may be isolated like *Ferroplasma acidiphilum*, an acidophilic, autotrophic, cell-wall-lacking, mesophilic new archea discovered by serial dilution of aqueous phase of a bioreactor of a pilot plant treating a gold-bearing arsenopyrite/pyrite concentrate (Golyshina *et al.*, 2000).

A trend in the results of the recent ecological studies is to say that *A. ferrooxidans* does not play the major role as originally asserted and all the effort of investigation in molecular biology on this micro-organism should be broadened to *L. ferrooxidans* and sulphur-oxidising bacteria (Rawlings, 1999). The fact is that these two organisms, namely *A. ferrooxidans* and *L. ferrooxidans*, which belong to very different phylogenetic branches, are found in the same ecosystems. If *A. ferrooxidans* can use ferrous and reduced forms of sulphur as substrate and *L. ferrooxidans* uses only ferrous, some persuasive description of the bacterial action (indistinguishable for one micro-organism to the other) shows these micro-organisms as controlling the catalytic oxidation of pyrite by the regeneration of ferric ion close to the sulfide surface, the ferric iron being the only electrochemical reagent dissolving pyrite (Crundwell, 2001). Actually, exopolymers from the bacteria could generate a layer, which constitutes a special, enlarged reaction space (Gehrke *et al.*, 1998; Kinzler *et al.*, 2001). However, *A. ferrooxidans* could have a different way of degrading pyrite from *L. ferrooxidans*; one could be more chemical by action of carrier bio-compounds and the other strictly electrochemical by ferric action (Tributsch, 1999), respectively. This would explain the differences of predominating areas of operating conditions of optimum growth for these species and would be a beginning of explanation of the symbiotic behaviour of species in mixed cultures. Kinetic modelling of pyrite oxidation, which takes into account differentiation of growth characteristics of the species, is restricted to consider their respective specific ferrous oxidation rates as an explanation to their predominance at the different stages of the pyrite degradation (Hansford, 1997).

It is noteworthy that cultures with mixed populations result in higher pyrite leaching efficiencies than cultures with individual pure bacterial components (Battaglia *et al.*, 1998; Hughes & Poole, 1989). Thus, if it is not demonstrated that there is symbiosis, at least there may be synergy to use a substrate.

The proportions of bacterial species in the BIOX® installation from Sao Bento (Brazil) was found 48% *L. ferrooxidans*, 34% *Acidithiobacillus thiooxidans* and 10% *A. ferrooxidans* as free unattached bacteria. The bacterial composition in the first tank of the Fairview Mine unit (South Africa) was near with *L. ferrooxidans* 57%, *A. thiooxidans* 30% and *A. ferrooxidans* 13%. Another observation by Dew *et al.* 1997 was that given the experimental errors there might be no difference between the proportions of free cells and attached-to-the-solids cells; however the detailed protocol and results of the measurements are not described. Beside, Coram and Rawlings (2002) discovered that the genus *Leptospirillum* consists in two different groups and the only one present in the commercial biooxidation tanks in South Africa was of a type proposed to be called *L. ferriphilum*.

Various techniques of molecular biology allow analysing and even quantifying the distribution of the species in a mixed culture like DGGE (Denaturing Gradient Gel Electrophoresis), FISH (Fluorescent In Situ Hybridization) (Jerez, 1997) and SSCP (Single Strand Conformation Polymorphism) (Foucher *et al.*, 2001).

2.2. Carbon Supply

The microbial consortia currently used at industrial scale are considered to be autotrophic, even if a number of species of heterotrophs, most of which belong to the genus *Acidiphilum*, grow in very close association with the obligate autotrophs (Rawlings, 1997). They do not use other carbon source than carbon dioxide and actually, other organic compounds are generally harmful to them.

Specific CO_2 supply is subject to discussion as an operating parameter in the industrial units of bioleaching. Actually, it is generally admitted that there is enough carbon dioxide in the air to feed continuous growth of bacteria in tank bioleach processes through air sparging to bring oxygen to the medium. At a time, Kelly and Jones (1977) said that for *A. ferrooxidans* oxidation could be "uncoupled" from assimilatory carbon metabolism and consequently, that abundant supply of CO_2 was not essential to the efficiency of the bioleach process. However, the possibility that bacterial growth could be amplified or reduced by an excess of carbon dioxide remains and may vary from one micro-organism to another (Norris, 1989). The fact is that bioleaching in continuous mode shows that oxygen uptake rate, measuring the extent of the sulfide oxidation, is correlated to carbon uptake rate (Oolman, 1993). At high bacterial growth rate, carbon uptake rate reaches levels such as CO_2 supply by the air injected is no more sufficient (Morin *et al.*, 1993) and carbon transfer rate becomes a limiting step of the sulfide oxidation rate (Acevedo *et al.*, 1997).

In real conditions of treatment, the question of supplying extra carbon dioxide is very often irrelevant because either the mineral contains carbonate that dissolves in acidic conditions of bioleaching to produce CO_2 in situ and/or limestone addition is used to control pH in the bioleach tanks. In the case where the buffering effect of the concentrate is due to another compound than a carbonate, like pyrrhotite, the question of extra carbon dioxide addition may be relevant.

It must be noted that as the in situ provision of CO_2 by addition of carbonate is likely to result in different growth rates of the different species of a mixed culture, it may be a parameter of pressure of selection on a culture. As such, this may play a role on the performances of bioleaching.

2.3. pH

Acidity level of bioleaching medium results from the balance of protons between net-consuming reactions (oxides/carbonates dissolution, arsenopyrite/pyrrhotite/iron oxidation, etc.) and reactions of sulphuric acid production and iron hydrolysis.

The optimal pH range is variable from one system to the other, one micro-organism to the other (Rawlings, 1997). On one hand, from the point of view of thermodynamics, higher the pH, faster the acid-producing reactions. However, a relative high pH, between 1.8 and 2, may lead to the precipitation of iron hydroxide in excess, raising the proportion of sterile surface interfering with the interaction of the bacteria with sulfides, increasing the slurry viscosity and making mixing and oxygen transfer less efficient. On the other hand, a low pH value, close to 1.0, of course is harmful to micro-organisms metabolism and can be very selective for acid-tolerant species making the biological system fragile. Other aspects can be also taken into consideration when pH is low. At low pH, ferric iron from iron sulfides becomes high in concentration in solution, which can reduce bacterial growth of species sensitive to this ion (Collinet & Morin, 1990). Less neutralising agent like limestone being added, less dioxide carbon is available in situ for stimulating bacterial growth and less gypsum is precipitated, which may have consequences on the retention time in neutralisation operation. In the case of refractory gold concentrates, it is observed that low pH reduces the risk of gold being encapsulated and less soluble during cyanidation treatment (Spencer, 2001). Another aspect is related to nutrients as some authors think that too much calcium could result in less phosphate being available in solution for the bacterial metabolism needs (Barrett *et al.*, 1993). This could be relevant for ammonium as well, because the formation of jarosite is enhanced by the addition of neutralising agent. It is true that one must be especially aware of nutrients availability in the primary stage of the bioleaching treatment in stirred tanks, where the major part of the bacterial growth occurs, much less in the following stages, it is why pH can be increased from the primary to the following stages.

A pH range of 1.4-1.7 is probably a good compromise between the risks mentioned above and the technical feasibility of controlling pH in huge stirred tanks. The fact is that considering the evident and unavoidable inefficiency of mixing systems at microscopic scale, the range of local pH in a tank is actually very large in any situation, from less than 1 to more than 7, respectively in zones where sulfuric acid is produced and in others where limestone is dissolved. The most important factor is to be able to maintain a stable bulk system in order to have the fastest bacterial growth, by the addition of limestone when required, and the resulting generation of carbon dioxide in situ and the lowest solids load.

2.4. Nutrients

The two essential components to provide for an optimal growth of the biomass are ammonium as nitrogen source and phosphate. Two others may be also required, which are potassium and magnesium, however they are often naturally available from the ore or concentrate. The required amounts of each nutrient is variable from one sulfide substrate to the other in the range of one to ten kg/t of sulfide concentrate and must optimised in continuous leaching conditions where water recycling is also considered.

For a long time, it has been considered that jarosite precipitated in proportion of the ammonium concentration in the slurry could be a cause of reduction of the bacterial activity by sulfide surface passivation. Some studies tend to say the opposite (Wiersma & Rimstidt, 1984; d'Hugues *et al.*, 1999); significant ammonium precipitation is beneficial, maybe precisely because it provides a micro-environment stimulating bacterial activity at the sulfide surface. This micro-environment created by the bacteria in order to displace the electrochemical equilibriums as suggested by Fowler (1999) would be efficient to oxidise pyrite for example as nutrients could insulate the sulfide surface.

Fertilisers can be used as sources of nitrogen and phosphorus: diammonium phosphate (DAP) and monoammonium phosphate (MAP). Ammonium sulphate may be required to complete the fertilisers in ammonium. Urea can replace ammonium sulphate but seems to be less useable to the bacteria, without being able to say whether it is due to the organic nature of this compound or to a less good availability of the ammonium (d'Hugues *et al.*, 1999). Nevertheless, urea produces much less precipitate in the medium than ammonium sulphate.

Eventually, the nutrient requirements are likely to change with time in an industrial unit (Dew *et al.*, 1997) and significant savings can result from regular check-up of this parameter.

2.5. Temperature and Heat Removal

Three temperature ranges are considered for present and future industrial-scale units of bioleaching in agitated tanks: the mesophiles typical range, which is 35-40 °C, the moderate thermophiles, 45-55 °C, and the extreme thermophiles, 60-85 °C. These ranges have no strict limits, and for example moderate thermophile micro-organisms like *Sulfobacillus thermosulfidoxidans* of which optimal growth temperature is 45-50 °C, can occur at 35 °C (Norris, 1997; Foucher *et al.*, 2001). Moreover, all these temperature ranges can occur in the same reacting system when heap, dump or in situ leaching techniques are considered and the inoculum for a bioheap should contain a diverse microflora adapted to those conditions (Brierley, 1997).

All the existing bioleach units in agitated tanks are at low temperature, near 40 °C, one used moderate thermophiles, near 50 °C, and several projects are in progress to use extreme thermophile (Rawlings *et al.*, 2003; Batty & Rorke, 2005) for the recovery of copper from chalcopyrite. The use of thermophilic bacteria follows the demonstration of the ability of *Sulfolobus*-like micro-organisms to dissolve copper from chalcopyrite up to 95% (Norris, 1997), whereas this mineral is refractory to complete degradation when using mesophilic bacteria. Temperature becomes not only a constraint in the design of installation but also a parameter for the selection of the most efficient biological system for dealing with a mineral in the most economical condition (Dew *et al.*, 1999).

As sulfide oxidation is exothermic heat removal is an important aspect of the design of the bioleach tanks. Higher the temperature in the tanks, lower the heat removal requirement, this may be important when water supply and energy for cooling would be too expensive. As an example, a reduction in cooling energy of 30% was calculated for bioleaching at 65 °C compared to bioleaching at 45 °C (Sandström *et al.,* 1997). However, it will not be omitted the fact that high temperature has its drawbacks. Corrosion of the containing materials may be higher (Ignatiadis and Morin, 2001) and water balance and handling of hot slurry are critical in terms of engineering design and control operability.

The heat produced by the complete oxidation of pyrite is 12,900 MJ/t and the heat removal that can reach more than 10 MW for a reasonably small plant throughput of 100 tpd concentrate is a function of the mineralogy and the degree of sulfide oxidation required. The heat generated by agitation is also an input to be accounted for in the design of the heat removal system, just as outputs due to heat absorption by incoming feed and in exit slurry product, to convective transfer, to moisture in the air leaving the system and evaporation. The heat removal is generally ensured by transfer to water circulating in internal reactor coils that can be more than a kilometre long in one tank. This piece of equipment is particularly critical as a leak of cooling water in a tank can reduce the bioleaching activity by dilution. By the way, this is because of this risk that no anti-bacterial or anti-scaling reagents are used in the cooling water. Radiation methods using UV light for example are preferred.

2.6. O_2 Transfer

Oxygen transfer is a key parameter for the design and the cost of a bioleaching installation.

In order to use relatively simple and not excessively expensive equipment in a way respecting the viability of the micro-organisms and copying their natural conditions of growth, oxygen is supplied by injection of air sparged and dispersed by mechanical means. The criteria that are used to fix the required air flow rate and air dispersion efficiency are the oxygen uptake rate (OUR), the oxygen utilisation rate and the minimum oxygen concentration. Gas hold-up is also a parameter that allows to make a coarse comparison between aeration systems and various operating conditions for a given agitation system.

The oxygen uptake rate is the flow of oxygen consumed by the oxidation reactions. It can be measured either by the balance on the gas in and out of the reactor or the oxidation rates of the concentrate sulfides. The slight difference between the two values is the oxygen consumed by the biomass for their metabolic activity, but it represents a very low proportion of the oxygen consumed.

The oxygen utilisation rate is the ratio of the amount of oxygen effectively consumed by the chemical reaction of sulfide oxidation over the amount of oxygen injected as air in the same period of time. It is directly dependent on the oxygen uptake rate, which means that it strictly depends on the biochemical oxygen demand potentially restricted to the physical capacity of transfer of the mixing system.

From the physical point of view of the mixing, the rate at which oxygen will dissolve is described by the following equation:

$$(1) \qquad d[O_2]/dt = k_l a \cdot ([O_2]_s - [O_2])$$

where:

- k_l is the mass transfer coefficient and a the interfacial area between gas and liquid media, $k_l a$ (s^{-1}) is then the overall mass transfer coefficient
- $([O_2]_s - [O_2])$ is the driving force of the transfer, $[O_2]_s$ being the oxygen concentration at saturation depending on the pressure of the gas phase, temperature and composition of the liquid phase and $[O_2]$ is the actual oxygen concentration

Two factors relative to the bacteria growth viability must be taken into consideration in the design of the agitation system. The first one is the lower oxygen concentration acceptable for the stable maintenance of aerobic conditions required by the bacterial population involved in such processes, which is generally agreed to be 1.5 $mg.l^{-1}$ (Dew & Miller, 1997), even if the system would still be efficient at as low as 0.1 $mg.l^{-1}$ (Ritchie & Barter, 1997). Obviously, the lowest the value of the acceptable oxygen concentration, the highest the driving force in oxygen concentration and the oxygen transfer rate. The second factor is to reduce to the minimum the risk of physical damage to the bacterial cell wall by shearing stress due to the rotation of a stirring mobile in the case of the use of a mechanical agitation system.

Many kinds of agitation systems have been designed in the ore dressing industry for various applications from large scale like in the phosphate processing and to very specific ones using oxygen as dissolved oxidant like for gold recovery by cyanidation in particular. Each has its own constraints. Obviously in the case of the bioleaching of sulfide concentrates, the energy consumed by aeration and mixing is related to the residence time and the concentrate or sulfide-to-oxidize concentration in the feed. Residence time is fixed by the kinetics of bioleaching and whatever the substrate, it is fitted to bacterial growth rate, i.e. in the range of several days to achieve complete sulfide oxidation. For the substrate concentration, it is demonstrated that oxygen transfer rate is a potential limiting factor (Bailey & Hansford, 1993).

There are two mixing systems currently in use at industrial scale for processing refractory gold sulfide concentrates.

The first system is a Rushton turbine, which is the traditional impeller used in processes requiring high gas dispersion rates. It is known that a Rushton turbine has a high power draw. It is also significantly sensitive to gas flow rate. Moreover, the radial flow can produce separate mixing zones above and below the horizontal axis where different solids suspension, heat transfer conditions, and chemical reactivity may exist, limiting the performances of the transfers.

In all the other installations dealing with refractory gold concentrates, it is the Lightnin A315® impeller. This system consists of a dual hydrofoil impeller producing air dispersion and axial mixing simultaneously (see illustration in Figure 2). Air dispersion results from sparging through a holes-riddled ring located below the impeller and air bubbles breaking by flow acceleration between the blades partly overlapped. As a condensed compromise between solid/liquid mixing and

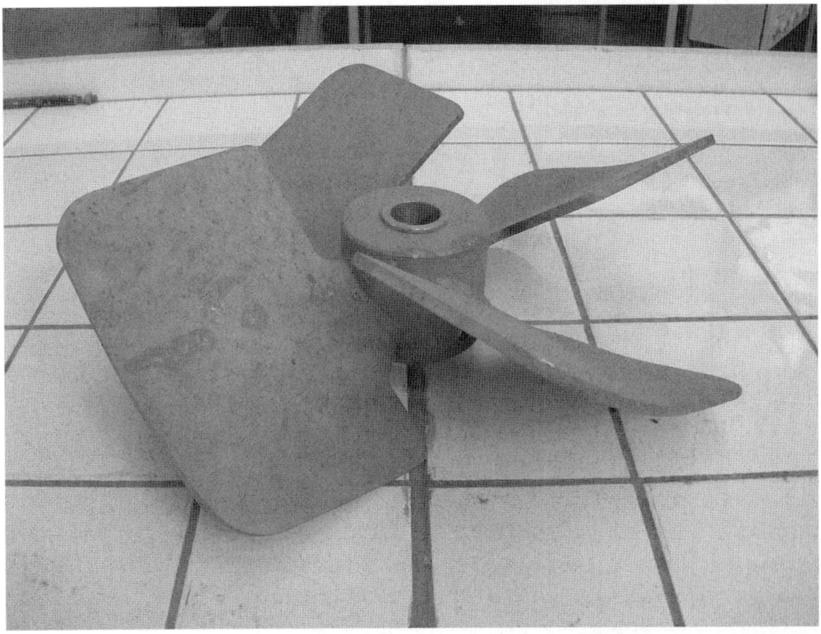

Figure 2. Lightnin A315 agitator at laboratory scale. In the demonstration BioCOP™ plant at Chuquicamata the word's largest A315 impeller used in each tank has a diameter close to 5m (Batty & Rorke, 2005)

gas dispersion, this system is particularly efficient in terms of power consumption but has in theory a relatively limited efficiency in air dispersion compared to the Rushton turbine.

The latest system operated at industrial scale for bioleaching in tanks and currently in use at the cobalt bioleaching plant in Kasese, Uganda, is the BROGIM® system, designed by Robin Industries (now MILTON ROY MIXING) (Bouquet & Morin, 2005).

BROGIM® consists of two different impellers mounted on the same rotating shaft (see Figure 3).

The lower is a flat blade disk turbine aimed at producing air dispersion according to the same principle as a Rushton turbine and one or two propellers located above the turbine, aimed at maintaining homogeneity of all the phases in the tank. Air is injected below the bottom turbine, through one or several open pipes. The number

Figure 3. Bottom part of the BROGIM agitation system in the KCC bioleach tank. The diameter of the bottom turbine is about 4.5m

of blades of the turbine (up to 36) and the size of these blades depend on the volume of the vessel and on the air dispersion requirement.

Actually, the main characteristic of the BROGIM® system is that the size of the propellers and the design of the turbine can be tailor-made according to process specifications so as to minimise capital and operating costs for both agitation and aeration.

The required air flow rates are very high and the supply of air is the greatest item of capital and running costs in a bioleach plant.

Agitation systems currently in use in mesophilic conditions have been scaled up with an objective of oxygen utilisation rate in the range between 30 to 40% while always keeping a safe oxygen concentration of at least 1.5 $mg.l^{-1}$, actually it is more than 40% at industrial scale whatever the system. It remains that the energy consumption for agitation and aeration in those conditions is as high as 1,500 to 2,000 kWh/tonne of sulphur oxidised.

Even if the scale-up predictions underestimated the system efficiencies at real scale such high energy consumption justifies looking for innovative ways of undertaking bioleaching at industrial scale. Concepts like external sparging by means of slurry pumping and venturi device (Pinches *et al.*, 1988), bubble column where mixing is provided by simultaneous upward motion of gas and recycled slurry (Foucher *et al.*, 2001) and air-lift reactor (Giaveno *et al.*, 2001) propose alternatives to conventional mechanical mixing. Apart from the high performances predicted for those systems, the question of their operability at large scale compared to the systems currently in practice at real scale is still to be answered.

Oxygen solubility in thermophilic systems is reduced to 1/3 of what could be expected at mesophile conditions. This fact suggested to BHP-Billiton that in their development work of chalcopyrite using thermophiles oxygen enrichment of the air injected would be a way to maintain a sufficient concentration of oxygen in the bioleach slurry (Batty & Rorke, 2005). An automated dissolved oxygen strategy was used to control the oxygen consumption and prevent the micro-organisms to suffer from their high sensitivity to this gas. Oxygen utilization up to 90% was reached and an oxygen concentration of 90% in the feed gas would be required to maximize the efficiency at industrial scale.

3. SOLIDS CONCENTRATION AND RESIDENCE TIME

Solids loading limit is generally admitted to be 20% (weight %) in the case of mesophiles and 12.5% in the case of thermophiles.

The general reasons for the limitation in solids concentration in bioleaching include oxygen and carbon dioxide availability, bacteria-to-solid ratio, mechanical trauma to bacteria and the build-up of toxic products (Bailey & Hansford, 1993).

The more reduced limit in solids concentration for the thermophiles is explained by the higher sensitivity of these micro-organisms to shearing as they have no cell wall but only a cell membrane. Batty and Rorke (2005) stress that this characteristic largely contributes to confirm that bioleaching at high temperature is a different technology than the mesophile one.

The microbial growth rates as the main kinetic parameter of the auto-catalytic phenomenon impose the kinetics of bioleaching. A complete degradation of the sulfides is generally in the range of 4 to 6 days. A partial bio-oxidation as achieved in the Sao Bento plant may require no more than two days retention time.

4. REACTORS DESIGN AND CONFIGURATION

4.1. Criteria for Design

The design of a unit of bioreactors from performance to practical points of view depends on many parameters of which the most important ones are:

- The maximum head pressure acceptable for the air injection to reach optimal efficiency and lower power consumption;
- The optimal maximum size of the agitation system in terms of power consumption vs. mechanical reliability;
- The optimal configuration of bioreactors as the best compromise between oxidation performances, power consumption and robustness of the system.

For the air injection blowers are preferred to compressors due to lower capital and operating costs for the former in the context of the very large air flow rates involved in these applications. The maximum air head pressure from blowers is a little more than 100 kPa, fixing the operating height of the bioreactors to about 10 meters in taking into account the specific density of the slurry and the gas hold-up.

A significant progress made in the conception of the bioleach tanks at industrial scale was in the effort to ensure an optimized control of the oxygen consumption, as required for the chalcopyrite bioleaching at high temperature (Batty & Rorke, 2005). Another one was to design an agitation system customized to the requirements in terms of power and oxygen transfer, as for the Kasese Cobalt Company plant (Bouquet & Morin, 2005).

4.2. Reactors Configuration

The design of a circuit of bioleach tanks at industrial scale is a compromise based on the main following criteria:

- To have the minimum number of tanks and the minimum overall volume of reaction;
- Correlatively to have the largest size of the tanks as possible, as the power consumption will decrease with the volume per tank by economy of scale;
- To have a longer retention time in the first stage of bioleaching (primary tanks) than in the next ones because the bio-oxidation rate is the highest at this stage and it is essential to ensure a complete and safe growth of the microbial consortia with all their components at this stage. A sufficient number of primary tanks will reduce the reduction of feed flow rate in case where a tank is bypassed for

maintenance or breakdown. With two or three primary tanks in parallel and the same volume of tanks as for the next stages that are in series, the retention time in the primary tanks is two or three times greater than in the downstream stages.

- To have a reasonable number of interchangeable tanks as breakdowns are unavoidable and it is essential to have the flexibility of modifying the circuit according to the circumstances. In a general way, the circuit must be designed to allow easy bypassing of tanks and derivation of flows.
- Secondary and tertiary stages in the series of tanks ensure the achievement of the bio-oxidation transformation. Air flow rates can be significantly reduced in those tanks compared to what is required in the primary tanks, but again flexibility must be ensured to cope with variations in the slurry flow.
- The transfer of slurry from one stage to the next one must be by gravity in order to avoid mechanical issues inherent to pumping.

4.3. Bioreactor Design and Corrosion Issues

When the tanks tend to be of large volume, the limitation in head pressure due to the use of blower for the air injection while taking care of the size of the agitation system results in shape ratios (height/diameter) close to 1.

The reactors and agitation systems in mesophilic processes are made of stainless steel (304L, 316L or SAF2205), or from mild steel with rubber lining for protection against corrosion and wear.

The internal coils that must be designed to be efficient in heat transfer and to be used as baffles for the mixing of the aerated slurry are also made of the same stainless steels. The reliability of their binding to the tank wall is critical as mixing forces and wearing may lead to serious incidents in case of failure.

Higher grades such as SAF2205 or 904L are required where water contains high chloride concentration (above 1,000 $mg.l^{-1}$ Cl^-) as experienced in some places in Western Australia (Dew *et al.*, 1997).

The high temperatures applied with thermophilic microbes result in higher constraints in corrosion and other materials are required (Batty & Rorke, 2005).

5. PROCESS MODELLING AND SIMULATION

Modelling of kinetics of bioleaching is mainly aimed at bringing more confidence to the scale-up of equipment and CAPEX estimates. Mathematical modelling provides data that feed simulation tools, which are used to design plants and to account for the conception or the modification of flowsheets.

In the case of bioleaching in agitated reactors, modelling also is an interesting scientific exercise as the apparently simplified and controlled conditions support a phenomenological approach integrating the fundamental mechanisms of interaction between micro-organisms and minerals.

The phenomenological approaches are much diversified depending on the physico-chemical aspects considered.

From the mineral point of view, the solubility of the sulfides and their reactivity to ferric ions, which is the main oxidizing agent of the medium, are essential to understand the nature of their degradation and how bacteria play a role in this transformation. Sand and co-workers (2001) and Tributsch (2001) have particularly detailed the chemical reactions that are thermodynamically susceptible to occur as primary source of energy to the bacterial growth. The role of ferric iron as energy vehicle explains the extent of the work to achieve a mathematical description of the kinetics of the oxidation of ferrous to ferric by oxygen catalysed by the bacterial activity (Dempers *et al.*, 2003). It is required to properly estimate the maximum yield of biomass production and the electrochemical conditions that lead to the oxidation of the sulfides. Ojumu and co-workers (2005) noted that some rate equations found in the literature include terms to account for the effects of temperature, pH, biomass concentration, ionic strength as well as inhibition due to arsenic but in general these effects have been studied in isolation and in ranges near the optimum. The fact is that few rate equations combine more than two effects and there is no clarity on how a comprehensive model to represent all effects should be constructed.

The interpretation that bioleaching mechanism is based on two sub processes of ferrous iron oxidation and chemical ferric leaching of the sulfide mineral gives a simple explanation to the predominance of the various bacterial species according to the sulfide minerals resulting from their ability of oxidation in various operating conditions (Hansford, 1997).

The first models of bioleaching in continuous-stirred-tank reactors broadly were either based on the principle of shrinking-particle or on the bacterial growth and attachment to solids, reflecting the debate of indirect or direct leaching mechanisms. Crundwell (1994) introduced the factor of particle-size distribution and the change in active surface due to the shrinkage in particle sizes and to the flow into and out of the reactors. Integrating the indirect leaching by ferric iron continuously generated by ferrous-oxidizing micro-organisms and the role of direct leaching by attached bacteria, the developed model could describe both batch and continuous kinetics of bioleaching in stirred tanks, after fitting of a reduced number of adjustable parameters. Reasonable predictions could be made to describe the kinetics of leaching of pyrite in a continuous pilot plant.

6. CONCLUSION

Bioleaching in continuous stirred reactors is now reaching the level of conventional practice after almost 20 years of existence for the pioneering installation of Fairview Mine in South Africa and about ten other industrial implementations. What seemed possible only for gold has now become a reality for cobalt (Hau *et al.*, 1997) and is on the way for nickel (Dew & Miller, 1997), zinc (Sandström *et al.*, 1997) and copper (d'Hugues *et al.*, 2001; Batty & Rorke, 2005). Henceforth, experience allows scaling up the equipment at industrial size and the performances to be predicted with more confidence and more reliability.

Bioleaching in continuous stirred tanks has some advantages compared to more conventional techniques (roasting/smelting and pressure leaching) of which some are the following:

- It is a very flexible technology, able to cope with various compositions of the sulfide concentrates with variable sulfide grades and potentially penalizing elements (As, Bi, Sb, etc.). It is also flexible in size and particularly appropriate to small installations for which there is no doubt that it is generally more competitive than conventional techniques.
- It is easy to operate and does not require a high level of qualification from the personnel in charge.
- The equipment used is not sophisticated and many parts can be found locally almost everywhere.
- All valuable metallic elements can be potentially recovered as pure metal or as intermediate products that can be commercialized to appropriate chemical end-users. A large range of hydrometallurgical alternatives exist to optimize the income from the treatment.
- The final wastes are confined to solids relatively stable and easy to handle.

After listing theses main positive characteristics, the following current limitations must not be omitted:

- It is a slow process and the solids concentration of the feed is limited, which means large pieces of equipment for high feed flow rates requiring sufficient ground surface and relatively high power consumption.
- The industrial experience in this field remains limited and the projects still seem to present a relatively high level of risk for investors, even if no industrial project has ever been directly subject to failure due to the bioleaching treatment only. The engineering background is shared by a limited number of experienced engineers and is not taught in school of engineers as such for example.
- The hydrometallurgical treatments that are required to purify the pregnant solution and allow the recovery of marketable products are sometimes costly because they are sophisticated and consume expensive reagents or power.

Some of these limitations are not totally crippling defaults and will just require time to find adequate response or improvements certified by the industrial experience.

7. REFERENCES

Acevedo F, Gentina JC, Garcia N. 1997. Effect of CO_2 supply in the biooxidation of As refractory gold concentrate. In: Proceedings of the International Biohydrometalurgy Symposium and Biomine' 97; August 4–6; Australian Mineral Foundation Inc., Sydney. Glenside, Australia, PM1.1-PM1.2.

Bailey AD, Hansford GS. 1993. Factors affecting bio-oxidation of sulfide minerals at high concentrations of solids: a review. Biotechnol Bioeng 42: 1164-1174.

Barrett J, Hughes MN, Karavaiko GI, Spencer PA. 1993. Metal extraction by bacterial oxidation of minerals, Ellis Horwood series in inorganic chemistry, New York.

Battaglia-Brunet F, d'Hugues P, Cabral T, Cezac P, Garcia JL, Morin D. 1998. The mutual effect of mixed thiobacilli and leptospirilli. Miner Eng 11: 195-205.

Batty JD, Rorke GV. 2005. Development and commercial demonstration of the BioCOP™ thermophile process. In: Harrison STL, Rawlings DE, Petersen J, eds. Proceedings of the 16th International

Biohydrometallurgy Symposium, September 25-29, Cape Town, South Africa. Produced by Compress www.compress. co.za, 153-161.

Brierley JA. 1997. Microbial population in mineral biooxidation processes. In: Proceedings of the International Biohydrometalurgy Symposium and Biomine' 97; August 4–6; Australian Mineral Foundation Inc., Sydney. Glenside, Australia, PSC1.1-PSC1-10.

Bouquet F, Morin D. 2005. BROGIM®: a new Three-Phase Mixing System Testwork and Scale-up. In: Harrison STL, Rawlings DE, Petersen J, eds. Proceedings of the 16th International Biohydrometallurgy Symposium, September 25–29, Cape Town, South Africa. Produced by Compress www.compress. co.za, 173-182.

Clark ME, Batty J, van Buuren C, Dew D, Eamon D. 2005. Biotechnology in Minerals processing: technological breakthroughs creating value. In: Harrison STL, Rawlings DE, Petersen J, eds. Proceedings of the 16th International Biohydrometallurgy Symposium, September 25–29, Cape Town, South Africa. Produced by Compress www.compress. co.za, 17-24.

Collinet MN, Morin D. 1990. Characterization of arsenopyrite oxidizing Thiobacillus. Tolerance to arsenite, arsenate, ferrous and ferric iron. Antonie van Leeuwenhoek 57: 237-244.

Coram NJ, Rawlings DE. 2002. Molecular Relationship between Two Groups of the Genus *Leptospirillum* and the Finding that *Leptospirillum ferriphilum* sp. nov. Dominates South African Commercial Biooxidation Tanks That Operate at 40°C. Appl Environ Microbiol 68: 838-845.

Crundwell FK. 1994. Modelling of bacterial leaching. Chem Eng J 54: 207-220.

Crundwell FK. 2001. How do bacteria interact with minerals? In: Ciminelli VST, Garcia Jr. O, eds. Biohydrometallurgy: Fundamentals, Technology and Sustainable Development. Elsevier Science BV, Amsterdam, Part A, 149-157.

Dempers CJN, Breed AW, Hansford GS. 2003. The effect of maintenance on the ferrous-iron oxidation kinetics of *Leptospirillum ferrooxidans* In: Tsezos M, Hatzikioseyian A, Remoudaki E, eds. BioHydrometallurgy: a sustainable technology in evolution. Eds National Technical University of Athens.

Dew DW, Lawson EN, Broadhurst JL. 1997. The BIOX® process for biooxidation of gold-bearing ores or concentrates. In: Rawlings DE, ed. Biomining: Theory, Microbes and Industrial Processes. Springer-Verlag, Berlin, 45-80.

Dew D, Miller G. 1997. The BioNIC process: bioleaching of mineral sulfide concentrates for recovery of nickel. In: Proceedings of the International Biohydrometalurgy Symposium and Biomine' 97; August 4–6; Australian Mineral Foundation Inc., Sydney. Glenside, Australia, M7.1.1-M7.1.9.

Dew DW, van Buuren C, McEwan K, Bowker C. 1999. Bioleaching of base meta sulfide concentrates: a comparison of mesophile and thermophile bacterial cultures. In: Amils R, Ballester A, eds. Biohydrometallurgy and the environment toward the mining of the 21st century, Proceedings of the International Biohydrometallurgy Symposium IBS-99, El Escorial, Spain. Elsevier, Amsterdam, Part A, 229-238.

d'Hugues P, Cezac P, Battaglia F, Morin D. 1999. Bioleaching of a cobaltiferous pyrite at 20% solids: a continuous laboratory-scale study. In: Amils R, Ballester A, eds. Biohydrometallurgy and the environment toward the mining of the 21st century, Proceedings of the International Biohydrometallurgy Symposium IBS-99, El Escorial, Spain. Elsevier, Amsterdam, Part A, 167-176.

d'Hugues P, Morin D, Foucher S. 2001. HIOX® Project: a bioleaching process for the treatment of chalcopyrite concentrates using extreme thermophiles. In: Ciminelli VST, Garcia Jr. O, eds. Biohydrometallurgy: Fundamentals, Technology and Sustainable Development. Elsevier Science BV, Amsterdam, Part A, 75-83.

Foucher S, Battaglia-Brunet F, d'Hugues P, Clarens M, Godon JJ, Morin D. 2001. Evolution of the bacterial population during the batch bioleaching of a cobaltiferous pyrite in a suspended-solids bubble column, and comparison with a mechanically-agitated reactor. In: Ciminelli VST, Garcia Jr. O, eds. Biohydrometallurgy: Fundamentals, Technology and Sustainable Development. Elsevier Science BV, Amsterdam, Part A, 3-11.

Fowler TA. 1999. Interactions in the bacterial leaching of pyrite and sphalerite by *Thiobacillus ferrooxidans*, Ph. D. thesis, University of the Witwatersrand, Johannesburg, RSA.

Gehrke T, Telegdi J, Thierry D, Sand W. 1998. Importance of extracellular polymeric substances from *Thiobacillus ferrooxidans* for bioleaching. Appl Environ Microbiol 64: 2743-2747.

Giaveno A, Chiacchiarini P, Lavalle L, Donati E. 2005. Reversed flow airflift reactor characterization for bioleaching applications. In: Amils R, Ballester A, eds. Biohydrometallurgy and the environment toward the mining of the 21st century, Proceedings of the International Biohydrometallurgy Symposium IBS-99, El Escorial, Spain. Elsevier, Amsterdam, Part A, 125-135.

Golyshina OV, Pivovarova TA, Karavaiko GI, Kondrateva TF, Moore ERB, Abraham WR, Lünsdorf H, Timmis KN, Yakimov MM, Golyshin PN. 2000. Ferroplasma acidiphilum gen.nov., sp.nov., an acidophilic, autotrophic, ferrous-iron-oxidizing, cell-wall-lacking, mesophilic member of the Ferroplasmaceae fam.nov., comprising a distinct lineage of the Archaea. Int J Sys Evol Microbiol 50: 997-1006.

Gomez JM, Cantero D, Johnson DB. 1999. Comparison of the effects of temperature and pH on iron oxidation and survival of *Thiobacillus ferrooxidans* (type strain) and a '*Leptospirillum ferrooxidans*'-like isolate. In: Amils R, Ballester A, eds. Biohydrometallurgy and the environment toward the mining of the 21st century, Proceedings of the International Biohydrometallurgy Symposium IBS-99, El Escorial, Spain. Elsevier, Amsterdam, Part A, 689-696.

Hansford GS. 1997. Recents development in modeling the kinetics of bioleaching. In In: Rawlings DE, ed. Biomining: Theory, Microbes and Industrial Processes. Springer-Verlag, Berlin, 151-175.

Hau JM, Morin D, Ollivier P, Aird J. 1997. Hydrometallurgical recovery of cobalt dissolved by bioleaching of a cobaltiferous pyrite concentrate. In: Hoberg H, von Blottnitz H, eds, Proceedings of the XX International Mineral Processing, Aachen, Germany. GDMB, Clausthal-Zellerfeld, Germany, 557-566.

Hughes MN, Poole RK. 1989. Metals and Micro-organisms, Chapman and Hall, London.

Ignatiadis I, Morin D. 2001. In: Ciminelli VST, Garcia Jr. O, eds. Biohydrometallurgy: Fundamentals, Technology and Sustainable Development. Elsevier Science BV, Amsterdam, Part A, 95-105.

Jerez C. 1997. Molecular methods for the identification and enumeration of bioleaching micro-organisms. In: Rawlings DE, ed. Biomining: Theory, Microbes and Industrial Processes. Springer-Verlag, Berlin, 281-297.

Johnson DB, Francisco FR. 1997. Heterotrophic acidophiles and their roles in the bioleahcing of sulfide minerals. In: Rawlings DE, ed. Biomining: Theory, Microbes and Industrial Processes. Springer-Verlag, Berlin, 259-279.

Kelly DP, Jones CA. 1977. Factors affecting metabolism and ferrous iron oxidation in suspensions and batch cultures of *Thiobacillus ferrooxidans*: relevance to ferric iron leach solution regeneration. In: Murr LE, Torma AE, Brierley JA, eds. Metallurgical applications of bacterial leaching and related phenomena. Academic Press, New York, 19-44.

Kinzler K, Gehrke T, Telegdi J, Sand W. 2001. Bioleaching – a result of interfacial processes caused by extracellular polymeric substances (EPS). In: Ciminelli VST, Garcia Jr. O, eds. Biohydrometallurgy: Fundamentals, Technology and Sustainable Development. Elsevier Science BV, Amsterdam, Part A, 191-197.

Morin D, Battaglia F, Ollivier P. 1993. Study of the bioleaching of a cobaltiferrous pyritic concentrate. In: Torma AE, Wey JE, Lakshamanan VL, eds, Biohydrometallurgical Technologies. The Minerals, Metals and Materials Society, Warrendale, Pennsylvania, 147-156.

Norris PR. 1989. Factors affecting bacterial mineral oxidation: the example of carbon dioxide in the context of bacterial diversity. In: Proceedings of the International Biohydrometallurgy Symposium'89, CANMET, Ottawa, 7-14.

Norris PR. 1997. Thermophiles and bioleaching. In: Rawlings DE, ed. Biomining: Theory, Microbes and Industrial Processes. Springer-Verlag, Berlin, 247-258.

Ojumu TV, Petersen J, Searby GE, Hansford GS. 2005. A review of rate equations proposed for microbial ferrous-iron oxidation with a view to application to heap leaching. In: Harrison STL, Rawlings DE, Petersen J, eds. Proceedings of the 16th International Biohydrometallurgy Symposium, September 25–29, Cape Town, South Africa. Produced by Compress www.compress. co.za,85-93.

Oolman T. 1993. Bioreactor Design and Scaleup – Apllications in Minerals Bioleaching. In: Torma AE, Wey JE, Lakshamanan VL, eds. Biohydrometallurgical Technologies. The Minerals, Metals and Materials Society, Warrendale, Pennsylvania, 401-416.

Pinches T, Chapman JT, te Riele WAM and van Staden M. 1988. The performances of bacterial leach reactors for the pre-oxidation of refractory gold-bearing sulfide concentrates. In: Norris PR, Kelly DP, eds. Biohydrometallurgy, Proceedings of the International Symposium, Warwick, Antony Rowe Ltd. Chippenham, Wiltshire, UK, 329-344.

Rawlings DE. 1997. Mesophilic, autotrophic bioleaching bacteria: description, physiology and role. Rawlings DE, ed. Biomining: Theory, Microbes and Industrial Processes. Springer-Verlag, Berlin, 229-245.

Rawlings DE. 1999. The molecular genetics of mesophilic, acidophilic, chemolithotrophic, iron or sulfide-oxidizing microorganisms. Amils R, Ballester A, eds. Biohydrometallurgy and the environment toward the mining of the 21st century, Proceedings of the International Biohydrometallurgy Symposium IBS-99, El Escorial, Spain. Elsevier, Amsterdam, Part B, 3-20.

Rawlings D, Dew D, du Plessis C. 2003. Biomineralization of metal-containing ores and concentrates. Trends Biotechnol 21: 38-44.

Ritchie I, Barter J. 1997. Process and engineering design aspects of biological oxidation plants. In: Proceedings of the International Biohydrometalurgy Symposium and Biomine' 97; August 4–6; Australian Mineral Foundation Inc., Sydney. Glenside, Australia, M14.4.4-M14.4.8.

Sand W, Gehrke T, Jozsa PG, Schippers A. 2001. (Bio)chemistry of bacterial leaching–direct vs. indirect bioleaching. Hydrometallurgy 59: 159-175.

Sandström A, Sundkvist JE, Petersson S. 1997. Bio-oxidation of a complex zinc sulfide ore: a study performed in continuous bench and pilot scale. In: Proceedings of the International Biohydrometalurgy Symposium and Biomine' 97; August 4–6; Australian Mineral Foundation Inc., Sydney. Glenside, Australia, M1.1.1-M1.1.11.

Spencer PA. 2001. Influence of bacterial culture selection on the operation of a plant treating refractory gold ore. Int J Miner Process 62: 217-229.

Tributsch H. 1999. Direct versus indirect bioleaching. Amils R, Ballester A, eds. Biohydrometallurgy and the environment toward the mining of the 21st century, Proceedings of the International Biohydrometallurgy Symposium IBS-99, El Escorial, Spain. Elsevier, Amsterdam, Part A, 51-60.

Tributsch H. 2001. Direct versus indirect bioleaching, Hydrometallurgy 59: 177-185.

Wiersma CL, Rimstidt JD. 1984. Rates of reaction of pyrite and marcasite with ferric iron at pH 2. Geochim Cosmochim Acta 48: 84-92.

CHAPTER 8

BIOREACTOR DESIGN FUNDAMENTALS AND THEIR APPLICATION TO GOLD MINING

FERNANDO ACEVEDO AND JUAN CARLOS GENTINA
School of Biochemical Engineering, Pontifical Catholic University of Valparaiso, Valparaiso, Chile

1. INTRODUCTION

Microbial growth and product formation can be considered the resultant of thousands of metabolic enzyme catalyzed chemical reactions and as such can be represented by an overall stoichiometric equation (Wang *et al.*, 1979; Acevedo, 1987; Nielsen *et al.*, 2003). For heterotrophs in aerobic growth the following equation can be written per mole of biomass:

$$1 \cdot C_aH_bO_c + m \cdot NH_3 + n \cdot O_2 \rightarrow CH_eO_fN_g + r \cdot CO_2 + t \cdot H_2O + u \cdot C_hH_iO_jN_k \quad (1)$$

in which $C_aH_bO_c$ represents the carbon and energy source, $CH_eO_fN_g$ the biomass and $C_hH_iO_jN_k$ an extracellular metabolite.

In bioleaching autotrophic microorganisms are involved. In this case carbon dioxide is used as carbon source and the energy is obtained from the oxidation of the mineral substrate that can de e.g. ferrous iron or reduced sulfur species:

$$1 \cdot CO_2 + m \cdot NH_3 + p \cdot (Fe^{2+} + H^+) + n \cdot O_2 \rightarrow CH_eO_fN_g + p \cdot Fe^{3+} + t \cdot H_2O \quad (2)$$

In the case of a sulfide mineral such as pyrite:

$$1 \cdot CO_2 + m \cdot NH_3 + n \cdot FeS_2 + q \cdot H_2O + r \cdot O_2 \rightarrow CH_eO_fN_g + w \cdot Fe_2(SO_4)_3 + z \cdot H_2SO_4 \quad (3)$$

In this context the vessel in which microbial growth and activity takes place, the traditional fermentor, can be regarded as a reactor, an important consideration when

E.R. Donati and W. Sand (eds.), Microbial Processing of Metal Sulfides, 151–168.

facing its design and operation. Being so, in principle the bioreactor can be of any of the types developed for chemical reactors: stirred tank, tubular, different types of columns (fixed bed, expanded bed, fluidized, bubbling column, air-lift column, plate column), and especial designs such as loop reactors, solid substrate reactors and heap reactors. Most of these designs are actually used as bioreactors, but clearly stirred tanks are by far the mostly used in industrial fermentations aimed to the production of commodities and fine biochemicals, while different type of columns are widely used in the biological treatment of solid, liquid, and gaseous effluents.

Equations (1) to (3) are mass balance equations and as such are not useful to formulate kinetic equations because they do not represent a very important characteristic of microbial growth: its autocatalytic nature (Atkinson, 1971, 1974; Acevedo, 2000; Rossi, 2001; González *et al.*, 2004). In microbial growth cells act as products and as reactants. This situation can be expressed as:

$$(4) \qquad S + X \rightarrow X + X$$

where S represents the nutrients or substrates and X the biomass or cells. In equations (1) to (3) the biomass term in the left member of equation (4) has been cancelled with an equivalent molar amount of cells in the right member, something licit to do in a mass balance equation. But if the interest is to write a reaction rate equation this cannot be done because every reactant must be considered (Levenspiel, 1999a).

The rates of cell growth and nutrient consumption are represented by (Pirt, 1975; Wang *et al.*, 1979; Esener *et al.*, 1983; Posten & Cooney, 1993):

$$(5) \qquad r_x = \frac{dX}{dt} = \mu \cdot X$$

$$(6) \qquad r_s = -\frac{dS}{dt} = \frac{1}{Y_x} \cdot \frac{dX}{dt}$$

These two equations contain three variables, X, S and μ, so a third equation is required:

$$(7) \qquad \mu = \mu_M \cdot \frac{S}{K_s + S}$$

Equation (7), the Monod equation (Monod, 1949), represents the most commonly used relation between substrate concentration and specific growth rate and implies that growth takes place in the absence of any type of inhibition. Several alternative, and better suited to the specific case, non-Monodian and inhibition models have been proposed (Levenspiel, 1980; Zwietering *et al.*, 1990; Blanch & Clark, 1996; Peleg, 1997; Nielsen *et al.*, 2003a).

2. SELECTION OF IDEAL BIOREACTORS

Although there are several types of bioreactors, from the standpoint of their kinetic behavior only two ideal models can be distinguished: the ideal continuous stirred tank (CSTR) and the plug flow tubular reactor (PFR). The CSTR model implies that its content is totally mixed, so no gradients are present; this condition can only be approached in non-ideal or real reactors. On the other hand the PFR model implies that the fluid inside the tubular reactor presents a plug flow regime which is defined as an orderly flow with no backmixing and no axial dispersion, so all linear velocities in the radial direction are the same. Again, this situation can only be approached in actual operation (Bailey & Ollis, 1986; Levenspiel, 1999b).

One interesting approach to reactor selection is based in the criterion of minimum reaction volume (equivalent to minimum residence time) for a defined limiting reactant conversion (Levenspiel, 1999c). Mass balances on the limiting reactant in a CSTR and in a PFR operated at steady state give:

CSTR:

$$F(S_0 - S) = r_s \cdot V \quad (8)$$

Defining the fractional conversion χ as $(S_0\text{-}S)/S_0$:

$$\frac{\tau}{S_0} = \frac{V}{F \cdot S_0} = \frac{\chi}{r_s} \quad (9)$$

PFR:

$$F \cdot dS = r_s \cdot dV \quad (10)$$

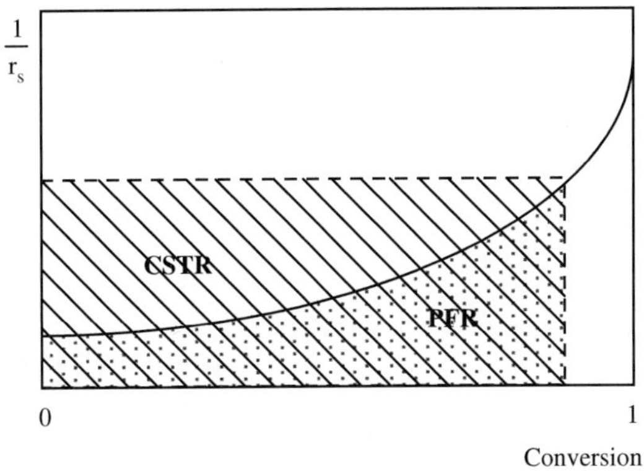

Figure 1. Comparison of the performance of continuous stirred tank and plug flow reactors for the case of regular kinetics

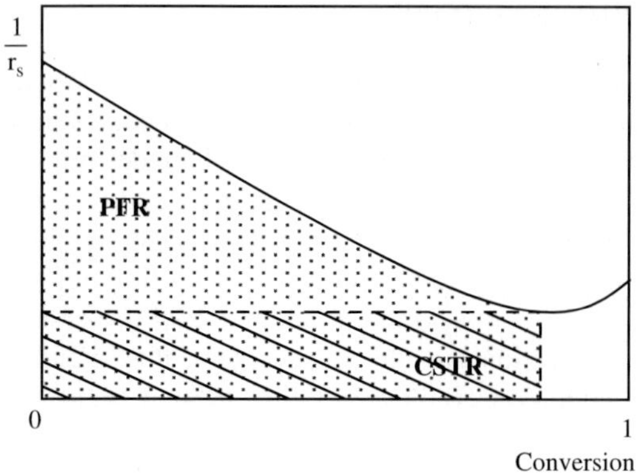

Figure 2. Comparison of the performance of continuous stirred tank and plug flow reactors for the case of an autocatalytic reaction

$$(11) \qquad \frac{\tau}{S_0} = \frac{V}{F \cdot S_0} = \int \frac{d\chi}{r_s}$$

In a χ vs $1/r_s$ plot, the residence time, equivalent to the reaction volume for a defined flow rate, divided by the inlet reactant concentration is represented by the area of the rectangle for a CSTR and by the area under the curve for a PFR (Figure 1). So for an elementary reaction a PFR requires less residence time and reaction volume than a CSTR. But this situation changes in the case of autocatalytic reactions such as microbial growth. It can be seen from Figure 2 that now a CSTR is a better choice than a PFR.

3. OPTIMAL REACTOR CONFIGURATION IN BIOLEACHING

As depicted in Figure 2, the χ vs $1/r_s$ curve for autocatalytic reactions presents a minimum which is also an inflection point. The position of this minimum can be determined as follows:

Using equations 6 and 7 and considering that at steady state $X = Y_X \cdot (S_0 - S)$ we can write:

$$(12) \qquad \frac{1}{r_s} = \frac{K_S + S}{\mu_M \cdot S \cdot (S_0 - S)}$$

Making:

$$(13) \qquad \frac{d\left(\frac{1}{r_s}\right)}{dS} = 0$$

We obtain:

$$(14) \qquad S^* = -K_s + \sqrt{K_s^2 + S_0 \cdot K_s}$$

and

$$\chi^* = \frac{S_0 - S^*}{S_0} \qquad (15)$$

Equations (13) and (14) imply that the position of the minimum is determined by the inlet reactant concentration and by the value of the saturation constant. In conventional industrial fermentations this minimum is located at high conversions and the selection of a CSTR holds. This is because culture media components are chosen considering their affinity with the cell population, so their K_S values are very low, in the order of grams per cubic meter. Nevertheless the situation in bioleaching is different: we cannot freely choose the energy source which is the mineral, and minerals usually present K_S values as high as of 3 to 6 kg/m^3, so the minimum in the curve is placed at relatively high S values and low conversions (Acevedo, 2000). In Figure 3 it can be seen that neither a CSTR nor a PFR is the optimum alternative. Minimum reaction volume is obtained with one CSTR followed by a PFR connected in series. In that way each type of reactor is used in the situation in which their performance is optimal. Because the actual operation of a plug flow tubular reactor for a three-phase system (liquid, gas and suspended solids) is not possible, the PFR behavior is simulated by a train of two to four CSTR's in series.

To illustrate the above, a 100 tonne/day plant pyritic gold concentrate operating at 20% w/v pulp density was considered (Soto *et al.*, 2002). The concentrate contained 67.6% pyrite and 15 g gold/tonne. Using the model proposed by Crundwell (2000/2001), Figure 4 was obtained.

From Figure 4 and a fractional conversion of 0.9 it can be established that the optimal reactor arrangement is one CSTR of 1160 m^3 followed by three CSTR's of 500 m^3 each, with a total volume of 2660 m^3, with an economy of 23.5% compared with the alternative of using a single 3479 m^3 CSTR.

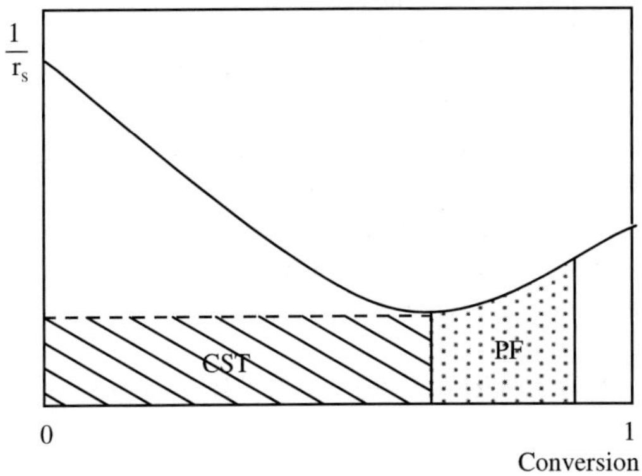

Figure 3. Comparison of the performance of continuous stirred tank and plug flow reactors for the case of biooxidation of a mineral concentrate of high saturation constant

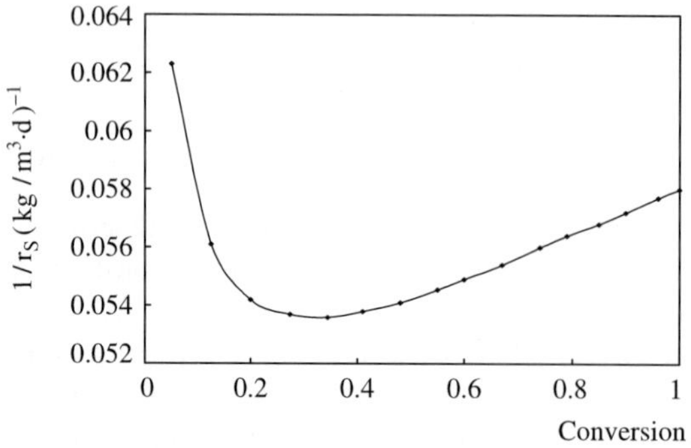

Figure 4. Inverse reaction rate as a function of conversion for the case of the biooxidation of a pyritic gold concentrate

Some additional considerations can be made. First, a tank of 1160 m^3can be considered too large because of construction difficulties and cost. Second, from the standpoint of maintenance and stock of spare parts it would be desirable to have equal volume tanks. So the reactor arrangement could consist of two CSTR's of 600 m^3 each in parallel acting as primary reactor and three 600 m^3 CSTR's in series as secondary reactors. This gives a total reaction volume of 3000 m^3, a 16,7% economy as compared to use a single primary rector represented by six 600 m^3 in parallel.

Actual pilot plant and commercial size plants for the biooxidation of gold and other base metal concentrates have commonly been designed with this bioreactor configuration (Dew, 1995; Dew *et al.*, 1997; Miller, 1997; Heinzle *et al.*, 1999; Ly, 2005).

4. OXYGEN AND CARBON DIOXIDE REQUIREMENTS DURING BIOOXIDATION OF SULFIDE MINERALS

It is known that microorganisms involved actively in biooxidation of minerals are autotrophs and have an aerobic energetic metabolism that uses molecular oxygen as final electron acceptor. Thus, the biooxidation of sulfide minerals faces the requirements of supplying sufficient oxygen and carbon dioxide in order to avoid being rate limited by them. In fact, both gases must be transferred not only in the right amount but also at the right rate (Acevedo & Gentina, 1989; Bailey & Hansford, 1993; Nagpal *et al.*, 1993).

In a general form, the microbial carbon dioxide uptake rate (CUR) is represented by the following equation:

$$(16) \qquad CUR = \frac{\mu X}{Y_{CO_2}}$$

where Y_{CO2} is the carbon dioxide yield coefficient in kg of cells/ kg of CO_2. Y_{CO2} will depend on the actual cell carbon content and its value should be around 0.55 kg of cell/ kg of CO_2 consumed.

On the other hand, the microbial oxygen uptake rate (OUR) is represented by:

$$(17) \qquad OUR = \frac{\mu\, X}{Y_{O_2}}$$

where Y_{O2} is the molecular oxygen yield coefficient in kg of cells/ kg of O_2, whose value depends on the type of energy source being used by the cells. In this sense, the lower the available free energy of the energy source, the lower will result the molecular oxygen yield (Madigan *et al.*, 1997). As a comparison, calculating Y_{O2} for a bacteria growing on glucose according to the expression presented by Mateles (1971) and for a bacteria oxidizing ferrous ion according to the relation proposed by Acevedo and Gentina (1989), results in 1.3 and 0.04 kg of cells/ kg of O_2 respectively, meaning that it is required 32 times more oxygen using ferrous ion than glucose to produce the same amount of biomass.

Both gases exhibit low solubility in aqueous solutions, making it necessary to transfer them continuously during the biooxidation process. Air has been the normal source for both gases, but since CO_2 and O_2 are not in the exact proportion they are needed, it normally happens that CO_2 becomes the limiting one (Acevedo *et al.*, 1998; Boon & Heijnen, 1998).

The gas transfer rate (GTR) to an aqueous media is given by the following relation:

$$(18) \qquad GTR = k_L a\, (C^* - C_L)$$

where $k_L a$ represents the volumetric gas transfer coefficient depending upon the geometry of the transfer system, the rheology of the liquid phase, the environmental conditions and most important the operating conditions. k_L stands for the mass-transfer coefficient in liquid film and a the gas-liquid interfacial area per unit volume of liquid. The difference C^* - C_L represents the driving force of the transfer phenomenon, being limited, as said earlier, by the small value of C^* for both O_2 and CO_2. The partial pressure of O_2 and CO_2 in air at 101 330 Pa are 21 279.3 and 30.4 Pa respectively, resulting values of C^* at 25 °C equal to 0.0085 kg/m^3 for oxygen and 0.00044 kg/m^3 for carbon dioxide (Hougen *et al.*, 1954).

In order to avoid gas limitation during biooxidation, the oxygen and carbon dioxide transfer rates must at least equal the oxygen and carbon dioxide consumption rates. If that is not possible, the microbial activity becomes limited and cell growth rate slows down affecting the kinetics of mineral solubilization.

To state the oxygen and carbon dioxide demands in a biooxidation process based on microbiological parameters it is not an easy task. There are severe experimental limitations to measure cell concentration and specific growth rate in a biooxidation reactor, without mentioning the likely presence of a mixed culture and multiple energy sources. Undoubtedly much more research efforts should be done in order to overcome these difficulties. There are alternative ways based on the overall reaction equation of the process including or not the presence of microbial cells. In the first case, in order to balance the equation, some microbiological parameters like the yield of the energy source (sulfide mineral) in biomass or the relation of solubilized base metal per unit of cell mass must be known (Acevedo, 2000). These equations are useful to establish quantitative relationships between the consumption of oxygen and carbon dioxide per unit mass of oxidized sulfide or grown biomass. Additionally, these types of relationships need a kinetic expression like the rate of mineral oxidation in order to finally calculate the required gas transfer rates.

As an example, considering just pyrite, the overall reaction is:

$$(19)\qquad 4\ FeS_2 + 15\ O_2 + 2\ H_2O \rightarrow 2\ Fe_2\,(SO_4)_3 + 2\ H_2SO_4$$

From equation (19) is possible to conclude that 1 kg of O_2 is required in order to oxidize 1 kg of pyrite.

The overall process reaction including the cell term shown in equation (3) allows for the evaluation of the CO_2 and O_2 required to oxidize 1 kg of pyrite, but the pyrite yield coefficient in kg of cells/ kg of FeS_2 or a related independent ratio must be previously known.

If an average oxidation rate of $1.39 \cdot 10^{-4}$ kg of $FeS_2/m^3 \cdot s$ is assumed (Boon & Heijnen, 1998a; Rossi, 1999; Acevedo *et al.*, 2004) the oxygen demand can be calculated as $1.39 \cdot 10^{-4}$ kg of $O_2/\ m^3 \cdot s$. The k_La to satisfy such a demand can be estimated with equation (18) assuming $C^* - C_L$ equal to 0.0075 kg/m^3, resulting a k_La value of $1.85 \cdot 10^{-2}\ s^{-1}$. To execute the same calculations for CO_2 is a much more difficult task, especially because there is not enough information about microbiological parameters.

Dividing equations (16) and (17), and writing equation (18) for O_2 and CO_2 and dividing them, the following expressions can be set:

$$(20)\qquad \frac{OUR}{CUR} = \frac{Y_{CO_2}}{Y_{O_2}}$$

$$(21)\qquad \frac{OTR}{CTR} = \frac{k_L a_{O_2}\left(C^*_{O_2} - C_{LO_2}\right)}{k_L a_{CO_2}\left(C^*_{CO_2} - C_{LCO_2}\right)}$$

A common design criteria is to define aeration condition as to satisfy the process oxygen demand, which is obtained setting OUR and OTR equals.

Using equations (20) and (21), it follows that:

$$(22)\qquad CUR = \frac{k_L a_{O_2}\left(C^*_{O_2} - C_{LO_2}\right) Y_{O_2}}{k_L a_{CO_2}\left(C^*_{CO_2} - C_{LCO_2}\right) Y_{CO_2}} CTR$$

The value of the proportional factor of equation (22) will define if the biooxidation process will be limited by CO_2 or not. As an example, at 25 ^oC the ratio of volumetric oxygen and carbon dioxide transfer coefficients is 1.176 (Boogerd *et al.*, 1990) and assuming Y_{O2} and Y_{CO2} equal to 0.04 and 0.55 respectively, it is found out that for a fix temperature, the relation among CUR and CTR depends upon the actual O_2 and CO_2 gradient concentrations:

$$\text{(23)} \qquad \text{CUR} = 0.0855 \frac{(C^*_{O_2} - C_{LO_2})}{(C^*_{CO_2} - C_{LCO_2})} \text{CTR}$$

According to Equation (23), if the gradient ratio results higher than 11.7 then CUR will be higher than CTR, meaning that the dissolved CO_2 concentration in the liquid medium limits cell activity. According to Myerson (1981) the critical oxygen concentration for the growth of *A. ferrooxidans* at 25 °C is 0.00041 kg/m^3. Assuming this value and C^*_{O2} equal to 0.0085 kg/m^3, the oxygen gradient becomes 0.0081 kg/m^3 which determine that the carbon dioxide gradient should be 0.00069 kg/m^3 in order not to limit the cell growth. This is impossible using air since the highest carbon dioxide gradient would approach C^*_{CO2} that at 25 °C is 0.00044 kg/m^3. To reverse this situation, a CO_2 enrichment or k_La increase policy should be applied.

5. AGITATION IN BIOOXIDATION REACTORS

Agitation in biooxidation reactors faces two important goals: one aimed to obtain a high degree of homogeneity of the gas – liquid – solid system and the other to enhance the oxygen and carbon dioxide transfer rates from the gas to the liquid phase and the heat transfer rate necessary to keep operation temperature within the recommended range. The work done by Gormely and Brannion (1989), Spencer *et al.* (1997), Greenhalgh and Ritchie (1999), Harvey *et al.* (1999), are a useful source of information on the different aspects related to the design requirements of a biooxidation reactor. Furthermore, the design of the agitation system is not only limited by cost considerations but also by the response of the microbial cells, which can suffer metabolic stress and damage by high shear stress and abrasive conditions generated during agitation (Toma *et al.*, 1991; Deveci, 2004).

The solids present in a biooxidation reactor are usually flotation concentrates with particle sizes under 75 μm and densities over 2000 kg/m^3. The former facilitates particle suspension and mixing, but the latter represent a serious impediment to get a reasonable homogeneity.

Mechanical and pneumatic agitation is commonly used in reactors. In mechanically stirred tanks the agitator consists of a motor, a shaft and one or more impellers. Pneumatically agitated reactors are well represented by the air-lift reactor and the Pachuca tank. The last one, commonly used in hydrometallurgical duties, differs from the air-lift in that it has a lower height to diameter ratio (Perry *et al.*, 1984; van't Riet & Tramper, 1991).

The disk or Rushton turbine, the classical impeller that has been used in microbial cell cultures, exhibits a high performance in the enhancement of the gas-liquid transfer rate, but at an elevated energy cost and high shear stress

(Spencer *et al.*, 1997). However, this type of impeller is not suited for systems operating with heavy solid particles since its radial flow pattern is not especially effective to generate a homogeneous suspension. Additionally, it can be affected by flooding phenomenon when high gas throughput is required (Lally, 1987). In comparison to common industrial fermentations, biooxidation demands higher mixing characteristics and lower oxygen transfer rates. As a consequence, impellers that generate axial flow patterns are preferred since create better suspension condition. González *et al.* (2003) compared the mixing characteristics of a pitch blade turbine and marine helix, concluding that although the first one require lower agitation rate, the marine propeller is preferable because requires less power and produces a more homogeneous suspension. Among the axial flow type impellers the hydrofoil design presents several advantages especially because its good mixing characteristics, low shear and gas transfer capabilities compatible with the bio-oxidation processes demands (Dew *et al.*, 1997; Acevedo, 2000). In order to satisfy situations of higher oxygen demand it has been proposed to use a mixed impeller arrangement consisting of a disk turbine at the bottom and a propeller(s) at the upper part of the shaft (Bouquet & Morin, 2005).

The gas transfer coefficient depends strongly on the operation conditions, especially on agitation and aeration. For mechanically agitated systems the usual correlations for k_La are of the form (Cooper *et al.*, 1944):

$$k_La = K\left(\frac{P_g}{V}\right)^{\alpha} v_S^{\beta} \qquad (24)$$

where P_g represents the gassed agitation power drawn by the liquid, V the liquid volume, v_S the superficial gas velocity, defined as the gas flow rate per unit of vessel cross-sectional area, and K, α and β are constants. The values of α and β reveal the relative influence of agitation power and aeration rate on the gas transfer coefficient. Some equations use the rotational speed of the agitator instead of the gassed power per unit volume (van't Riet, 1979) or add it as a third term (Aiba *et al.*, 1973).

In the case of pneumatically agitated bioreactors, the exponent α of equation (24) is 0, so k_La depends only on the superficial gas velocity (Canales *et al.*, 2002; van't Riet & Tramper, 1991). In other correlations β is set equal to 0 and the gassed agitation power replaced by aeration power (Acevedo *et al.*, 1988). For both types of agitation systems, some authors have incorporated a term in the k_La expression related to the pulp density, which has a negative effect (Canales *et al.*, 2002; Mills *et al.*, 1987). For instance Canales *et al.* (2002) correlated the k_La with the operation conditions in the biooxidation of an enargite-pyrite refractory gold concentrate carried on in a laboratory aerated column:

$$k_La = 24.89V_S^{(0.171-0.011\Phi)} \qquad (25)$$

where v_S is in cm/s, Φ as percentage (w/v) and k_La in h^{-1}. It was stated that at constant v_S the hold up, defined as the volume fraction of bubbles in the liquid, decreases as solids concentration increases, implying that the inverse relation between $k_L a$ and Φ is a consequence of hold up reduction and therefore interfacial gas-liquid area reduction.

Mechanically agitated reactors have limitations with respect their sizes, because cost and mechanical complexity related to agitation increases at higher rates as volume goes over 500 m^3. In this sense, pneumatically agitated reactors could play an important role when high reaction volumes are needed, especially because their construction and operation costs should be lower due to the absence of an agitator.

Also this type of reactors offer comparable gas transfer coefficients and lower shear stress, generating an adequate environment for cell activity.

Furthermore, other types of reactors have been proposed for biooxidation applications that can also provide the necessary mixing and gas transfer capabilities. That is the case of the Biorotor (Loi *et al.*, 1995), the Aerated Trough Bioreactor and the Delft Inclined Plate Bioreactor (Rossi, 1999, 2001).

6. HEAT TRANSFER IN BIOOXIDATION REACTORS

The process temperature control is a necessary condition ~~in~~ order to keep the highest rate of biooxidation. Microbial cells impose temperature conditions, since they have an optimum temperature for their metabolic activity. Temperatures over the optimum will accelerate cell death and at the opposite direction will slow down metabolic reaction kinetics. The several heat fluxes present in the biooxidation operation must be balanced; otherwise a temperature shift with negative consequences for the process could take place.

Under normal operating conditions, the following general energy balance can be applied to a typical aerated and agitated continuous bioreactor:

$$(26) \qquad Q_{acc} = Q_{met} + Q_{ag} + Q_{aer} + Q_{sens} - Q_{evap} - Q_{surr} - Q_{hxcr}$$

where Q_{acc} stands for the rate of heat accumulation per unit volume unit in the reactant system, being the other terms rates of heat of generation/removal and the subscripts met, ag, aer, sens, evap, surr and hxcr stand for metabolic, agitation, aeration, sensible, evaporation, surrounding and heat exchanger respectively.

Equation (26) is the base of the design of heat exchange system required to balance heat fluxes to keep a constant operation temperature inside the reactor (Wang *et al.*, 1979). Constant temperature implies that Q_{acc} is zero and in order to establish the heat flux that has to be handled by the heat exchanger system all others terms of equation (26) must be evaluated.

In general Q_{met} and Q_{ag} are the most significant heat fluxes (Cooney, 1981), but also it is necessary to pay attention to the other terms. The metabolic heat generation Q_{met} is the net result of the metabolic reactions carried on by cells, which depend on the organism itself, the energy source being used, cell amount and the specific cell growth rate. Based on a stoichiometric equation describing the biooxidation reactions, the Q_{met} can be evaluated as the difference between the combustion or formation heats of the products, including the cells, and the reactants. In the biooxidation of minerals the heat generated by the oxidation of the reduced sulfur compound (i.e. pyrite has a standard heat of reaction equal to -1481 kJ/mol FeS_2) predominates over the energy expenditure for cell synthesis (low cell mass and growth rate), therefore the actual heat of reaction should be slightly lower than the amount released by the biooxidation of the energy source. In the case of heterotrophic microorganisms a simple relationship has been developed to estimate Q_{met} as a function of the oxygen consumption rate (Cooney *et al.*, 1968), a relationship that would be worth to adjust and validate for quimiolito-autotrophic cells.

Heat of agitation Q_{ag} is a consequence of the energy imparted to the fluid, a fraction of which is dissipated as heat. The amount of heat generation is a function of the agitation system, reactor geometry, operation conditions and fluid physical characteristics. The agitation systems reviewed in the preceding section that expend lower power inputs for a given duty will generate less agitation heat.

Aeration heat Q_{aer} is associated to the energy dissipated due to the isothermal expansion of the gas inside the vessel and depends mainly on the hydrostatic pressure in the reactor. In this sense, the aeration heat can be significant for industrial scale agitated reactors and pneumatically agitated reactors.

The sensible heat Q_{sens} results of the difference in temperatures between a feeding stream and the reactor. Biooxidation process has gaseous, liquid and solid feeding streams, which can be conditioned to the reactor temperature before get inside in order to neglect their influence in the energy balance.

The evaporation heat Q_{evap} comes from the bubbling of unsaturated air through the vessel. This effect can be avoided in small size reactors by saturating the incoming air with water, something that is not always practical at industrial scale. This loss of heat becomes significant for reactors with high volumetric air throughput, causing in addition broth volume reduction if an efficient condenser is not installed in the exhaust line.

Heat losses to the surroundings Q_{surr} represents heat flux through the reactor walls. The bigger the reactor volume the less significant is this term in the overall energy balance.

Heat exchange Q_{hxcr} corresponds to the heat flux through the exchange system of the reactor in order to equilibrate the energy balance and to maintain the reaction temperature constant.

Making $Q_{acc} = 0$, the general energy balance becomes:

$$Q_{hxcr} = Q_{met} + Q_{ag} + Q_{aer} + Q_{sens} - Q_{evap} - Q_{surr} \tag{27}$$

The exchange systems of reactors are coils or hollow baffles that expose the necessary surface area for addition/removal of heat.

The design equation for the exchange system has the form (McCabe & Smith, 1956):

$$Q_{hxcr} = U\ A\ \Delta_{ln}T \tag{28}$$

where U represents the overall heat-transfer coefficient, A the exchange area, and $\Delta_{ln}T$ the logarithmic mean temperature difference given by the internal reactor temperature and in and out temperatures of circulating fluid inside the exchange system.

Current biooxidation processes of refractory gold ores utilize mesophilic microbial consortia growing at 40–50 °C with a net heat generation, being necessary to cool down the reactor (Dew *et al.*, 1997; Ly, 2005). The task of heat removal is facilitated as the reactor temperature is set at higher levels, as in the case of using hypertermophilic microorganisms, due to a more favorable temperature gradient.

7. USE OF BIOREACTORS IN GOLD ORE PROCESSING

The use of bioreactors in gold mining is a direct consequence of the lower occurrence in nature of easier ore forms to extract gold. This situation has forced the exploitation of the so called refractory gold ores, especially those where the submicron gold particles are occluded by insoluble sulfides as pyrite, arsenopyrite, pyrrhotite, enargite and the like (Olson, 1994). In these cases the cyanidation process of gold extraction becomes extremely ineffective because the sulfur compounds constitutes a strong barrier between cyanide reagent and gold particles, allowing figures as poor as 50% gold recovery or less (Rawlings & Silver, 1995). To reverse this situation it is necessary to remove part of these compounds in order to get 80% or more of gold extraction (Neale *et al.*, 2000). Figure 5 shows the process line for refractory gold ores. Several technological alternatives exist for the pre-treatment step, namely pressure oxidation, chemical oxidation, roasting and biooxidation. The last one has been the most popular and applied successfully at commercial scale starting from the late 80's. The biooxidation technology was developed for high value gold concentrates making economically feasible the use of aerated and agitated continuous bioreactors. Very well known technologies are offered by Gold Fields (BIOX®) (Dew, 1995; Dew *et al.*, 1997) and Mintek-BacTech (Neale *et al.*, 2000). The main consequence of performing biooxidation in stirred tanks is the strict control on several operation conditions, like pH, temperature, aeration, agitation, nutrients, suspension homogeneity, which affect favorably the process kinetics and yields.

Finely ground mineral concentrate (normally less than 75 μm) is fed to the first of three or more continuous stirred tanks reactors in series, together with nitrogen and phosphorous sources essential for the microorganisms. The main microbial cells present belongs to the *Acidithiobacillus* and *Leptospirillum* genera, being possible also to find *Sulfobacillus*, *Acidimicrobium* and species of the archeon *Ferroplasma* genus (Rawlings *et al.*, 2003). Typical operation conditions are pH between 1.2-1.8, dissolved oxygen concentrations above 0.002 kg/m^3, CO_2 enrichment, temperature in the range of 40–45 °C, 4 to 5 days of residence time and 20% of pulp density (Dew *et al.*, 1997; Ly, 2005).

One major constrain is that pulp density has to be no more than 20%. Higher values attempt against suspension homogeneity and more important, affects negatively the activity and viability of microorganisms (Rawlings *et al.*, 2003;

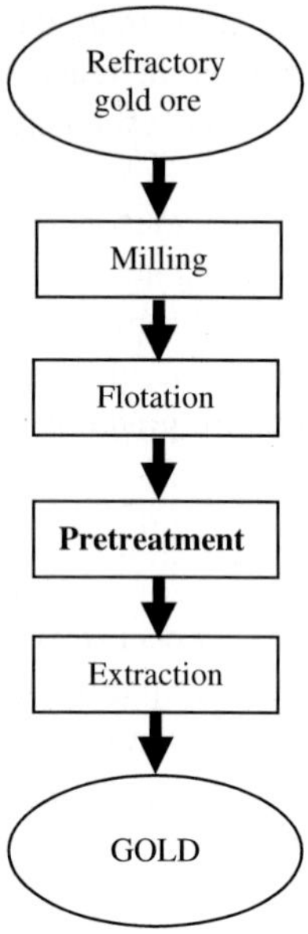

Figure 5. Process recovery of gold from refractory ores

Deveci, 2004). However, it is thought that higher pulp densities can be possible if proper adaptation of the cell population is done (Astudillo *et al.*, 2004).

Pretreatment of low value ores also has been developed to be performed in heaps (Brierley & Brierley, 1999). The total cost is reduced, but as the operation conditions can not be controlled in the same way as in a tank reactor, the solid residence time becomes much more extensive.

Table 1 gives a list of large-scale commercial biooxidation plants that have been established all over the world.

Most of the stirred tank processes listed in Table 1 correspond to BIOX® process with the exceptions of Youanmi plant, which used BacTech technology and Beaconsfield plus Laizhou that use Mintek-BacTech technology. The Harbour Lights plant was operated until gold ore was depleted.

All commercial existing plants utilize mesophilic or moderately thermophilic microorganisms. In the near future is expected that hypertermophilic archaea could be used to biooxidize highly refractory sulfides accompanying gold ores, like enargite, which are poorly solubilized inclusively at moderately thermophilic temperatures. This situation will originate new challenges to reactor design because of the lower solubility of gases at those temperatures, making harder to satisfy oxygen and carbon dioxide demands. Additionally, evaporation rates are high and sturdy materials will be required to construct the reactors in order to withstand the highly corrosive and acidic conditions. Some efforts have been made in this sense

Table 1. Commercial biooxidation plants[a]

Plant	Year commissioned	Capacity (tonne/day)	Total reactor volume (m^3)
Fairview, South Africa	1986	55	1415
Harbour Lights, Australia	1991	40	980
São Bento, Brazil	1991	300	1487
Wiluma, Australia	1993	154	4230
Youanmi, Australia	1994	120	3000
Ashanti-Sansu, Ghana	1994	960	21600
Tamboraque, Peru	1998	60	1570
Beaconsfield, Australia	1999	68	2310
Laizhou, China	2001	100	4050

[a]Adapted from Rawlings *et al.* (2003).

by means of the joint venture between BHP Billiton and Codelco Chile, who built and currently operate the plant in Chuquicamata, Chile, carrying on the BioCOP™ thermophile process (Batty & Rorke, 2005). This is a 20 000 tpa of copper cathodes facility that turned to be the first one in the world to extract copper by bioleaching of primary copper sulfide concentrates.

8. NOMENCLATURE

A	heat exchange area	t	time
C^*	equilibrium gas concentration	T	temperature
C_L	gas concentration in the liquid	U	overall heat transfer coefficient
E_z	axial dispersion coefficient	V	reactor volume
F	volumetric flow rate	v	linear velocity
$k_L a$	volumetric gas transfer coefficient	v_S	superficial gas velocity
K_S	saturation constant	X	cell concentration
L	tubular reactor length	Y_i	nutrient cell yields
Pe	Peclet number		
P_g	gassed agitation power	α, β	exponents
Q_i	heat generation/removal fluxes	Φ	solids concentration
r_s	volumetric rate of nutrient consumption	χ	conversion
		μ	specific growth rate
r_x	Volumetric rate of biomass growth	τ	residence time
S	nutrient concentration in exit line	χ^*	conversion at the higher rate of reaction
S^*	nutrient concentration that gives the higher rate of reaction	μ_M	maximum specific growth rates
S_0	nutrient concentration in inlet line	$\Delta_{ln}T$	logarithmic mean temperature difference

9. REFERENCES

Acevedo F. 1987. Mass balancing: an effective tool for fermentation process optimization. Crit Rev Biotechnol 6: 309-322.

Acevedo F. 2000. The use of reactors in biomining processes. Electronic J Biotechnol 3: 184-194. Available from www.ejbiotechnology.info.

Acevedo F, Cacciuttolo MA, Gentina JC. 1988. Comparative Performance of Stirred and Pachuca Tanks in the Bioleaching of a Copper Concentrate. In: Norris PR, Kelly DP, eds. Biohydrometallurgy, Kew Surrey.

Acevedo F, Gentina JC, Valencia P. 2004. Optimization of pulp density and particle size in the biooxidation of a pyritic gold concentrate by *Sulfolobus metallicus*. World J Microbiol Biotechnol 20: 865-869.

Acevedo F, Gentina JC. 1989. Process engineering aspects of the bioleaching of copper ores. Bioprocess Eng 4: 223-229.

Acevedo F, Gentina JC, García N. 1998. CO_2 supply in the biooxidation of an enargite-pyrite gold concentrate. Biotechnol Lett 20: 257-259.

Aiba S, Humphrey AE, Millis NF. 1973. Biochemical Engineering, 2nd ed., Academic Press, New York.

Astudillo C, González P, Gentina JC, Acevedo F. 2004. Adaptation of *Sulfolobus metallicus* to high pulp densities. In: Proceedings of the 12th International Biotechnology Symposium; Santiago, Chile.

Atkinson B. 1971. Biochemical Reaction Engineering. In: Richardson JF, Peacock DG, eds. Chemical Engineering, vol. 3, Pergamon Press, Oxford.

Atkinson B. 1974. Biochemical Reactors, chapter 3, Pion Ltd., London.

Bailey AD, Hansford GS. 1993. Factors affecting bio-oxidation of sulfide minerals at high concentrations of solids: a review. Biotechnol Bioeng 42: 1164-1174.

Bailey JE, Ollis DF. 1986. Biochemical Engineering Fundamentals, 2nd ed., chapter 9. McGraw-Hill Book Co., New York.

Batty JD, Rorke GV. 2005. Development and commercial demonstration of the BioCOPTM thermophile process. In: Harrison STL, Rawlings DE, Petersen J, eds. Proceedings of the 16th International Biohydrometallurgy Symposium, September 25 - 29, Cape Town, South Africa. Produced by Compress www.compress.co.za.

Blanch HW, Clark DS. 1996. Biochemical Engineering, chapter 3. Marcel Dekker Inc., New York.

Boogerd FC, Bos P, Kuenen JG, Heijnen JJ, van der Lans RGJM. 1990. Oxygen and carbon dioxide mass transfer and the aerobic, autotrophic cultivation of moderate and extreme termophiles: a case study related to the microbial desulfurization of coal. Biotechnol Bioeng 35: 1111-1119.

Boon M, Heijnen JJ. 1998. Gas-liquid mass transfer phenomena in bio-oxidation experiments of sulfide minerals: a critical review of literature data. Hydrometallurgy 48: 187-204.

Boon M, Heijnen JJ. 1998a. Chemical oxidation kinetics of pyrite in bioleaching processes. Hydrometallurgy 48: 27-41.

Bouquet F, Morin D. 2005. BROGIM®: a new three-phase mixing system – testwork and scale-up. In: Harrison STL, Rawlings DE, Petersen J, eds. Proceedings of the 16th International Biohydrometallurgy Symposium, September 25 – 29, Cape Town, South Africa. Produced by Compress www.compress.co.za.

Brierley JA, Brierley CL. 1999. Present and Future Commercial Applications of Biohydrometallurgy. In: Amils R, Ballester A, eds. Biohydrometallurgy and the environment toward the mining of the 21st century, Proceedings of the International Biohydrometallurgy Symposium IBS-99, El Escorial, Spain. Elsevier, Amsterdam, Part A, 81-90.

Canales C, Gentina JC, Acevedo F. 2002. Efecto de la aireación y la densidad de pulpa en el coeficiente de transferencia de oxígeno y la retención de aire en una columna de burbujeo para la biooxidación de concentrados de oro. In: Proceedings of the XV Chilean Congress of Chemical Engineering, October 22 – 24, Punta Arenas, Chile.

Cooney CL, Wang DIC, Mateles RI. 1968. Measurement of heat evolution and correlation with oxygen consumption during microbial growth. Biotechnol Bioeng 11: 269-281.

Cooney CL. 1981. Growth of Microorganisms. In: Rehm H-J, Reed G, eds. Biotechnology, vol.1, Verlag Chemie, Weinheim.

Cooper CM, Fernstrom GA, Miller SA. 1944. Performance of agitated gas-liquid contactors. Ind Eng Chem 36: 504-509.

Crundwell FK. 2001. Modeling, simulation, and optimization of bacterial leaching reactors. Biotechnol Bioeng 71: 255-265.

Deveci H. 2004. Effect of particle size and shape of solids on the viability of acidophilic bacteria during mixing in stirred tank reactors. Hydrometallurgy 71: 385-396.

Dew DW. 1995. Comparison of Performance for Continuous Bio-oxidation of Refractory Gold ore Flotation Concentrates. In: Jerez JCA, Vargas T, Toledo H, Wiertz JV, eds. Biohydrometallurgical processing, Proceedings of the International Biohydrometallurgy Symposium IBS-95, Vina del Mar, Chile. University of Chile, Santiago de Chile, Vol. 1, 239-252.

Dew DW, Lawson EN, Broadhurst JL. 1997. The BIOX® Process for Biooxidation of Gold-Bearing Ores or Concentrates. In: Rawlings DE, ed. Biomining: Theory, Microbes, and Industrial Processes, Springer-Verlag, Berlin, 45-80.

Esener AA, Roels JA, Kossen NWF. 1983. Theory and applications of unstructured growth models: kinetic and energetic aspects. Biotechnol Bioeng 25: 2803-2841.

González R, Gentina JC, Acevedo F. 2004. Biooxidation of a gold concentrate in a continuous stirred tank reactor: mathematical model and optimal configuration. Biochem Eng J 19: 33-42.

González R, Gentina JC, Acevedo F. 2003. Optimisation of the solids suspension condition in a continuous stirred tank reactor for the biooxidation of refractory gold concentrates. Electronic J Biotechnol 6: 233-240. Available from www.ejbiotechnology.info.

Gormely LS, Brannion RMR. 1989. Engineering design of microbiological leaching reactors. In: Proceedings of the International Biohydrometallurgy Symposium, August 13–18, Jackson Hole, Wyoming.

Greenhalgh P, Ritchie I. 1999. Advancing reactor designs for the gold bioleach process. In: Proceedings of the Biomine '99 and Water Management in Metallurgical Operations, August 23–24, Australian Mineral Foundation Inc, Perth. Glenside, Australia.

Harvey PI, Batty JD, Dew DW, Slabert W, van Buuren C. 1999. Engineering considerations in bioleach reactor design. In: Proceedings of the Biomine '99 and Water Management in Metallurgical Operations, August 23–24, Australian Mineral Foundation Inc, Perth. Glenside, Australia.

Heinzle T, Miller D, Nagel V. 1999. Results of an integrated pilot plant operation using the BioNIC? process to produce nickel metal. In: Proceedings of the Biomine '99 and Water Management in Metallurgical Operations, August 23–24, Australian Mineral Foundation Inc, Perth. Glenside, Australia.

Hougen OA, Watson KM, Ragatz RA. 1954. Chemical Process Principles, 2nd ed, John Wiley & Sons, New York.

Lally KS. 1987. A-315 axial flow impeller for gas dispersion. Lightnin Technical Report 144.00.

Levenspiel O. 1980. The Monod equation: a revisit and a generalization to product inhibition situations. Biotechnol Bioeng 22: 1671-1687.

Levenspiel O. 1999a. Chemical Reaction Engineering, 3rd ed, chapter 2, John Wiley & Sons, New York.

Levenspiel O. 1999b. Chemical Reaction Engineering, 3rd ed, chapter 5, John Wiley & Sons, New York.

Levenspiel O. 1999c. Chemical Reaction Engineering, 3rd ed., chapter 6, John Wiley & Sons, New York.

Loi G, Trois P, Rossi G. 1995. Biorotor®: a New Development for Biohydrometallurgical Processing. In: Jerez JCA, Vargas T, Toledo H, Wiertz JV, eds. Biohydrometallurgical processing, Proceedings of the International Biohydrometallurgy Symposium IBS-95, Vina del Mar, Chile. University of Chile, Santiago de Chile, Vol. 1, 263-272.

Ly ME. 2005. Biooxidación de Concentrados Refractarios de Arsenopirita en Tanques Agitados. In: Acevedo F, Gentina JC, eds. Fundamentos y Perspectivas de las Tecnologías Biomineras. Ediciones Universitarias de Valparaíso, Valparaíso, Chile.

Madigan MT, Martinko JM, Parker J. 1997. Biología de los Microorganismos, 8th ed. Prentice Hall Iberia, Madrid, Spain.

Mateles RI. 1971. Calculation of the oxygen required for cell production. Biotechnol Bioeng 13: 581-582.

McCabe WL, Smith JC. 1956. Unit Operations of Chemical Engineering, Kogakusha Company Ltd, Tokyo, Japan.

Miller PC. 1997. The Design and Operating Practice of Bacterial Oxidation Plant Using Moderate Thermophiles. Rawlings DE, ed. Biomining: Theory, Microbes, and Industrial Processes, Springer-Verlag, Berlin, 81-102.

Mills DB, Bar R, Kirwan DJ. 1987. Effect of solids on oxygen transfer in agitated three-phase systems. AICHE J 33: 1542-1549.

Monod J. 1949. The growth of bacterial cultures. Ann Rev Microbiol 3: 371-394.

Myerson AS. 1981. Oxygen mass transfer requirements during the growth of *Thiobacillus ferrooxidans* on iron pyrite. Biotechnol Bioeng 23: 1413-1416.

Nagpal S, Dahlstrom D, Oolman T. 1993. Effect of carbon dioxide concentration on the bioleaching of a pyrite-arsenopyrite ore concentrate. Biotechnol Bioeng 41: 459-464.

Neale JW, Pinches A, Deeplaul V. 2000. Mintek-BacTech's bacterial-oxidation technology for refractory gold concentrates: Beaconsfield and beyond. J South African Inst Min Metall 100: 415-421.

Nielsen J, Villadsen J, Lidén G. 2003. Bioreactor Engineering Principles, 2nd ed, chapter 3, Kluwer Academic/Plenum Publishers, New York.

Nielsen J, Villadsen J, Lidén G. 2003a. Bioreactor Engineering Principles, 2nd ed, chapter 7, Kluwer Academic/Plenum Publishers, New York.

Olson JO. 1994. Microbial oxidation of gold ores and gold bioleaching. FEMS Microbiol Lett 119: 1-6.

Peleg M. 1997. Modeling microbial populations with the original and modified versions of the continuous and discrete logistic equation. Crit Rev Food Sc 37: 471-490.

Perry RH, Green DW, Maloney JO. 1984. Perry's Chemical Engineers' Handbook, 6th ed, McGraw-Hill Book Company, New York, 19-50.

Pirt SJ. 1975 Principles of Microbe and Cell Cultivation, chapters 2, 4, 21, 24. Blackwell Scientific Publications, Oxford.

Posten CH, Cooney CL. 1993. Growth of Microorganisms. In: Rehm H-J, Reed G, eds. Biotechnology, 2nd ed., vol. 3, Verlag Chemie, Weinheim.

Rawlings DE, Silver S. 1995. Mining with microbes. Bio-technol 13: 773-778.

Rawlings DE, Dew D, du Plessis C. 2003. Biomineralization of metal-containing ores and concentrates. Trends Biotechnol 21: 38-44.

Rossi G. 2001. The design of bioreactors. Hydrometallurgy 59: 217-231.

Rossi G. 1999. The Design of Bioreactors. In: Amils R, Ballester A, eds. Biohydrometallurgy and the environment toward the mining of the 21st century, Proceedings of the International Biohydrometallurgy Symposium IBS-99, El Escorial, Spain. Elsevier, Amsterdam, Part A, 61-80.

Soto L, Gentina JC, Acevedo F. 2002. Optimización del sistema de reacción para la biooxidación continua de un concentrado refractario de oro. In: Proceedings of the XV Chilean Congress of Chemical Engineering, October 22–25, Punta Arenas, Chile.

Spencer PA, Satalic DM, Baxter KG, Pinches T. 1997. Key aspects in the design of a bacterial oxidation reactor. In: Proceedings of the International Biohydrometalurgy Symposium and Biomine' 97; August 4–6; Australian Mineral Foundation Inc., Sydney. Glenside, Australia.

Toma MK, Ruklisha MP, Vanags JJ, Zeltina MO, Leite MP, Galinina NI, Viesturs UE, Tengerdy RP. 1991. Inhibition of microbial growth and metabolism by excess turbulence. Biotechnol Bioeng 38: 552-556.

van't Riet K. 1979. Review of measuring methods and results in nonviscous gas-liquid mass transfer in stirred vessels. Ind Eng Chem Proc Des Dev 3: 357-364.

van't Riet K, Tramper J. 1991. Basic Bioreactor Design, Marcel Dekker Inc., New York.

Wang DIC, Cooney CL., Demain AL, Dunnill P, Humphrey AE, Lilly MD. 1979. Fermentation and Enzyme Technology, chapter 6, John Wiley & Sons, New York.

Zwietering MH, Jongenburger I, Rombouts FM, van't Riet K. 1990. Modeling the bacterial growth curve. Appl Environ Microbiol 56: 1875-1881.

CHAPTER 9

AIRLIFT REACTORS: CHARACTERIZATION AND APPLICATIONS IN BIOHYDROMETALLURGY

ALEJANDRA GIAVENO[1], LAURA LAVALLE[1], PATRICIA CHIACCHIARINI[1] AND EDGARDO DONATI[2]

[1]*Laboratorio de Biolixiviación, Facultad de Ingeniería, Universidad Nacional del Comahue, Neuquén, Argentina*

[2]*Centro de Investigación y Desarrollo en Fermentaciones Industriales (CINDEFI-CONICET), Facultad de Ciencias Exactas, Universidad Nacional de La Plata, La Plata, Argentina*

1. INTRODUCTION

Over the last decade biooxidation for the pretreatment of refractory sulfidic gold concentrates and the bioleaching of copper have been applied with increasing frequency. In addition to heap-leach facilities for the bioleaching of copper, several biotank oxidation plants are also operational for the pretreatment stage in the processing of refractory gold ores in large-scale operations (Brierley, 1997; Acevedo, 2000; Rossi, 2001). The application of commercial biooxidation plants using bioreactors started in South Africa and expanded first into Africa and then Australia, South America and, more recently, into China (Morin *et al.*, 2005). These commercial operations show that bioleaching/biooxidation processes can be viable options for the mining industry. Moreover, there is currently a great deal of interest in improving tank technologies for use in the recovery of other metals, such as Cu, Ni, Zn and Co (Brierley & Brierley, 1999; Okibe *et al.*, 2003; Morin *et al.*, 2005; Acevedo & Gentina, 2005; Sand & Gehrke, 2006).

The reactors most commonly employed in biohydrometallurgy are the Stirred Tank Reactor (STR), the Bubble Column (BCR) and the Airlift Reactor (ALR). Additionally, cylindrical vessels with a conical bottom known as Pachuca tanks have traditionally been used in biomining. Furthermore, new reactor designs have recently been developed: the Low Energy Bioreactor, the Delft Inclined Plate Bioreactor and the Revolving Drum Bioreactor (Biorotor) (Rossi, 2001).

Numerous different papers have been published on bioleaching-biooxidation processes using ALRs and Pachuca tanks (Atkins *et al.*, 1986; Acevedo *et al.*, 1987; Roy *et al.*, 2000; Acevedo, 2000; Rossi, 2001; Mousavi *et al.*, 2005; Shi & Fang, 2005); however, there is limited knowledge on the applications and characterization

E.R. Donati and W. Sand (eds.), Microbial Processing of Metal Sulfides, 169–191.

of airlift bioreactors. In an effort to better understand the performance of ALRs, in this chapter we describe a reverse flow airlift reactor designed by our research group and outline the hydrodynamic characterization tests. Moreover, it includes the modeling of the iron (II) oxidation kinetic by a bacterial strain and some biohydrometallurgical applications tests using ores from the north Patagonia region of Argentina.

2. AIRLIFT REACTOR

2.1. General Description of ALRs

ALRs and BCRs are the two main types of pneumatically agitated reactors. Both of these systems can be gas-liquid or gas-liquid-solid contacting devices. BCRs are simply cylindrical vessels that are aerated at the bottom where the air produces a random flow pattern. ALRs are distinguished by fluid circulation in a well-defined and clear cyclic pattern through channels providing a loop for recycling the liquid. These reactors usually consist of four sections: the riser, the gas separator, the downcomer and the base. In the riser both the gas and liquid flow upwards. The gas is injected at the bottom of this section. In the gas separator the gas disengages, totally or partially, from the liquid. In the downcomer the liquid flows down from the top to the bottom of the reactor. The different volumes of gas retained in the riser and the downcomer create a pressure difference that forces the fluid from the bottom of the downcomer towards the riser, a process that determines the liquid circulating (Onken & Weiland, 1983; Merchuk, 1986; Chisti, 1998). Three types of conventional ALRs are predominately used in research and in the industry, two of them presents internal loop configuration: the concentric (draft) tube and the split vessel (either circular or rectangular), and the third kind is the external loop (Siegel *et al.*, 1988; Chisti & Moo-Young, 2002).

ALRs have been designed and constructed on a large scale for a few specific processes such as waste-water treatment, single-cell protein production and fuel ethanol production (Merchuk, 1986). In the hydrometallurgical industry Pachuca tanks are used for the production of non-ferrous metals, including gold, uranium, zinc and copper (Clark, 1984; Merchuk & Siegel, 1988; Roy & Sheckhar, 1996). ALRs have also been used in the bacterial leaching of metals from sewage sludge (Couillard & Mercier, 1991; Tyagi *et al.*, 1991). Additionally, ALRs and Pachuca tanks have been used in coal biodesulfurization (Bos *et al.*, 1986; Beyer *et al.*, 1987; Ryu *et al.*, 1993).

ALRs present several advantages such as high gas dispersion efficiency, good heat transfer characteristics, rapid mixing, simplicity of construction and a low probability for the loss of sterility (Merchuk & Siegel, 1988; Chisti, 1989; Chisti, 1998; Schügerl & Lübbert, 1995). In conventional STRs and BCRs the energy required for the movement of the fluids is introduced focally at one point in the reactor. Consequently, energy dissipation is very high in the vicinity of the stirrer or the sparger that produces a heterogeneous shear field. On the other hand, the homogeneity of the stress forces is the main advantage offered by ALRs (Merchuk, 1990; Merchuk *et al.*, 1994; Merchuk & Berzin, 1995). Therefore the hydrodynamic environment developed is more suitable for cells that are susceptible to physical damage by mechanical agitation or fluid turbulence (Chisti, 1998). Moreover, the low shear rates facilitate the growth of biofilms, which can increase the reaction rate (Contreras *et al.*, 1999). Another advantage is that more sensitive biomasses like thermophiles can survive (Siegel *et al.*, 1988). Compared to STRs, ALRs require less power to attain a given rate of gas–liquid mass transfer. In the same way, a desk study reported by Ruitenberg *et al.* (2001) has shown that the investment costs of ALRs are significantly lower as compared to STRs. Furthermore, the operational costs for ALRs are the same or lower as compare to conventional mechanically agitated reactors. In comparison to bubble columns, ALRs have the additional advantages of loop reactors, such as increased heat and mass transfer capacity and a reduction in the energy consumption for mixing (Siegel & Robinson, 1992). Additionally, ALRs outperform BCRs in suspending solids (Chisti, 2005).

However, the application of ALRs in industry is still limited for several reasons. ALRs are less flexible to changes in process requirements (Siegel & Robinson, 1992). STRs, though subject to failures, have two operational degrees of freedom: agitation speed and aeration rate. Moreover, if a performance problem occurs, the stirrer motor and the type, number, location or diameter of the impeller can be modified relatively easily to overcome the problem. There is little tolerance for design flaws in ALRs because they do not have such modifiable features (Chisti, 1998). Furthermre, ALRs are limited in handling high-viscosity fluids because the high energy dissipation from wall friction leads to a very low circulation velocity and inefficient mixing (Merchuk, 1986).

On the other hand, there is a lack of appropriate methods to determine primary hydrodynamic parameters (e.g., linear circulation velocity, gas hold-up, gas–liquid mass transfer) and appropriate hydrodynamic models that take into account the reactor scale (Roy & Sheckhar, 1996; Roy *et al.*, 2000; Blažej *et al.*, 2004). For these reasons, an accurate description of the performance of airlift bioreactors is still difficult (Znada *et al.*, 2004).

2.2. Description of the Reactor used for the Experiments

The airlift reactor was constructed in transparent acrylic with a thickness of 6 mm. The vessel had an external diameter of 0.115 m and was 0.875 m high, containing a 0.700 m high concentric tube with a 4.484×10^{-3} m^2 riser area and a downcomer area of 7.473×10^{-3} m^2 ($A_r/A_d = 0.6$). A schematic diagram of the airlift bioreactor used is shown in Figure 1. The gas was supplied through a uniformly perforated ring, located in the annulus, at the bottom of the reactor. A reverse flow, which offers a better heat-transference coefficient and lower level of foam formation, was obtained in this way (Aprea *et al.*, 1997). A personal computer with a Pentium® processor, a PCL-818HG I/O card and data acquisition and control software (Genie® 2.0 from

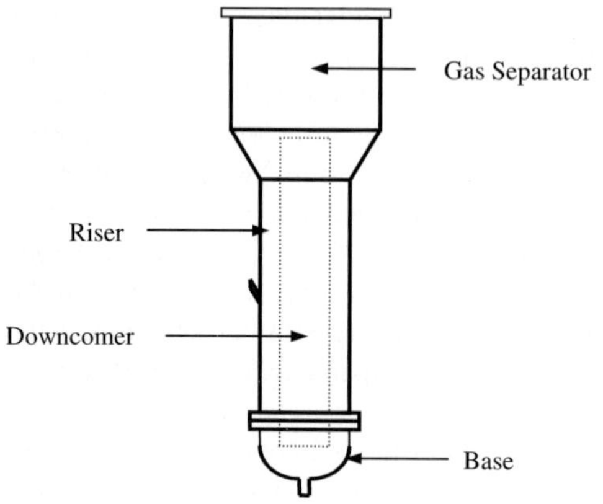

Figure 1. Schematic diagram of the airlift reactor

Advantech Corp.) were used for data acquisition. For all the tests, tap water and air were the liquid and gas phases, respectively. The temperature was kept constant at 25 °C.

2.3. Estimation of the Most Typical Characterization Parameters: T_M, ε, k_La and P_w

In the characterization of ALRs, the mixing parameters are very important for predicting the reactor behavior. Such mixing characteristics are usually presented in terms of the mixing time of the whole system (Onken & Weiland, 1983; Choi *et al.*, 1995). Gas hold-up is another parameter that is widely studied due to its influence on the gas residence time, liquid velocity and mixing characteristics of the reactor, but its main importance lies in its influence on the gas-liquid interfacial contacting area, which is available for mass transfer (Wu & Wu, 1990). The transfer of gas in the gas-liquid system is the limiting step for the growth rate of the microorganisms and product formation. Therefore, the determination of k_La, i.e. the overall mass transfer coefficient, plays an important role in the characterization of bioreactors (Koide *et al.*, 1983).

In our experiments, the volumetric gas flow rate was varied up to $0.016 \times 10^{-3} m^3 s^{-1}$, which corresponds to superficial gas velocities J_{Gr} (based on the riser cross-sectional area) of up to 0.0037 m s^{-1}. This parameter was calculated using the following expression:

$$J_{Gr} = F/A_r \qquad (1)$$

where F is the volumetric gas flow rate and A_r is the cross-sectional area of the riser.

The use of solid particles as flow followers of neutral buoyancy was applied here to visualize flow regimes and mixing characteristics of the reactor (Choi *et al.*,

1995; Schügerl & Lübbert, 1995). This technique was employed to evaluate the mean liquid circulation time using colored PVC particles as flow followers. The time required for every solid particle to travel through the reactor in a single loop was measured. At low superficial gas velocities, a uniform distribution of bubbles into the annulus was observed and the movement of flow followers showed a very stable liquid circulation. At high superficial gas velocities, it was observed that a proportion of the flow followers remained trapped in eddies. Additionally, the number of bubbles that entered into the downcomer and the level of bubble coalescence in the riser increased on increasing the superficial gas velocity (J_{Gr}). A histogram of circulation times obtained for a superficial gas velocity of 0.015 m s^{-1} is shown in Figure 2A.

Due to the high dispersion of the measured values, the mean liquid circulation time was computed from an average of 113 measurements and a value of 15.4 s was obtained. The calculated time gives an idea of the movement of the particles within the reactor and the mixing behavior as that both properties are related to the size and density of the solids and also the superficial gas velocity used. The frequency of the liquid flow patterns shows a long tail, which was considered to be due to trajectory excursions and intermittent entrapment of particles by eddies (Merchuk & Siegel, 1988).

Mixing characteristics are expressed in terms of mixing time (T_M), which was determined by a signal response technique using an acid tracer and a pH electrode detector (Merchuk *et al.*, 1998). The pH electrode was located at the top of the riser and the tracer (1 ml of concentrated HCl) was injected at the top of the downcomer. The mixing time was calculated from the response curves as the time required to achieve 99% of the final tracer concentration. Higher degrees of mixing give rise to shorter mixing times. As can be seen in Figure 2B, at low gas velocities the mixing time is longer due to a less turbulent flow and this is improved with increasing

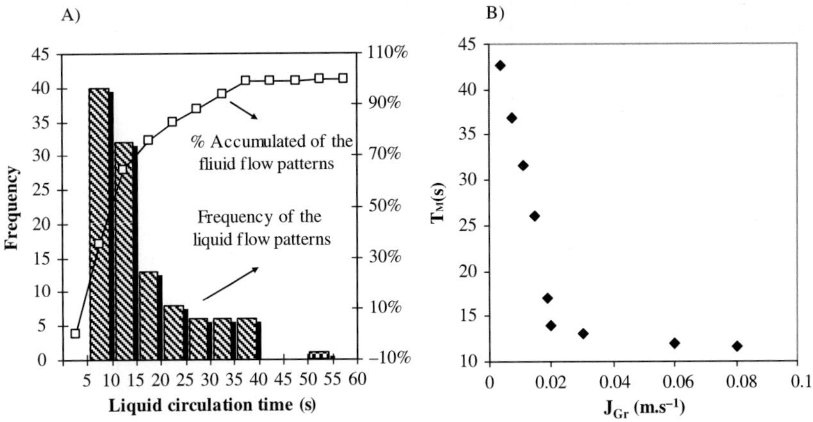

Figure 2. Flow regime histogram at J_{Gr} = 0.015 m s^{-1} and B) effect of superficial gas velocity (J_{Gr}) on mixing time (T_M)

superficial gas velocity. At superficial gas velocities higher than 0.02 m s^{-1}, the mixing time was practically independent of this parameter. This effect is similar to that found by others authors (Merchuk *et al.*, 1998; Sánchez Mirón *et al.*, 2004). The mixing behavior was characterized by very rapid homogenization and an excellent mixing capability.

For short reactors, such as the one described in this work, the determination of the gas volumetric fraction requires precise measurements of very small pressure drops to distinguish between riser and downcomer hold-up. Therefore, only the overall gas hold-up (ε) was determined. The ε values were determined by the volume expansion method (Wu & Wu, 1990). Overall gas hold-up is given by the following equation:

$$(2) \qquad \varepsilon = V_S/(V_L + V_S)$$

where V_L is the initial volume of the liquid in the reactor without aeration and V_S is the spread volume when the reactor is aerated.

The experimental data for overall gas hold-up (ε) as a function of the superficial gas velocity (J_{Gr}) is shown in Figure 3A. It is possible to identify two regions, each separated by a threshold superficial gas velocity value close to 0.02 m s^{-1}. A good correspondence was found with the power law for each part of the curve, which can be expressed in a general way as follows:

$$(3) \qquad \varepsilon = a \cdot J_{Gr}^{b}$$

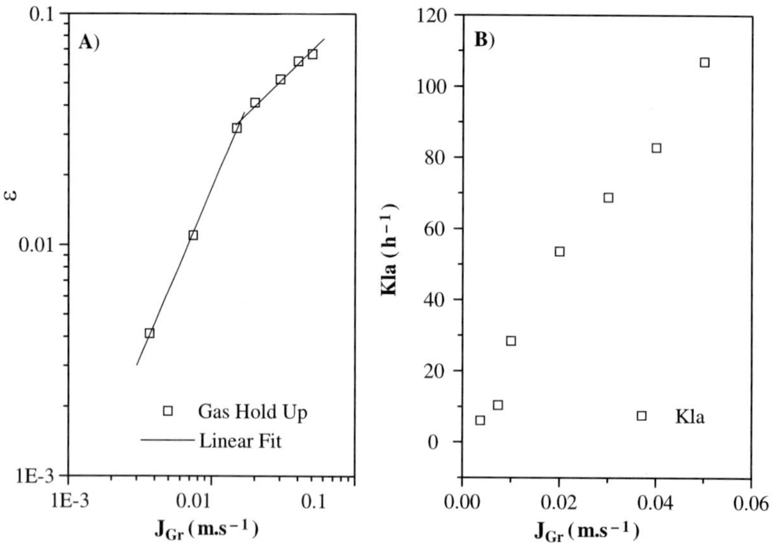

Figure 3. A) The overall gas hold-up (ε) and B) the mass transfer coefficient (k_{La}) as a function of the superficial gas velocity

The parameters *a* and *b* depend on the flow regime and the properties of the fluid. In the first region, the gas hold-up increases quickly with the superficial velocity, in agreement with an exponent *b* equal to 1.46. Above the velocity threshold, a slight increase in the gas hold-up, corresponding to an exponent *b* equal to 0.54 can be observed.

Each part of the curve in Figure 3A represents well-differentiated mixing behavior. At low superficial gas velocities, a low level of bubble recirculation was observed. The downcomer and the riser can be clearly distinguished and this allows an ordered movement of bubbles to be seen, a situation that corresponds to a uniform bubble flow (Merchuk *et al.*, 1998). At gas velocities over 0.02 m s^{-1} a transition towards turbulent flow takes place and this is characterized by an erratic movement and high coalescence of bubbles. It is important to highlight that the similar threshold value of superficial gas velocity was obtained for the mixing time (Figure 2B).

The overall volumetric gas-liquid mass transfer coefficient ($k_L a$) was determined using the static method Koide *et al.* (1983). The k_L term is the true mass transfer coefficient and *a* is the gas-liquid interfacial area per unit liquid volume. The dissolved oxygen concentration was measured with a polarographic sensor. The corresponding mass balance for the oxygen dissolving process can be written as:

$$k_L a = -\frac{2.303(1-\varepsilon)}{t}\log\frac{C_i - C}{C_i - C_o} \tag{4}$$

where C_o and C_i are the initial and saturated concentrations of dissolved oxygen in the liquid, respectively, while C is the same concentration measured at time t. Finally, the value of $k_L a$ was calculated as the slope of the curve obtained in the plot of log $(C_i\text{-}C)/(C_i\text{-}C_o)$ versus time.

It can be seen from Figure 3B that the overall mass transfer coefficient increased with the increasing superficial gas velocity. The obtained values are higher than those reported by Wu and Wu (1990) for an airlift with gas dispersion into the tube and are similar to those reported by Koide *et al.* (1983) for a system with gas dispersion into the annulus. Comparison of this performance with that of mechanically agitated reactors used in bioleaching shows that our airlift has similar $k_L a$ values to the STR and Delft Inclined Plate Bioreactor but slightly lower values than an aerated trough bioreactor (Loi *et al.*, 1995; Rossi, 2001). However, the $k_L a$ values reported by Rossi (2001) for the revolving drum bioreactor (Biorotor) are at least one order of magnitude higher than those for the other reactors mentioned above.

The power input in a reactor controls most of its performance parameters and affects the economics of operation. The specific power input is often used as the basis for comparing the performance of bioreactors and it can be calculated using the method proposed by Chisti (1989). When our reactor was operated at J_{Gr} values between 0.015 and 0.08 m s^{-1}, the specific power input (Pw) was in the range 0.09 to 0.5 kW m^{-3}. These values are in agreement with power inputs data reported for cell culture applications and they are lower than the typical maximum values in pneumatically agitated bioreactors (Loi *et al.*, 1995; Chisti, 1998).

2.4. New Model Based on Kriging Interpolation Proposed to Monitor the Reactor Performance

Modeling is an interesting way to interpret the mechanisms and internal processes carried out in the biotechnology field. The models are often used for purposes like (i) reactor design, operation and optimization; and (ii) monitoring and control of the variables. This knowledge enables us to develop a model, related to bioleaching processes, that is useful to evaluate the ALR performance.

Taking into account the wide scope for the application of modeling, numerous research models have been published that describe the bioleaching processes. Different kinds of models can be found in the literature but the majority of them are "knowledge-based models" (Shrihari *et al.*, 1990; Hadaddin *et al.*, 1995; Crundwell, 1997; Nemati & Webb, 1997, Ojumu *et al.*, 2005). This kind of models represents an interesting and useful way to analyze and validate biological mechanisms and kinetics laws. However, it is difficult to obtain such theoretical models because the key variables are generally non-linear functions that are related in a complex way. Moreover, they require the estimation of a large number of parameters, which are often difficult to identify due to identificability and distinguishability problems (Chappell & Godfrey, 1990).

On the other hand, some of the model approaches are based on the ability to match input-output mapping rules of unmodeled nonlinear systems, called "black-box" or input/output model types. Usually, a good input/output model instead of the knowledge-based one is good enough for use in most engineering applications. Models using artificial neural networks (ANN) or fuzzy logic and interpolation based models have been developed. In the last decade, ANN has become the most widely reported procedure to obtain this kind of black-box models (Kurtanjek, 1994). However, issues such as the selection of the structure and dimensions of the ANN, convergence to local minima and a rigorous model validation remain open questions. Moreover, since ANN is based on the identification of a large number of parameters, it often requires a large amount of data. In addition, fuzzy logic control (FLC) has its own limitations. The analytical study of fuzzy logic (De Gloria *et al.*, 1999) is still trailing its implementation and a great deal of work still lies ahead, particularly in the area of stability and performance analysis. Furthermore, as far as solutions to practical problems are concerned, fuzzy logic control design is an issue dependent on the adaptation of an exiting fuzzy logic controller to a different control problem – a goal that is not straightforward.

The aim is to obtain a "black-box" model capable of representing ferrous iron oxidation by *Acidithiobacillus ferrooxidans* and monitoring the ALR performance. At the same time, it was hoped to follow the growth of these bacteria in a batch system, thus overcoming the problems associated with ANN and FLC systems.

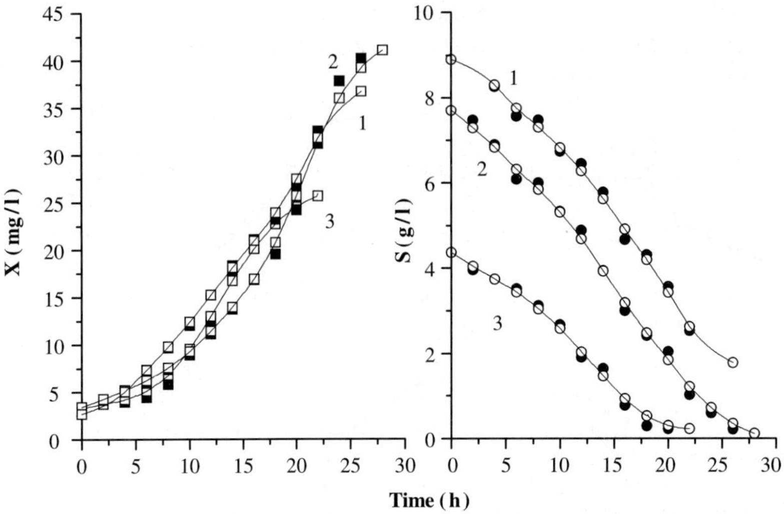

Figure 4. Kinetics of biomass growth and substrate consumption. (•) Experimental samples of S, (∘) Smoothed samples of S, (■) Experimental samples of X, (□) Smoothed samples of X, Continue line: Interpolated samples using Kriging

The use of a popular method known as Kriging, which is based on the spatial interpolation of measured data, was proposed (Gatti *et al.*, 2003).

In order to collect the experimental data kinetic growth tests were carried out using 9K medium (Silverman & Lundgren, 1959) and *A. ferrooxidans* DSM 11477 (Deutsche Sammlung von Mikroorganisem) as inoculum. The pH and temperature were controlled throughout the experiments at pH = 2.0 and T = 30±1°C. The bioreactor was operated at $J_{Gr} = 0.015\ m\ s^{-1}$. The evolution of the key variables (X and S) for three kinetic tests carried out under different initial conditions is represented in Figure 4.

From the interpolated data, the time derivative of X(t) and S(t) was obtained using Euler algorithm and a set of values of $a_X(\zeta_k)$ and $a_S(\zeta_k)$ with k=1,2,….,96 was obtained. Using these N = 96 sampled data the semivariogram was calculated and then surfaces of $\hat{a}_X(\zeta)$ and $\hat{a}_S(\zeta)$ for different values of ζ were obtained by means of ordinary and universal Kriging interpolation.

Two other growth kinetics of *A. ferrooxidans* were run with the aim of comparing ordinary and universal Kriging interpolation methods and experimental data. The results are shown in Figure 5.

Significant differences were not found when the surfaces were estimated using either ordinary or universal Kriging methods. Therefore, it was possible to use the ordinary Kriging method to determine $\hat{a}_X$ and $\hat{a}_S$. In order to validate the model, the last two kinetics were compared to the variables estimated by the model. The simulated model predictions are shown in Figure 6 together with experimental data and error bound.

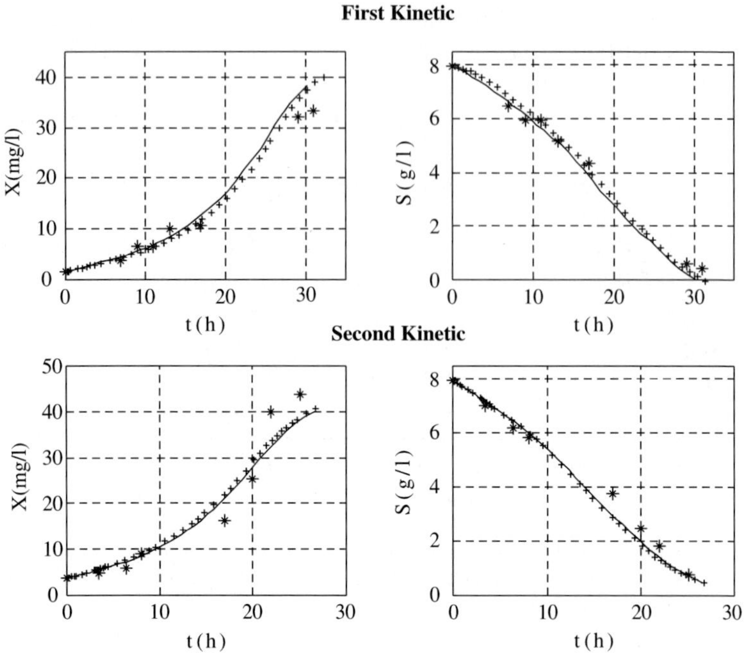

Figure 5. Comparison between ordinary and universal Kriging interpolation methods. (*) experimental data, (+) model estimation using universal Kriging, (-) model estimation using ordinary Kriging

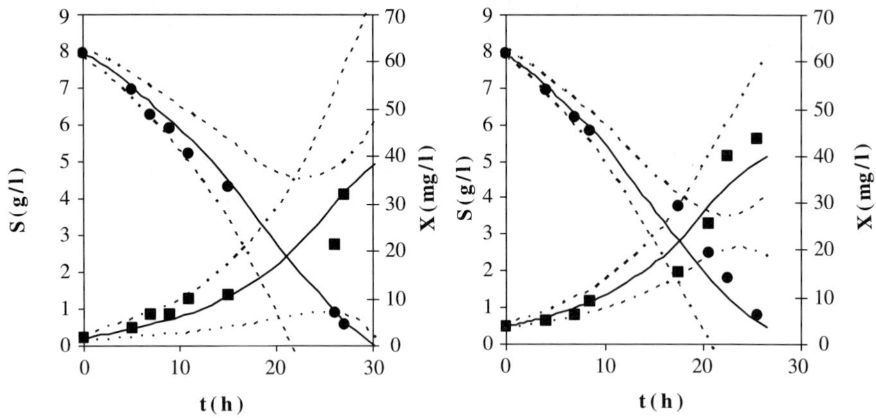

Figure 6. Kinetics of growth of *A. ferrooxidans* used for model validation. (•) Experimental samples of S. (■) Experimental samples of X. (—) Model Estimation. (- - -) Error bound for 67% of probability with $\sigma_{ex}(0)$=0.3(mg/l) and $\sigma_{es}(0)$=0.1 (g/l)

The predicted output variables of the model match very satisfactorily the experimental data within the bounds of one standard deviation, which means that the representation of the global model of two dimensions is correct.

It should be highlighted that the specific growth rate $\mu_X(\zeta)=a_X(\zeta)/X$ and the specific substrate consumption rate $\mu_S(\zeta)=a_S(\zeta)/X$ were used instead of $a_X(\zeta)$ and $a_S(\zeta)$, since these have a physical sense and they can easily be compared with other reported values easily.

Values of $\mu_X(\zeta)$ were calculated from the surfaces obtained using ordinary Kriging interpolation. Although data for maximum specific growth rate (μ_{XM}) were not found in the literature for the growth of *A. ferrooxidans* in an airlift reactor, our data were compared to those reported for other batch cultures. The μ_{XM} obtained was $0.19 \pm 0.01\ h^{-1}$ and this is consistent with the values reported by Shrihari *et al.* (1990), Gómez *et al.* (1996), and Lavalle *et al.* (2005).

The biological ferrous iron oxidation rate obtained in our ALR using different energy source concentrations was in the range 0.3 to 0.5 $g\ l^{-1}h^{-1}$ and these are the same order of magnitude as those listed in the reports by Nemati *et al.* (1998) and Park *et al.* (2005) for free cells.

It can be concluded that the model based on Kriging interpolation can be applied not only to represent ferrous iron oxidation by *A. ferrooxidans* but also to evaluate the performance of the ALR. In this way the operational parameters could be changed easily through setting another, more convenient superficial gas velocity in order to obtain the optimal operation conditions.

3. APPLICATIONS IN BIOHYDROMETALLURGY: BIOLEACHING AND BIOOXIDATION OF ORES FROM THE NORTH PATAGONIA REGION OF ARGENTINA

A survey of the literature on biohydrometallurgical topics revealed that the papers devoted to bioreactors represent less than 5% of the total (Rossi, 2001). In this area, a great deal of attention has been paid to stirred tank bioreactors followed by bubble columns machines and the characteristic Pachuca tanks. Airlift bioreactors are mentioned very little in the literature in terms of biohydrometallurgical processes. In many cases the papers refer mainly to the analysis of the performance and design of bioreactors.

In 2005 Shi and Fang reported the bioleaching of marmatite flotation concentrate by adapted strains of *A. ferrooxidans* and *L. ferrooxidans* in mixed culture with a pulp density of 15% in a 10-L airlift reactor. It was found that the addition of Fe^{2+} ion was favorable for extracting zinc (50 $g\ L^{-1}$ in leach liquor), demonstrating the technological feasibility of the marmatite bioleaching using adapted mixed cultures.

In a recent study Mousavi *et al.* (2005) compared the ability of *A. ferrooxidans* and a thermophilic *Sulfobacillus* in copper recovery from chalcopyrite concentrate in an airlift bioreactor. It was found that maximum copper dissolution (85%) was achieved with *Sulfobacillus* at 10% w/v pulp density after 10 days.

With the aim of evaluating the performance of our airlift reactor, several bioleaching and biooxidation tests were carried out. Although it is well known that bioprocessing of run-of-mine ores is not justified on economic grounds, especially when the ore is low grade, the effect of pulp density on the recovery of copper from Campana Mahuida ores was studied. In addition, the biological pretreatment of a refractory gold mineral from Andacollo ores was analyzed. Other tests concerning the bioprocessing of polymetallic ores conducted in the ALR have been described previously (Giaveno *et al.*, 2005). All of the mineral samples were obtained from reservoirs located in Neuquén, Argentina.

Bearing in mind the results of the characterization experiments, it was decided to operate the reactor at a superficial gas velocity $J_{Gr} = 0.015\ m\ s^{-1}$. This value enabled the system to work in the range of the uniform bubble flow regime. In this way, the bubble recirculation into the downcomer was sufficient to maintain satisfactory absorption of the air components to support bacterial growth. Moreover, this superficial gas velocity kept all solid particles suspended as it was determined by visual observation (Koide *et al.*, 1984).

3.1. Bioleaching of Campana Mahuida Ores using a Collection Strain of *Acidithiobacillus Ferrooxidans*

Samples of copper ores were obtained from the Tres Puntas peak at the Campana Mahuida reservoir (Giaveno *et al.*, 1995). The ore sample used was prepared from 57 selected portions, taken at different depths from six drill holes ranging from 37 m to 88 m, containing the following sulfide minerals: chalcocite 0.1%, covellite 0.3%, pyrite 2.2% and traces of chalcopyrite. Additionally, it was possible to identify in the samples other oxidized copper and iron mineralogical species like malachite, cuprite, azurite, hematite and goethite. The chemical composition of these two metals was: copper 0.75% and iron 3.15%. The sample was ground to below 74 μm. The strain used in this study was *A. ferrooxidans* DSM 11477, which was grown in 9K medium pH 1.8 without adaptation to the mineral.

The airlift reactor was operated with a final volume of 8 dm^3 of iron free 9K medium pH 2.0 inoculated at 10% v/v. The initial bacterial concentration was $1x10^7$ bact ml^{-1}. The initial pulp density values were 3, 5 and 10% w/v. All experiments were carried out at 30^oC and air was supplied at a rate of $0.07x10^{-3}m^3\ s^{-1}$ (J_{Gr}= 0.015 m s^{-1}). Sulfuric acid was added during the first few days in order to stabilize the pH. Periodically, 10 ml samples were withdrawn from the reactor and centrifuged to remove the solids from leaching solutions. The supernatant was analyzed for pH, Eh, ferrous and total iron, copper and bacterial concentrations. Sterile control tests were carried out by adding 2% w/v thymol in methanol instead of the inoculum.

Atomic absorption spectrophotometry was used routinely for the determination of copper and total iron in aqueous solution. The ferrous ion concentration was quantified by permanganate titration. The bacterial population in the liquid phase was determined by direct counting in a Petroff-Haüsser chamber using a phase-contrasting microscope.

The representative behavior of Eh and pH values during bioleaching experiments is shown in Figure 7. An increase of total iron (data not shown) was detected up to 20 days; after that point a strong precipitation was observed and this is consistent with the decrease in pH (iron hydrolysis). The increase in Eh indicates iron(II) oxidation by bacterial activity (this phenomenon was not observed in sterile controls) and is in agreement with the increase in the bacterial population in the suspension.

Copper recoveries in bioleaching experiments for three different pulp densities are shown in Figure 8. Results for the 3% pulp density sterile control have also been included. The copper extraction in that control was slightly higher than 30% which in turn is much lower than the copper recoveries in inoculated systems. Results for the other sterile control were very similar. The highest copper extraction value (83%) was obtained at 3% w/v pulp density. Final copper extractions at 5% and 10% solid concentrations were 75% and 70%, respectively. The net process was acid consuming even though acid production took place in the later phases of the experiments. The acid consumption average was 36 g H_2SO_4/Kg ore (6 g H_2SO_4/g Cu leached) including the initial pH adjustments.

The low final extraction rate can be explained in terms of the formation of a product layer on the mineral surface, which creates a diffusion barrier to the

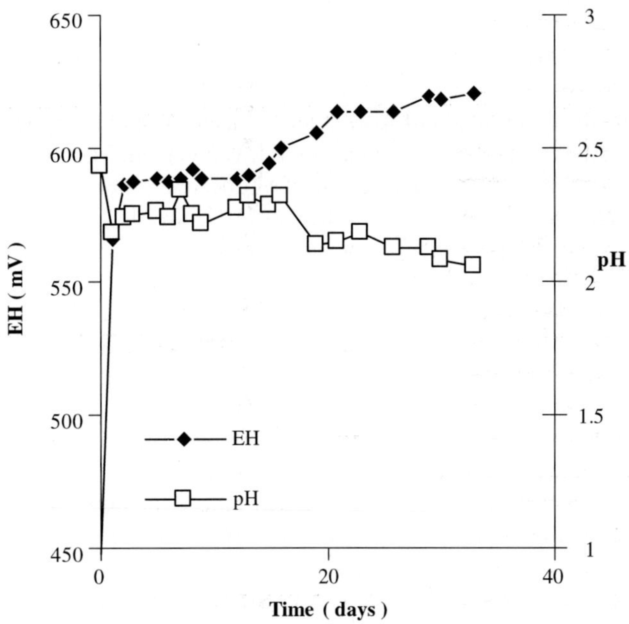

Figure 7. Redox potential and pH for a 3% of pulp density

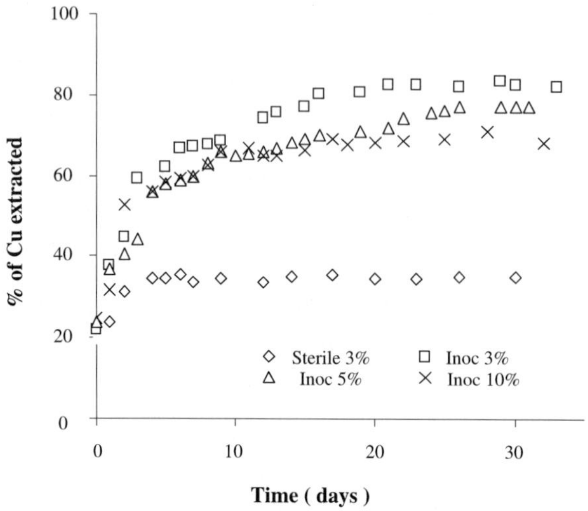

Figure 8. Copper extraction from Campana Mahuida ore for three pulp densities and the sterile control

interfacial fluxes of reactants and products (Chaudhury & Das, 1987). If the bioleaching kinetic was controlled by diffusion through these products, following the model of shrinking core-product layer diffusion, the kinetics may be correlated graphically using Equation (5), where K_p = the parabolic rate constant (days^{-1}), t= time (days) and X = fraction of copper reacted (Viñals, 2000).

$$(5) \qquad K_p t = 1 - 2/3X - (1 - X)^{2/3}$$

Based on the experimental data presented in Figure 8 for Campana Mahuida copper bioleaching, a plot of $(1 - 2/3\ X - (1 - X)^{2/3})$ versus time was obtained. The linear fits for the sterile experiment and for the inoculated ones at different pulp densities are shown in Figure 9. All experimental values were used to plot the linear fit although it was not observed the period of colonization of the ore by bacteria, reported by Lizama, 2003.

These results clearly indicate that copper leaching in the inoculated systems obeys the proposed model. On the other hand, the lack of fitting in the abiotic test, where copper is only solubilized by the action of acid, could indicate that the process in the sterile system is controlled by chemical reactions. Our results show that the solids concentration had a significant impact on the bioleaching efficiency and a high pulp density resulted in lower copper extractions. Similar results were obtained by Mousavi *et al.* (2005) working with an airlift bioreactor but using a concentrate of chalcopyrite. Similarly, a decrease in metal extraction was reported at higher pulp densities when a flotation concentrate of marmatite was bioleached by adapted cultures in an airlift reactor (Shi & Fang, 2005). It is believed that significant damage to the microorganisms and insufficient oxygen transfer mass are the reasons that preclude the use of a high solids concentration (with about

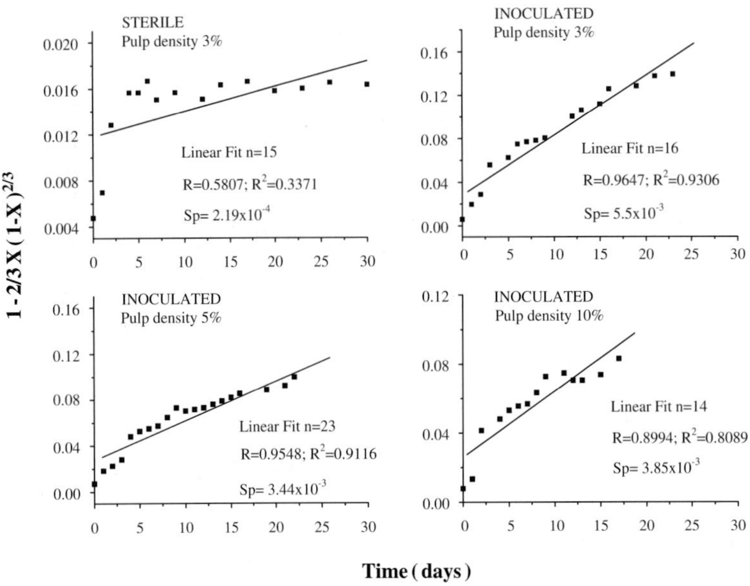

Figure 9. $1 - 2/3\ X - (1 - X)^{2/3}$ versus time from the copper bioleaching data in Fig. 8

20% being the limit) in STRs and Pachuca tanks (Bailey & Hansford, 1993; Rossi, 2001). The particle attrition due to the high solids concentration and fine particle sizing could cause inhibitory effects on bacterial metabolism (Escobar *et al.*, 1995; Deveci, 2004).

3.2. Biooxidation of Andacollo Ores using Collection Strains of *Acidithiobacillus Ferrooxidans* and *Acidithiobacillus Thiooxidans*

Gold occurs commonly in association with base metals, where the ore is frequently refractory because gold is trapped in the matrix of metallic sulfides, mainly pyrite and arsenopyrite. In such cases the conventional methods for liberating gold are roasting or oxidation at high pressure. The residues are then treated by cyanidation. However, these treatments are environmentally unacceptable because of the resulting atmospheric pollution by sulfur dioxide. Bacterial pretreatment leads to the destruction of the sulfide matrix in order to liberate the gold physically and to make the ore amenable to cyanidation (Brierley, 1997). Metals dissolution due to the pretreatment helps to diminish the cyanide consumption during the later process of gold recovery.

In this section we describe the biooxidation process on a refractory gold ore from Andacollo using mixed cultures of *A. ferrooxidans* (DSM 11477) and *A. thiooxidans* (DSM 11478). The conditions for bacterial activity were selected from previous experiments in shake flasks (data not shown). The increase in the level of gold

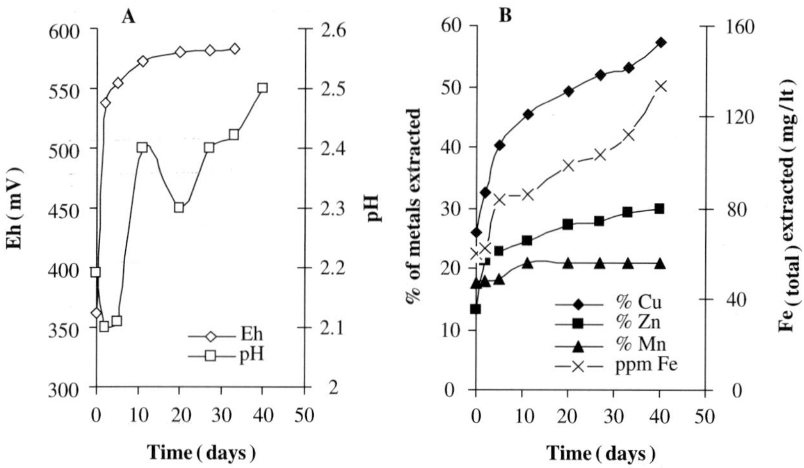

Figure 10. A) Eh and pH evolution. B) Extraction of Cu, Zn, Mn and Fe(total) concentration

recovery by cyanidation on using the biooxidation pretreatment and the other relevant metals were evaluated.

The mineralogical composition was sphalerite 0.71%, galena 0.76%, pyrite 9.51%, chalcopyrite 0.16%, arsenopyrite 0.08% and iron oxides 0.13%. The chemical composition was 10 g/t Au, 238 g/t Cu, 2975 g/t Zn, 0.39 % MnO and 6.53 % Fe_2O_3. The mineral size was less than 74μm and the material was added at 5% of pulp density to give a final volume of 8 dm^3 of iron-free 9K medium. Air was supplied at a rate of $0.07x10^{-3}m^3\ s^{-1}$ (J_{Gr}= 0.015 m s^{-1}). The experiment was carried out at 30°C and the initial pH was 2.0 (not controlled). The methods used were the same as those mentioned in Section 3.1. *A. thiooxidans* was cultivated routinely in iron-free 9K medium with sulfur as the energy source. After the pH stabilization had been achieved, the inoculation with the mixed culture was carried out.

The initial sulfuric acid consumption before inoculation was 67.5 g H_2SO_4/kg ore average, which indicates the presence of high alkaline gangue contents such as carbonate minerals. The evolution of pH and Eh as a function of time is shown in Figure 10A. The Eh rapidly increased during the initial days of the experiment and reached a stable value of 585 mV after 40 days. This indicates an adequate level of activity for the *A. ferrooxidans* culture. The final bacterial population in suspension was $7x10^7$ bact ml^{-1}. The amount of acid generated by the microorganisms could not be enough to maintain the initial pH 2.0.

The copper, zinc, manganese and iron evolution profiles during the 40 days of the biooxidation process are shown in Figure 10B. The extraction percentages reached were 57.2% of Cu, 30% of Zn and 21% of Mn. These values are the same as those obtained in previous shake flask experiments, under same experimental conditions. Metal extraction percentages in the sterile control were 20% of Cu, 11%

of Zn and 23% of Mn. Thus, microbial action allowed to increase significantly the extractions of Cu and Zn which were essentially as sulfides in the ore. However, bacterial action did not enhance Mn recovery. Probably, it is due to manganese was as carbonate in the ore and this compound is easily solubilized in acid media even in the absence of microorganisms.

The presence of iron in solution indicates that *A. ferrooxidans* contributed to pyrite dissolution (Sand *et al.*, 1999). In addition, arsenopyrite was also degraded by bacterial action (arsenic dissolution was 2.3% whereas in the sterile control it was negligible). This low extraction of arsenic is in agreement with XRD analysis, which shows the presence of arsenopyrite and a small amount of pyrite in the biooxidation residues. According to the XRD analysis, galena was also oxidized and the lead released from the sulfide appeared in the residues as lead jarosite $PbFe_6(SO_4)_4(OH)_{12}$. Gold extraction through cyanidation was 74% in the mineral previously treated by biooxidation. This extraction was 20% higher than that detected in the mineral without pretreatment. The extraction level was similar to others reported in a heap biooxidation process (Whitlock, 1997) but lower than those described in some reports where another type of reactor was used (Jones & Hackl, 1999). This is probably due to the particle size used being smaller than the optimum particle size.

4. NOMENCLATURE

A	Coefficient
A_r	riser area, m^2
A_d	downcomer area, m^2
B	Exponent
C	oxygen concentration in the liquid at time t, mg dm^{-3}
C_0	initial concentration of dissolved oxygen, mg dm^{-3}
C_i	saturated concentration of dissolved oxygen, mg dm^{-3}
F	volumetric gas flow rate, $m^3\ s^{-1}$
J_{Gr}	superficial gas velocity in the riser, m s^{-1}
$k_L a$	volumetric gas transfer coefficient, h^{-1}
K_P	parabolic rate constant, day^{-1}
S	substrate concentration, g l^{-1}
t	time, s
T_M	mixing time, s
V_S	spread volume of the liquid with aeration, dm^3
V_L	initial volume of the liquid without aeration, dm^3
X	fraction reacted of copper
X	cell concentration, mg l^{-1}
ε	gas hold up
μ_X	specific growth rate, h^{-1}
μ_S	specific substrate consume rate, h^{-1}
μ_{XM}	maximum specific growth rate, h^{-1}

5. APPENDIX

Kriging is a stochastic method, which guarantees the error variance between the real and the estimated values of the variables to be within a minimum upper bound.

The problem was formulated in the following way: the large numbers of phenomena that affect the kinetic processes in a bioleaching batch system were described by a set of non-linear differential equations, which were expressed using a general formula:

$$\dot{\zeta}(t) = A(\zeta(t)) \tag{6}$$

Where $\zeta(t)$ is the vector of the variables in time (state variables) (e.g., biomass, substrate, product, pH temperature) and $A(\zeta(t))$ is a vector formed with elements $a(\zeta(t))$, which are scalar non-linear functions of the state variables. Taking into account the following assumptions: (i) pH and temperature were constant; (ii) O_2 and $CO_2$are not limiting substrates and (iii) the inhibitory effect of product P is negligible in the range of concentrations of Fe^{2+} tested (Nemati & Webb, 1997), it was possible to obtain a system (*S*) with two variables (X and S) and a model (*M*) could be formulated in two dimensions.

$$S: \begin{bmatrix} \dot{X}(t) \\ \dot{S}(t) \end{bmatrix} = \begin{bmatrix} a_X(\zeta(t)) \\ a_S(\zeta(t)) \end{bmatrix} \tag{7}$$

$$M: \begin{bmatrix} \dot{\hat{X}}(t) \\ \dot{\hat{S}}(t) \end{bmatrix} = \begin{bmatrix} \hat{a}_X(\hat{\zeta}(t)) \\ \hat{a}_S(\hat{\zeta}(t)) \end{bmatrix} \tag{8}$$

If we consider a two variable model each $a(\zeta_0)$ element can be seen as a surface with an unique scalar value $a(\zeta_0)$ at a geographical position ζ_0 as can be seen in Figure 11.

Considering that the non-linear functions were unknown, the goal of this work was to obtain a model (*M*) based on a set of sampled values so that the error between the system and the model variables was bounded (Figure 12).

We assumed that the standard deviation of the error was restricted to an upper bound. Moreover, this upper bound was the minimum as represented by the following equations:

$$\begin{bmatrix} [Var(e_X)]^{1/2} \\ [Var(e_S)]^{1/2} \end{bmatrix} \leqslant \begin{bmatrix} J(\hat{a}_X) \\ J(\hat{a}_S) \end{bmatrix} = J(A) \tag{9}$$

$$\min_{\hat{a}} \{||J(A)||\} \tag{10}$$

Given N measured data of $a_X(X,S)$ and $a_S(X,S)$, the minimum of $||J(A)||$ was obtained using Kriging Interpolation. In order to achieve this, the estimated values of a_X and a_S should be calculated using the following equation:

$$\hat{a}(\zeta_0) = YG^{-1}g \tag{11}$$

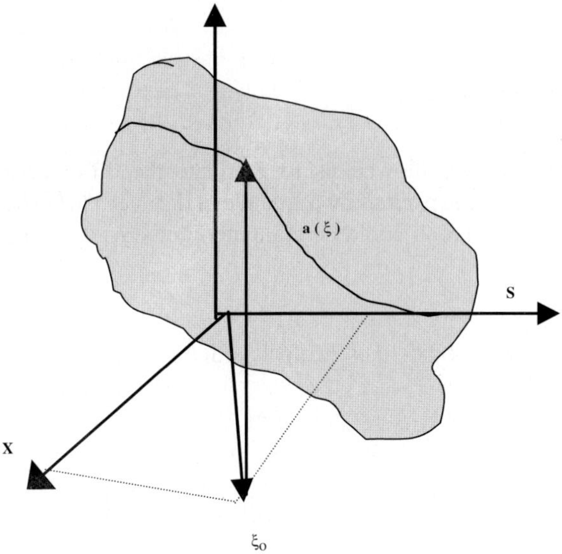

Figure 11. Schematic representation of a response surface obtained from the model variables

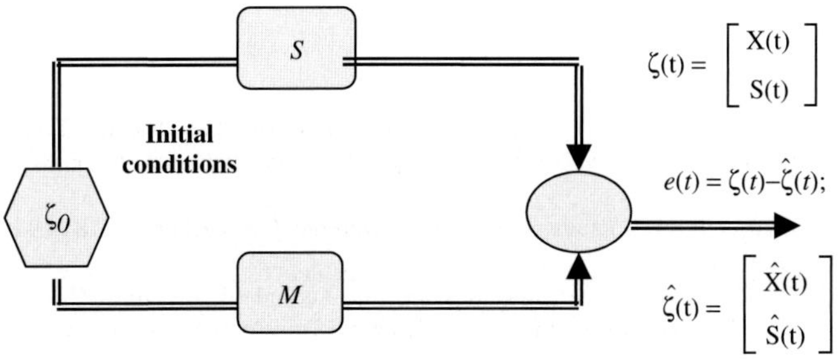

Figure 12. Minimum error between variables measured in the system and output model values

Where Y, G and g are defined as follows:

(12) $$Y = [a(\zeta_1)\ldots a(\zeta_0)0];$$

(13) $$G = \begin{bmatrix} \gamma(d_{11}) & \cdots & \gamma(d_{N1}) & 1 \\ \vdots & \ddots & \vdots & \vdots \\ \gamma(d_{1N}) & \cdots & \gamma(d_N) & 1 \\ 1 & \cdots & 1 & 0 \end{bmatrix}; \quad g = \left[\gamma(d_{01})\ldots\gamma(d_{0N})\ 1\right]^T;$$

And the semivariogram $\gamma(d)$ is:

$$(14) \qquad \gamma(d) = \frac{1}{n(d)} \sum_{k=1}^{N} \left(a\left(\zeta_k\right) - a\left(\zeta_{k+d}\right)\right)^2$$

The semivariogram was a statistic that allowed us to calculate the differences between two sampled data, which were separated by a Euclidean distance: "d". The upper bound of the error standard deviation could be computed numerically using the following inequality:

$$(15) \qquad \sigma_e(t) \leq \sigma_e(0) + \beta\sqrt{\frac{\pi}{2}} \int_0^t \sigma_e(s)ds + \int_0^t \sigma_\Delta(\zeta(s))ds = J(\hat{A})$$

where

$$(16) \qquad \sigma_e(t) = \begin{bmatrix} \sigma_{eX}(t) \\ \sigma_{eS}(t) \end{bmatrix}; \sigma_e(0) = \begin{bmatrix} \sigma_{eX}(0) \\ \sigma_{eS}(0) \end{bmatrix}; \text{and} \sigma_\Delta(\zeta) = \begin{bmatrix} \sigma_{\Delta X}(\zeta) \\ \sigma_{\Delta S}(\zeta) \end{bmatrix};$$

Note: $\sigma_e(t)$=standard deviation $X_{(t)}$ and $S_{(t)}$; $\sigma_e(0)$= standard deviation at initial conditions; $\sigma_\Delta(\zeta)$= standard deviation of the function $a(\zeta)$ and β is a matrix with the system constants derived from the semivariogram.

REFERENCES

Acevedo F, Cacciuttolo MA, Gentina JC. 1988. Comparative Performance of Stirred and Pachuca Tanks in the Bioleaching of a Copper Concentrate. In: Norris PR, Kelly DP, eds. Biohydrometallurgy, Kew Surrey.

Acevedo F. 2000. The use of reactors in biomining processes. Electronic J Biotechnol 3: 184-194. Available from: www.ejbiotechnology.info

Acevedo F, Gentina JC. 2005. Biolixiviación de minerales de oro. Acevedo F, Gentina JC, eds, Fundamentos y perspectivas de las tecnologías biomineras. Ediciones Universitarias de Valparaíso, Valparaíso, Chile, 79-91.

Aprea JL, Giaveno MA, Chiacchiarini PA, Lavalle TL, de la Fuente MV. 1997. Air lift reactor implementation for bioleaching of low grade copper ores. In: Proceedings of the International Biohydrometallurgy Symposium, Biomine IBS-97, August 4-6. Australian Mineral Foundation Inc., Sydney. Glenside, Australia.

Atkins A, Pooley F, Townsley C. 1986. Comparative mineral sulphide leaching in shake flasks, percolation columns, and Pachuca reactors using *Thiobacillus ferrooxidans.* Process Biochem 21: 3-10.

Bailey AD, Hansford GS. 1993. Factors affecting bio-oxidation of sulphide minerals at high concentrations of solids: a review. Biotechnol Bioeng 42: 1164-1174.

Beyer M, Ebner H, Assenmacher H, Frigge J. 1987. Elemental sulphur in microbiologically desulphurized coals Fuel 66: 551-555.

Blažej M, Kiša M, Markoš J. 2004. Scale influence on the hydrodynamics of an internal loop airlift reactor. Chem Eng Proces 43: 1519–1527.

Bos P, Huber TF, Kos CH, Ras C, Kuenen JG. 1986. A Dutch feasibility study on microbial coal desulphurization. In Lawrence RD, Branion RMR, Ebner HG, Ritcey GM. Process Metallurgy. Fundamental and Applied Biohydrometallurgy. Elsevier, Amsterdam, 385-394.

Brierley CL. 1997. Heap leaching of Gold-Bearing Deposits: Theory and Operational Description. In: Rawlings DE, ed. Biomining: Theory, Microbes and Industrial Processes. Springer-Verlag, Berlin, 103-115.

Brierley JA, Brierley CL. 1999. Present and Future Commercial Applications of Biohydrometallurgy. In: Amils R, Ballester A, eds. Biohydrometallurgy and the environment toward the mining of the 21st century, Proceedings of the International Biohydrometallurgy Symposium IBS-99, El Escorial, Spain. Elsevier, Amsterdam, Part A, 81-90.

Chappell MJ, Godfrey KR. 1990. Global identificability of the parameters of Nonlinear Systems with specified inputs: a comparison of methods. Mathematical Biosciences 102: 41-73.

Chaudhury GR, Das RP. 1987. Bacterial leaching- complex sulfides of copper, lead and zinc. Int J Min Proces 21: 57-64.

Chisti MY. 1989. Airlift Bioreactors. Elsevier Applied Science. New York.

Chisti MY. 1998. Pneumatically agitated bioreactors in industrial and environmental bioprocessing: Hydrodynamics, hydraulics and transport phenomena. Appl Mechanics Rev 51: 33-112.

Chisti MY, Moo-Young M. 2002. Bioreactors. Meyers, RA, ed. Encyclopedia of Physical Science and Technology. Academic Press, San Diego, vol. 2, 247-271.

Chisti MY. 2005. Mass transfer. Kirk-Othmer Eds. Kirk-Othmer Encyclopedia of Chemical Technology, 5th edition. Wiley, New York, vol. 15, 1–75.

Choi K, Chisti MY, Moo-Young M. 1995. Split-channel rectangular airlift reactors: Enhancement of performance by geometric modifications. Chem Eng Commun 138: 171-181.

Clark NN. 1984. Predicting the Circulation Rate in Pachuca Tanks with Full Height Draft Tubes. Minerals and Metallurgical Processing 226-231.

Contreras A, García F, Molina E, Merchuk J. 1999. Influence of sparger on energy dissipation, shear rate, and mass transfer to sea water in a concentric-tube airlift bioreactor. Enzyme Microb Technol 25: 820–830.

Couillard D, Mercier G. 1991. Optimum residence time (in CSTR and Airlift Reactor) for bacterial leaching metals from anaerobic sewage sludge. Water Res. 25: 211-218.

Crundwell F. 1997. The kinetics of the chemiosmotic proton circuit of the iron-oxiding bacterium *Thiobacillus ferrooxidans*. Bioelectrochem Bioenerg 43: 115-122.

De Gloria P, Ferrari D, Grosso M, Olivieri M, Puglisi I., 1999.Implementation Techniques and Their Applications Ed. by Leondes C. Fuzzy Theory Systems, Techniques and Applications. Academic Press, USA, chapter 1, 19-50.

Deveci H. 2004. Effect of particle size and shape of solids on the viability of acidophilic bacteria during mixing in stirred tank reactors. Hydrometallurgy 71: 385-396.

Escobar B, Godoy Y, Pardo E, Wiertz JV. 1995. Bioleaching of copper by *Thiobacillus ferrooxidans* at incresing pulp density. In: Jerez JCA, Vargas T, Toledo H, Wiertz JV, eds. Biohydrometallurgical processing, Proceedings of the International Biohydrometallurgy Symposium IBS-95, Vina del Mar, Chile. University of Chile, Santiago de Chile, Vol 1, 273-281.

Gatti M, Milocco R, Giaveno A. 2003. Modeling the bacterial oxidation of ferrous iron with *Acidithiobacillus ferrooxidans* using Kriging interpolation. Hydrometallurgy 71: 89-96.

Giaveno MA, Chiacchiarini PA, Lavalle TL, de la Fuente MV, Aprea JL. 1995. Bioleaching of patagonic copper ores in flask and percolating columns. In: Jerez JCA, Vargas T, Toledo H, Wiertz JV, eds. Biohydrometallurgical processing, Proceedings of the International Biohydrometallurgy Symposium IBS-95, Vina del Mar, Chile. University of Chile, Santiago de Chile, Vol 1, 343-350.

Giaveno MA, Chiacchiarini PA, Lavalle TL, Donati ER. 2005. Reversed flow airlift reactor characterization for bioleaching applications. In: Harrison STL, Rawlings DE, Petersen J, eds. Proceedings of the 16th International Biohydrometallurgy Symposium, September 25–29, Cape Town, South Africa. Produced by Compress www.compress.co.za, 83-85.

Gómez JM, Caro I, Cantero D. 1996. Kinetic equation for growth of *Thiobacillus ferrooxidans* in submerged culture over aqueous ferrous sulphate solutions. J Biotechnol 48: 147-152.

Hadaddin J, Dagot C, Fick M. 1995. Models of bacterial leaching. Enzyme Microb Technol 17: 290-305.

Jones L, Hackl R. 1999. Sources of high cyanide consumption for a biooxidized refractory gold concentrate. In: Amils R, Ballester A, eds. Biohydrometallurgy and the environment toward the mining of the 21st century, Proceedings of the International Biohydrometallurgy Symposium IBS-99, El Escorial, Spain. Elsevier, Amsterdam, Part A, 337-346.

Koide K, Horibe K, Kawabata H, Ito S. 1984. Critical gas velocity required for complete suspension of solid particles in solid-suspended bubble column with draught tube. J Chem Eng Japan 17: 368-374.

Koide K, Sato H, Iwamoto S. 1983. Gas holdup and volumetric liquid-phase mass transfer coefficient in bubble column with draught tube and with gas dispersion into annulus. J Chem Eng Japan 16: 407-413.

Kurtanjek Ž. 1994. Modeling and Control by Artificial Neural Network in Biotechnology. Comp Chem Eng 18: S627-S631.

Lavalle TL, Chiacchiarini PA, Pogliani C, Donati ER. 2005. Isolation and characterization of acidophilic bacteria from Patagonia, Argentina. Process Biochem 40: 1095-1099.

Lizama HM, Fairweather MJ, Dai Z, Allegretto TD. 2003. How does bioleaching start? Hydrometallurgy 69: 109-116.

Loi G, Trois P, Rossi G. 1995. Biorotor®: a New Development for Biohydrometallurgical Processing. In: Jerez JCA, Vargas T, Toledo H, Wiertz JV, eds. Biohydrometallurgical processing, Proceedings of the International Biohydrometallurgy Symposium IBS-95, Vina del Mar, Chile. University of Chile, Santiago de Chile,Vol 1, 263-272.

Merchuk JC. 1986. Hydrodynamics and hold-up in air-lift reactors. In: Cheremisinoff NP, ed. Encyclopedia of Fluid Mechanics. NP Gulf Publishing Company, Houston, Texas, USA, 1487-1511.

Merchuk J, Siegel M. 1988. Air-Lift reactors in chemical and biological technology. J Chem Technol Biotechnol 41: 105-120.

Merchuk J. 1990. Why use airlift bioreactors? Tibtech 8: 66-71

Merchuk J, Ben-Zvi S, Keskhavan N. 1994. Why use bubble-column bioreactors? Tibtech 12: 501-511

Merchuk J, Berzin I. 1995. Distribution of energy dissipation in airlift reactor. Chem Eng Sci 50: 2225-2233

Merchuk JC, Contreras A, García F, Molina E. 1998. Studies of mixing in a concentric tube airlift bioreactor with different spargers. Chem Eng Sci 53: 709-719.

Morin D, Lips A, Pinches A, Huisman J, Frias C, Norberg A, Forssberg E. 2005. BioMinE integrated project for the development of biotechnology for metal-bearing materials in Europe. In: Harrison STL, Rawlings DE, Petersen J, eds. Proceedings of the 16th International Biohydrometallurgy Symposium, September 25–29, Cape Town, South Africa. Produced by Compress www.compress.co.za, 23-40.

Mousavi S, Yaghmaei S, Vossoughi M, Jafari A, Hoseini S. 2005. Comparison of bioleaching ability of two native mesophilic and thermophilic bacteria on copper recovery from chalcopyrite concentrate in an airlift biorreactor. Hydrometallurgy 80: 139-144.

Nemati M, Webb CA. 1997. Kinetic Model for Biological Oxidation of Ferrous Iron by *Thiobacillus ferrooxidans*. Biotechnol Bioeng 53: 478-486.

Nemati M, Harrison S, Hansford G, Webb C. 1998. Biological oxidation of ferrous sulfate by *Thiobacillus ferrooxidans:* a review on the kinetic aspects. Biochem Eng J 1: 171-190.

Ojumu T, Petersen J, Searby G, Hansford G. 2005.A review of rate equations proposed for microbial ferrous-iron oxidation with a view to application to heap bioleaching. In: Harrison STL, Rawlings DE, Petersen J, eds. Proceedings of the 16th International Biohydrometallurgy Symposium, September 25–29, Cape Town, South Africa. Produced by Compress www.compress.co.za, 85-93.

Okibe N, Gericke M, Hallberg K, Johnson B. 2003. Enumeration and Characterization of Acidophilic Microorganisms Isolated from a Pilot Plant Stirred-Tank Bioleaching Operation. Appl Environ Microbiol 69: 1936–1943.

Onken U, Weiland P. 1983. Airlift fermenters: Construction, Behavior and Uses. Mizrahi A, van Wezel A, eds. Advances in Biotechnological Processes. Alan R Liss Inc, New York, 67-95.

Park D, Lee D, Joung J, Park J. 2005. Comparison of different bioreactor systems for indirect H_2S removal using iron-oxidizing bacteria. Process Biochem 40: 1464-1467.

Rawlings DE. 1997. Mesophilic, autotrophic bioleaching bacteria: Description, physiology and role. In: Rawlings DE, ed. Biomining: Theory, Microbes and Industrial Processes. Springer-Verlag, Berlin, 229-245.

Rawlings DE, Tributsch H, Hansford GS. 1999. Reasons why "*Leptospirillum*" –like species rather than *Thiobacillus ferrooxidans* are dominant iron-oxidizing bacteria in many commercial processes for the biooxidation of pytite and related ores. Microbiology 145: 5-13.

Rossi G. 2001. The design of bioreactors. Hydrometallurgy 59: 217-231.

Roy GG, Sheckhar R. 1996. Oxygen mass transfer in air-agitated Pachuca tanks -Part 1: Laboratory-scale experimental measurements. Part 2: mathematical modelling of mass-transfer coefficients. Transactions of the Institution of Miningand Metallurgy 105 - Section C: 9-21.

Roy CG, Bera A, Mankar JH. 2000. Effect of design and operating parameters on gas hold up in Pachuca (air agitated) tanks, Transactions of the Institution of Mining and Metallurgy 109 - Section C: 90-96.

Ruitenberg R, Schultz C, Buisman C. 2001. Bio-oxidation of minerals in air-lift loop bioreactors. Int J Min Proces 62: 271-278.

Ryu H, Chang Y, Kim S. 1993. Microbial coal desulfurization in an airlift bioreactor by sulfur-oxidizing bacterium *Thiobacillus ferrooxidans.* Fuel Proces Technol 36: 267-275.

Sánchez Mirón A, Cerón García MC, García Camacho F, Molina Grima E, Chisti Y. 2004. Mixing in bubble column and airlift reactors. Chem Eng Res Design 82: 1367–1374.

Sand W, Gehrke T, Jozsa PG, Schippers A. 1999. Direct versus indirect bioleaching. In: Amils R, Ballester A, eds. Biohydrometallurgy and the environment toward the mining of the 21st century, Proceedings of the International Biohydrometallurgy Symposium IBS-99, El Escorial, Spain. Elsevier, Amsterdam, Part A, 27-49.

Sand W, Gehrke T. 2006. Extracellular polymeric substances mediate bioleaching/biocorrosion processes involving iron (III) ions and acidophilic bacteria. Res Microbiol 157: 49-56.

Shrihari SRB, Kumar R, Gandhi KS. 1990. Modelling of Fe^{+2}oxidation by *Thiobacillus ferrooxidans.* Appl Microbiol Biotechnol 33: 524-528.

Siegel M, Hallaile M, Merchuck J. 1988. Air-Lift reactors: Design, operation and applications. Adv Biotechnol Processes 7: 79-124.

Siegel M, Robinson C. 1992. Applications of airlift gas-liquid-solid reactors in biotechnology. Chem Eng Sci 47: 3215-3229.

Silverman M, Lundgren D. 1959. Studies on the chemoautotrophic iron bacterium *Ferrobacillus ferrooxidans.* I .An improved medium and a harvesting procedure for securing high cell yields. J Bacteriol 77: 642–647.

Shi S, Fang Z. 2005. Bioleaching of marmatite flotation concentrate by adapted mixed cultures in an air-lift reactor. Min Proces 76: 3-12.

Schügerl K, Lübbert A. 1995. Pneumatically agitated bioreactors. Asenjo J, Merchuck J, eds. Bioreactor system design. Marcel Dekker Inc, 257-303.

Tipre D, Vora S, Dave S. 2003. Comparison of air lift and stirred tank batch and semicontinuous bioleaching of polymetallic bulk concentrate. Tsezos M, Hatzikioseyian A, Remoudaki E, eds. Proceedings of the International Biohydrometallurgy Symposium, IBS-03, Athens, Greece, 21-29.

Tyagi R, Couilliard D, Tran F. 1991. Comparative study of bacterial leaching of metals from sewage sludge in continuous stirred tank and air-lift reactors. Process Biochem 26: 47-54.

Viñals J. 2000. Cinética de las reacciones heterogeneas. Ballester A, Verdeja L, Sancho J, eds, Metalurgia Extractiva. Editorial Síntesis, España, Vol. 1, 169-198.

Whitlock JL. 1997. Biooxidation of Refractory Gold Ores (The Geobiotics Process). In: Rawlings DE, ed. Biomining: Theory, Microbes and Industrial Processes. Springer-Verlag, Berlin, 117-127.

Wu W, Wu J. 1990. Airlift Reactor with Net Draught Tube. J Ferment Bioeng 70: 359-361.

Znada H, Báleš V, Markoš J, Kawase Y. 2004. Modeling and simulation of airlift bioreactors. Biochem Eng J 21: 73–81.

CHAPTER 10

PRINCIPLES, MECHANISMS AND DYNAMICS OF CHALCOCITE HEAP BIOLEACHING

JOCHEN PETERSEN[1] AND DAVID G. DIXON[2]

[1] *Department of Chemical Engineering, University of Cape Town, Private Bag, Rondebosch 7701, South Africa. E-mail: jochen.petersen@uct.ac.za*

[2] *Department of Materials Engineering, University of British Columbia, 309-6350 Stores Road, Vancouver, B.C., V6T 1Z4, Canada. E-mail: dixon@interchange.ubc.ca*

1. INTRODUCTION

Heap leaching is a well established mining technology that involves stacking crushed ore into piles constructed on an impermeable layer fitted with a solution drainage system, or arranged on a slope to facilitate drainage. In many cases the ore is agglomerated through tumbling with acid and/or irrigation solution prior to stacking.

Solution is irrigated (continuously or intermittently) over the top surface of the heap and left to seep through the ore bed where it can react with the target minerals. Dissolved metals are then transported with the flowing solution to the bottom of the heap from where they are removed via the drainage system into a collection pond as the pregnant leach solution (PLS). The target metal is removed from the PLS through a suitable technology (by solvent extraction, cementation or adsorption), and the barren solution (usually referred to as the raffinate) returned to the top surface of the heap.

The technology has been in use since the 1960's for the acid leaching of copper oxide minerals, and since the 1970's for the cyanide leaching of gold and silver. A related technology, dump leaching, is based on the same principles as heap leaching, but usually uses uncrushed run-of-mine (ROM) material, usually containing sulfide minerals, or spent residue from leached heaps with a mineral grade too low to warrant further processing. Any additional metal liberated in a dump contributes to additional revenue of the mining operation, but dump leaching is rarely the primary method of metal extraction.

It was in copper sulfide dump leach operations that the potential for bioleaching micro-organisms was first realized (Brierley, 1997; Olson *et al.*, 2003). As the organisms involved occur naturally, no external intervention is needed to sustain

E.R. Donati and W. Sand (eds.), Microbial Processing of Metal Sulfides, 193–218.

their growth. Subsequently, heap bioleaching of copper sulfide minerals as a primary method for metal extraction was introduced first at Lo Aguirre in Chile in 1980 and then in numerous other operations from the early 1990's (Brierley, 1997). In the same vein as conventional oxide heap leaching, crushed and agglomerated ore is stacked into piles and irrigated with recycled acid solution. In addition, air is blown through pipes placed underneath the heap to ensure ready supply of oxygen and CO_2 throughout. In some cases inoculation is practiced and the microbial environment is enhanced by addition of nutrient material to the feed solution, but in general microbial growth in bio heaps proceeds naturally with little external intervention (Schnell, 1997).

Heap bioleaching results in the oxidative dissolution of sulfide minerals. Industrially this is practiced in two main applications, the dissolution of pyrite and arsenopyrite in refractory gold ores to liberate cyanide leachable gold, and the dissolution of copper from sulfide minerals. Biological dissolution of copper sulfide minerals occurs also in dumps, although the lack of artificial aeration tends to limit the extent of this. Furthermore, acid rock drainage (ARD) can, in a sense, be seen as a natural dump bioleaching process.

2. CHALCOCITE HEAP BIOLEACHING IN PRACTICE

Heap bioleaching of chalcocite (Cu_2S) originated from acid leaching of copper oxide minerals, which, due to the geological nature of many copper ore bodies, usually contain a certain proportion of copper sulfides. Many copper deposits consist of layers or zones containing predominantly oxides (brochantite, antlerite, chrysocolla, cuprite, *etc.*), underlain by zones enriched with secondary copper sulfides (chalcocite, covellite, bornite, *etc.*), and finally by zones containing primarily chalcopyrite. These layers show a continuous transition and hence oxide ores always have a certain portion of secondary sulfides present. Likewise, chalcocite ore treated in heap bioleaching is rarely 'pure' but may also contain copper oxides and chalcopyrite to varying degrees.

As stated above, chalcocite heap bioleaching as a stand-alone extraction technology began to develop in the early 1990's primarily in Chile, where it is still dominant today. The technology is now also practiced in Peru, Australia, the USA, China and Myanmar and is being explored for numerous other locations (Olson *et al.*, 2003; Rawlings *et al.*, 2003; Ruan *et al.*, 2005). Thus, in terms of metal tonnage produced and revenue generated, chalcocite heap bioleaching is by far the largest application of bioleaching technology at present.

Operational practice varies considerably, with heaps built as discrete units on pads or by continuous addition, with lifts stacked on top of one another. Heap/lift height varies from 6 to 12 m. The ore material is typically crushed to –25 mm and agglomerated with acid and/or raffinate, either on conveyor belts or in agglomeration drums, and stacked either by truck or conveyor belt. Aeration through pipes laid out underneath the heap is now standard, but aeration rates vary from as little as 0.08 to as much as 2 $m^3/m^2/h$. Irrigation is typically done through drip lines spread over the heap surface, with drip point spacing ranging from 30 to 100 cm. Irrigation rates range from 5 to 20 L/m^2/h. Irrigation-rest cycles are practiced in some operations. Chalcocite heaps generally demonstrate only limited heat generation, with effluent and subsurface temperatures rarely exceeding 25 °C, unless there is a significant presence of liberated pyrite which can be oxidized. In operations at high altitude (*e.g.* in the Chilean Andes) the heap surface is sometimes covered with plastic sheets as radiation shields for heat retention during the night. Heap cycle times vary with operational practice, mineral grade and composition, but usually are in the order of 8 months to 2 years. Copper extractions achieved are usually around 80% for ores containing predominantly chalcocite.

Active inoculation with bacteria does not appear to be common practice, but occurs usually through the agglomeration process with bacteria contained in the raffinate, or alternatively bacterial populations begin to thrive naturally in the heap after some time. Little, if any, microbiological maintenance is practiced. Characterization of the microbial ecology in chalcocite heaps has revealed the presence of *Acidithiobacillus ferrooxidans*, *Acidithiobacillus caldus* and *Leptospirillum ferriphilum* as well as certain types of archaea (Demergasso *et al.*, 2003; Hawkes *et al.*, 2006).

As is well known, the principal mechanism of bioleaching is the oxidation of ferrous to ferric iron (with concomitant reduction of oxygen) which serves as the metabolic energy source for certain types of micro-organisms, such as *Acidithiobacillus ferrooxidans,* or *Leptospirillum* species:

(1) $$4Fe^{2+} + O_2 + 4H^+ \rightarrow 4Fe^{3+} + 2H_2O$$

The ferric iron, in turn, effects the oxidative dissolution of certain sulfide minerals, including chalcocite. Leaching of chalcocite by ferric iron in sulfate solution proceeds by a two-stage mechanism, whereby the first stage:

(2) $$Cu_2S + 1.6Fe^{3+} \rightarrow 0.8Cu^{2+} + 1.6Fe^{2+} + (Cu_{1.2}S)$$

releases about 40% of the initial copper and produces a covellite-like material (Bolorunduro, 1999). However, this is not a stoichiometric compound, as the reaction goes through a number of intermediates from $Cu_{1.97}S$ to $Cu_{1.12}S$ (Schnell, 1997), during which a dramatic change in density and significant disintegration of the mineral grains occurs. The "covellite" then reacts further:

(3) $$(Cu_{1.2}S) + 2.4Fe^{3+} \rightarrow 1.2Cu^{2+} + 2.4Fe^{2+} + S^0$$

The kinetics of the two stages are quite different, with stage 1 proceeding rapidly even at room temperature, exhibiting a relatively low activation energy (values from 4 to 25 kJ/mol have been reported). Rate laws describe the reaction roughly as first order in ferric iron concentration (Bolorunduro, 1999).

The rate of stage 2 leaching, on the other hand, is slow at ambient temperature, and has an activation energy as high as 100 kJ/mol, which results in dramatic acceleration of the reaction at higher temperatures. Rate equations take reaction (3) as approximately half order with respect to the ratio of ferric to ferrous concentrations.

The kinetics of the bio-oxidation reaction (1) have been well researched (see, for example, Ojumu *et al.*, 2006). The oxygen required for this reaction comes from heap aeration. The required acid, on the other hand, must be supplied either with the feed solution, or else through the bio-oxidation of pyrite or elemental sulfur:

(4) $$2FeS_2 + 15O_2 + 2H_2O \rightarrow 2Fe_2(SO_4)_3 + 2H_2SO_4$$

(5) $$2S^0 + 3O_2 + 2H_2O \rightarrow 2H_2SO_4$$

Although the literature concerning the kinetics of bioleaching is extensive, most work has focused on tank leaching processes maintained under constant conditions. In heap leaching, on the other hand, many parameters such as temperature, pH, and O_2 and CO_2 availability are not well controlled and may in fact limit effective bio-oxidation. Very little fundamental study has been directed at determining bio-oxidation kinetics under heap conditions (Ojumu *et al.*, 2006).

3. FUNDAMENTAL PRINCIPLES OF HEAP BIOLEACHING

Although heap leaching appears to be a very simple process in concept, the sub-processes taking place within the ore bed as the solution trickles through are rather complex and their interactions not yet fully understood. One can distinguish between processes ranging from the macro- to the micro-scale as is illustrated in Figure 1 (Dixon & Petersen, 2003).

At the macro scale, i.e. at the scale of the entire heap, leaching is governed primarily by transport of mass and energy into, through, and out of the heap structure. More specifically these include:

- Solution flow: In unsaturated, coarsely granular packed beds solution generally flows along tortuous pathways but remains stagnant in pores and crevices between particles. This strongly influences heap performance in terms of reagent delivery and product removal from the reaction sites within the ore particles.
- Heat flow: Heat of reaction, which is significant in sulfide leaching, is transported through the heap downward as sensible heat with the flowing solution and upward as latent heat with the flow of humid air. Depending on air and solution flow rates, heaps can assume certain temperature profiles, and judicious manipulation of these variables may even allow a certain degree of control.
- Gas flow: Although gas flow is usually well distributed in aerated heaps, ensuring ample supply of oxygen throughout, the supply of CO_2 may be limited under certain conditions. In unaerated heaps O_2 availability may also be limited, and gas distribution patterns are complex.

Level	*Sub-processes*	*Illustration*
Heap Scale	• Solution flow through packed bed • Gas advection • Water vapour transport • Heat balance	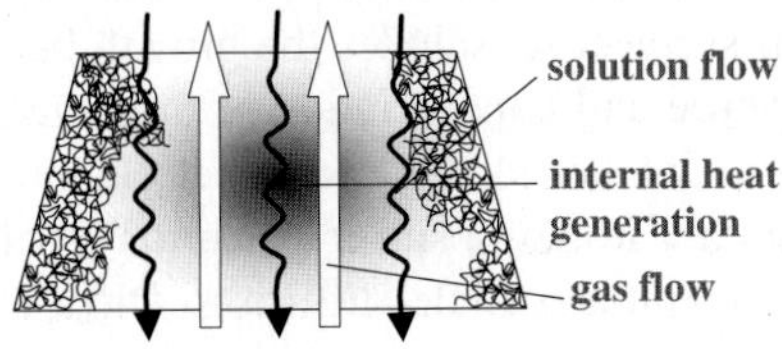
Cluster Scale	• Gas adsorption • Particle diffusion • Microbial growth • Microbial attachment • Microbial oxidation	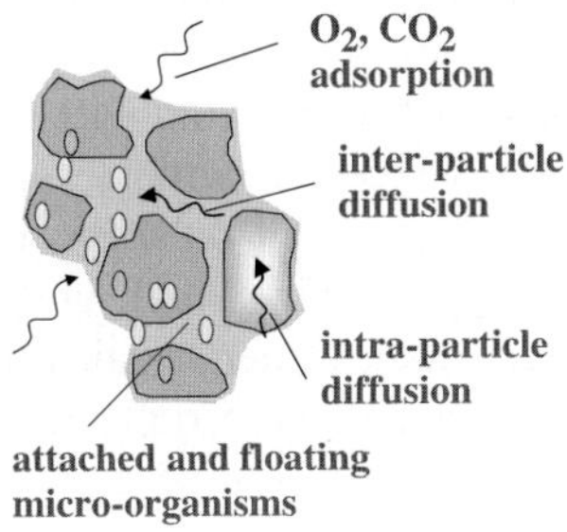
Particle Scale	• Topological effects • Intra-particle diffusion • Particle and grain size distribution	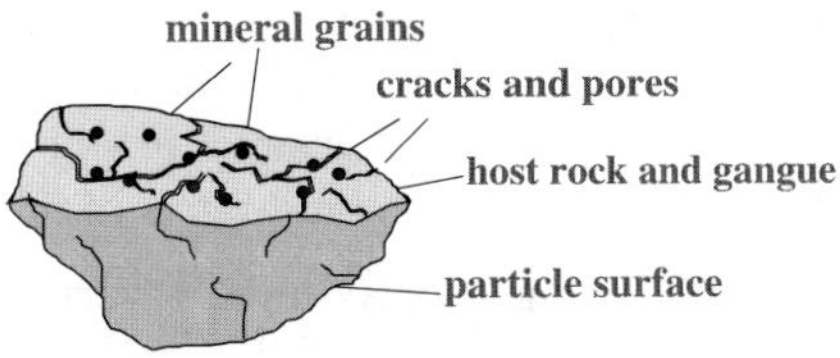
Grain Scale	• Ferric/ferrous reduction • Mineral oxidation • Sulfur oxidation • Surface processes	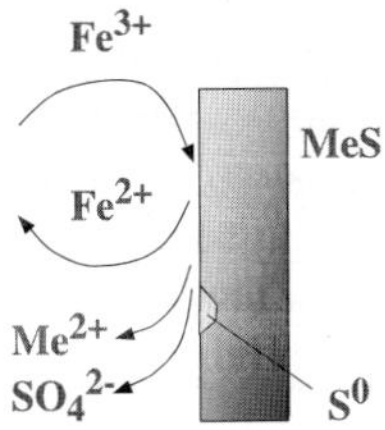

Figure 1. Schematic representation of sub-processes in heap bioleaching (adapted from Dixon & Petersen, 2003)

The next level, at the meso-scale, represents a cluster of particles. Here, three processes or groups of processes contribute to the overall rate of leaching:

- Gas (O_2 and CO_2) uptake from the gas into the liquid phase: This is a simple mass-transfer step, the rate of which is determined primarily by temperature, but which requires measurement of a system-specific mass-transfer coefficient.

- Bacterial growth, propagation and oxidation. This encompasses the complex interactions of different types of micro-organisms in the liquid phase and on the mineral surface. It includes the growth behaviour of each strain as a function of temperature and concentration of dissolved constituents (acid, ferrous and ferric iron, O_2, CO_2 as carbon source, etc.), and any synergies between these and the concomitant iron and sulfur oxidation reactions.
- Intra- and inter-particle diffusion. Diffusion is the main mode of transport of dissolved constituents from and to the moving solution into pore spaces between particles, and into cracks and fissures within particles. The effect of pore diffusion on overall kinetics is determined by the length of the diffusion path, which can be significant for systems with poor solution distribution.

At the level of a whole ore particle, topological effects present another degree of complexity. Topology refers to the way in which mineral grains (or, more precisely, their surfaces) are present in a single ore particle, and how they are distributed across different particle sizes. Mineral grains may occur as anything from free grains to highly localized spots at the particle surface to bands of grains permeating an inert matrix. Grain distribution and accessibility within particles directly determine the leachability of the target mineral. Furthermore, in low grade ores the mineralogy of the gangue matrix is also of some significance, as it can interfere with mineral leaching and biological phenomena.

Finally, at the level of a single mineral grain, leaching is governed by the chemical and electrochemical interactions at the grain surface. Sulfide minerals generally leach by the oxidative action of ferric ions in solution, and reaction kinetics is a function primarily of temperature (as characterized by the activation energy) and the concentration of ferric (and sometimes also ferrous) ions. The principal mechanisms of such reactions are well understood, except perhaps for chalcopyrite, but exact values of the critical parameters for each specific case remain subject to measurement. Further complication may arise when different minerals are galvanically coupled, as is the case with pyrite being present in many base metal sulfide ores. Direct biological interactions with exposed mineral surface, as stipulated in the contact leaching mechanism (Sand *et al.*, 2001) are also important at this level, although their significance in heap bioleaching (as opposed to concentrate leaching in tanks) has not yet been established.

The many sub-processes tend to interact with each other in patterns which in most cases are too complex to unravel intuitively. Any one of the sub-processes listed can become overall rate controlling under certain circumstances, and control can switch from one domain to another in the course of one particular leach scenario. With so many layers of complexity, diagnosing the effects of certain variables on overall heap performance becomes virtually impossible without a sophisticated modeling tool.

Heap leach models have been developed over the years as the technology evolved, first focused to acid leaching, then cyanide leaching, and finally bioleaching (Dixon, 2003). ARD models have been developed in parallel. The complexity of these models has continuously increased. The HeapSim model, developed by the authors

(Petersen & Dixon, 2003b, 2006a; Dixon & Petersen, 2003; Ogbonna *et al.*, 2005), takes into account all the various sub-processes and complexities outlined above without making any assumptions *a priori* as to which are rate determining. A similar degree of complexity is represented by the Phelps Dodge Copper Stockpile Model (Bennet *et al.*, 2003). HeapSim has been applied in the modeling of heap bioleaching of chalcocite (Dixon & Petersen, 2003a), sphalerite (Dixon & Petersen, 2004; Petersen & Dixon, 2006b), pyrite (Bouffard, 2003) and recently also chalcopyrite ores.

Models are usually calibrated against column tests of the ore in question. In such tests ore is filled into vertical tubes, usually 10-100 cm in diameter and 0.5 to 10 m in height, and irrigated with solution in similar fashion to full scale heap operations. Column tests are generally used to make decisions on the operating parameters of full-scale heaps, such as irrigation rates, crush sizes, temperature, inoculation mode, *etc*. The results of such tests are then used for scale-up to the heap operation, with or without the use of a model. However, the general experience is that column tests tend to over-predict rate and extent of metal recovery as compared with what is achieved in heaps designed on that basis, although this is at least partially due to incorrect scaling up, as is explained later. Therefore, the scope of column tests to inform the design of full-scale processes is limited at best.

4. COLUMN TESTWORK

The following gives a summary description of an extensive experimental column study of heap bioleaching of a typical low-grade chalcocite ore. Some of this work has been reported previously (Petersen & Dixon, 2003a, 2004).

4.1. Experimental

The experimental study was conducted in short columns, made from PVC pipe, 10 cm in diameter and with a bed height of 30 cm. Each column was immersed in a water bath for temperature control. In several experiments a number of these columns were run in series, whereby solution from one column would be pumped, via a collection cup, to the next column, using a multi-channel peristaltic pump. Tall column tests were conducted in 6 m vertical pipes. All columns were aerated from the bottom and the effluent drained through a looped pipe to avoid gas short-circuiting. Solution was fed from a single drip point at the top of the column, which was open to the atmosphere.

The sample ore, sourced from a heap bioleach operation in northern Chile, had an average Cu grade of 1.45%, approximately 25–30% reporting as acid soluble brochantite and the rest mostly as chalcocite. The pyrite content was around 2.5%. Top particle size was 20 mm and the <100 micron fines content was 10%. Prior to charging into the columns the ore was agglomerated with a mixture of 70 g water and 8 g concentrated sulfuric acid per kg of ore, resulting in a final moisture content of 7.5%.

Table 1. Column leach experiments conducted

Experiment	Height	Duration	Final Cu Extraction
Single Column Standard	30 cm	63 days	87.1%
4 Column Standard	1.2 m	69 days	81.8%
8 Column Standard	2.4 m	21 days	58.5%
Tall Column Standard	6 m	113 days	87.2%
Tall Column Low air & CO_2	6 m	113 days	80.8 %
4 Column Raffinate Solution	1.2 m	63 days	70.6 %

All experiments were fed solution on a once-through basis with the standard feed solution containing 7.5 g/L H_2SO_4, 0.7 g/L Fe(III) as sulfate and 1.3 g/L Fe(II) as sulfate (similar to conditions reported from the industrial operation). The standard experimental irrigation rate was 40 mL/h (corresponding to 5 L/m^2/h). Air was introduced at a standard rate of 260 NmL/min (2 Nm^3/m^2/h) of air enriched with 1% v/v CO_2 (to ensure that there was no limitation of carbon supply as the ore did not contain any carbonaceous minerals).

The mixed laboratory culture used was originally derived from a blend of pure cultures of *Acidithiobacillus ferrooxidans*, *Acidithiobacillus thiooxidans* and *Leptospirillum ferrooxidans*. This was maintained in 9K culture medium in an incubator shaking at 150 rpm at 30 °C, with weekly transfers. Columns were rinsed with acid solution containing no iron for one day, and subsequently inoculated by feeding 800 mL of inoculum solution before switching to the standard irrigation solution. Table 1 lists the column studies which are discussed in the following sections.

4.2. Chalcocite Leaching In Standard Laboratory Columns

Figure 2 shows Cu extraction versus time for the four standard experiments. In all curves alike, the first 30% of Cu extraction, associated with the dissolution and rinsing of brochantite, proceeds rapidly. Extraction then continues at a slower but more or less constant rate until about 60% extraction has been achieved and then gradually declines. All curves tend towards a total copper extraction of around 90%. The effective column height has only a minor effect on the rate of leaching, with the 6 m tall column taking only about twice as long to extract 80% Cu as the single 30 cm column. This indicates that the overall rate of Cu extraction is determined by the kinetics of the chemical reactions and is thus independent of the reactor size.

Although the focus of chalcocite leaching is on copper production, it is important to look also at the leaching and eventual fate of iron, especially that which is released through the competing bio-oxidation of pyrite as represented in Equation (4). Figure 3 shows the concentration in the effluent of total Fe, solution potential and pH for the 6 m column experiment. The effluent concentration of Fe drops below the feed concentration at 2 g/L only in the intermediate phase of the leach,

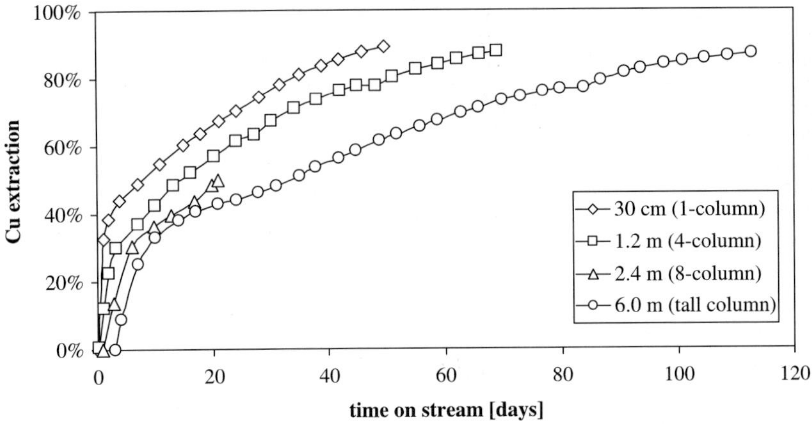

Figure 2. Cu extraction *vs.* time for columns of different length

but rapidly increases to net Fe production once the solution potential begins to increase. This is a strong indication that some pyrite leaching occurs in parallel with chalcocite leaching. Further evidence can be found in the declining pH at the later stages of the experiment, approaching the feed pH of 1.2. Although there is not net acid generation, this trend is nonetheless a clear indication that acid is produced in the column, counter-balancing the bio-oxidation reaction (1), which consumes acid at all times. This acid is either generated from secondary oxidation of elemental sulfur (produced from the stage-2 chalcocite reaction (3)), or from pyrite oxidation, or both. All other standard column experiments exhibited similar behavior, with pH in some cases eventually falling below the feed level.

The 8-column standard experiment was set up specifically to investigate the progression of dissolved constituents (acid, Fe(II), Fe(III) and micro-organisms) in the solution as it flows down the column through analyzing samples in the interim cups. The profiles of pH, solution potential (*vs.* Ag/AgCl) and microbial counts are shown in Figure 4. The experiment was stopped after 21 days, just as the wave of changes in solution potential began to break through the last column, to allow analysis of the solid residues.

The pH profiles clearly show two waves, one early peak moving through the columns during days 3–7 and one much broader wave passing through between days 8 and 20, with peak levels increasing from column to column (pH 1.4 in column 1 to 2.5 in column 8). In each column, solution potentials linger at low levels around 380 mV *vs.* Ag/AgCl for some time before rapidly rising to very high levels around 700 mV *vs.* Ag/AgCl. This potential rise progresses in a wave through the columns at a rate of about 30 cm (i.e. one column) every 2 days. Bacterial counts also indicate that the microbial population progresses through the columns in a front, with bacterial numbers rising to levels considered thriving (>10^7 active cells/mL) in the upper columns first before they proceed further down.

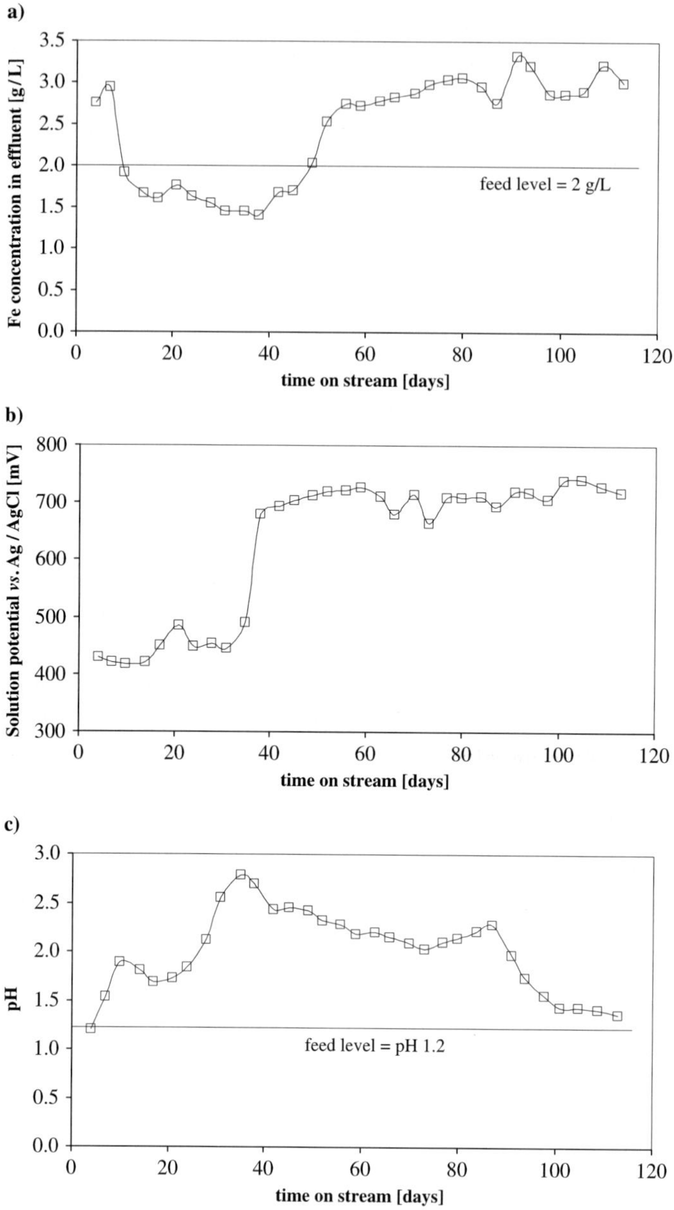

Figure 3. Results from 6m standard column experiment: a) Fe concentration, b) solution potential, and c) pH in column effluent

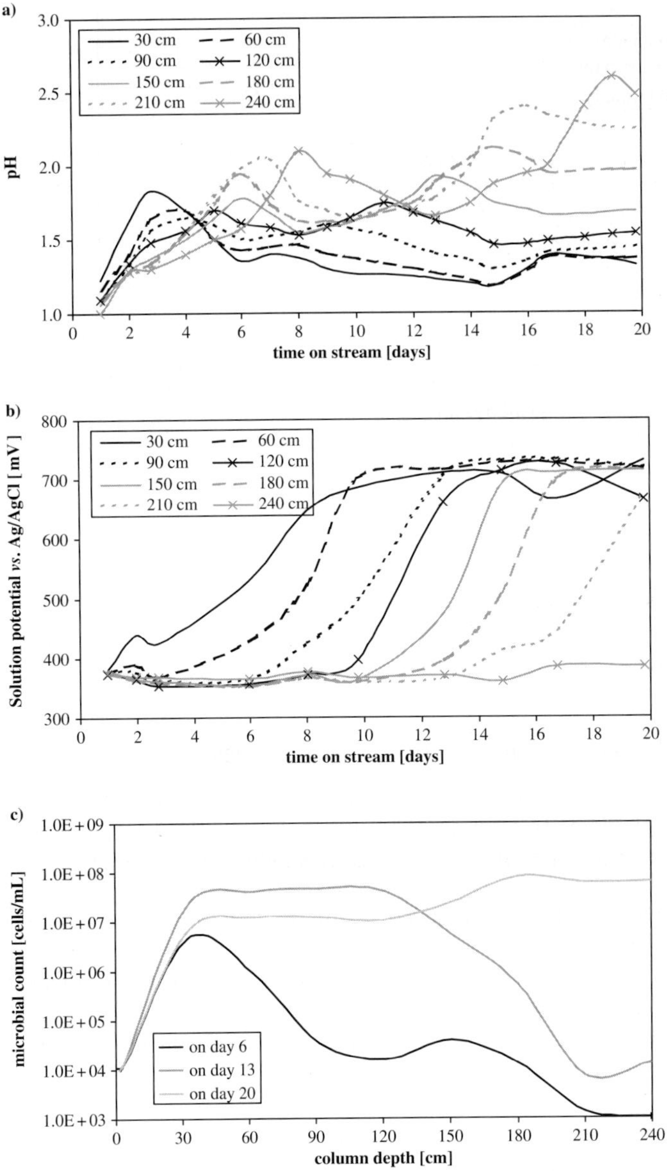

Figure 4. Results from 8-column standard experiment: a) pH progression; b) solution potential progression; c) progression of microbial counts in solution

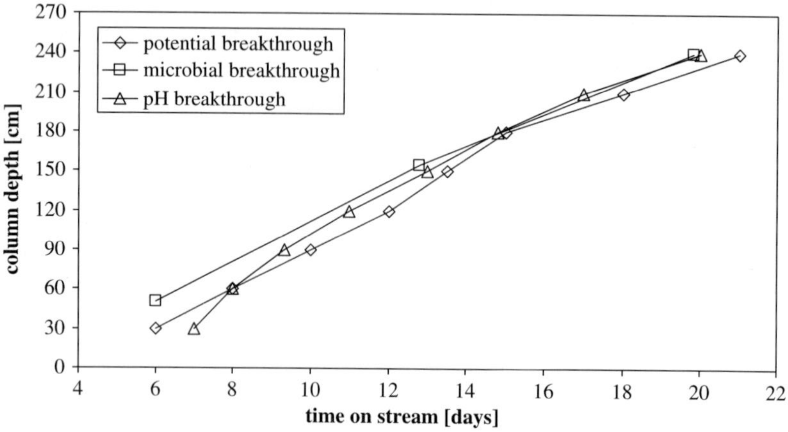

Figure 5. Microbial, pH and potential progression in experiment Z5

The progression of pH, potential and micro-organisms are clearly related. Figure 5 plots the progression with depth and time of the peak pH in the second wave, the points where the solution potential exceeds 500 mV vs. Ag/AgCl, and the points where the bacterial population in solution exceeds 10^7 cells/mL. The plot clearly shows that the three phenomena progress through the columns at more or less the same rate.

4.3. Discussion of Standard Experiments

Approximately 25–30% of the copper in the material reports as acid-soluble brochantite. Upon agglomeration with acid this is rapidly dissolved into the pore solution of the agglomerate material or forms readily soluble copper sulfate:

$$Cu_4SO_4(OH)_6 + 3H_2SO_4 \rightarrow 4CuSO_4 + 6H_2O \quad (6)$$

Upon rinsing the bed with feed solution, the acid-soluble copper is simply flushed from the column at a rate commensurate with the irrigation rate and total ore mass in the column. Once the soluble copper is flushed from the bed, all further copper released originates from oxidation of chalcocite.

As discussed above, ferric iron leaching of chalcocite proceeds in two stages. In the standard column experiments all ferric iron produced by the bacteria is initially consumed during first-stage chalcocite leaching. This reaction is so rapid that all ferric iron is consumed as soon as it is generated. As a consequence, no ferric iron can build up in solution and the redox potential remains quite low at or near the rest potential of chalcocite (approximately 370 mV *vs.* Ag/AgCl as observed in Figure 4b). This effect is often mistaken as an indicator for low microbial activity, although in fact the opposite is true.

The rate at which ferric iron is generated during this period is governed by the rate at which the bacteria can oxidize ferrous iron. This in turn is influenced by a large number of factors, such as temperature, nutrient and substrate concentrations, availability of oxygen and CO_2, *etc.* In the standard column experiments the rate limiting factor is the supply of acid to the reaction sites. According to Equation (1) the oxidation of ferrous to ferric iron requires acid. However, because the first stage of chalcocite leaching produces no elemental sulfur, and because pyrite leaching is inhibited at such low solution potentials, the only source of acid during this stage is the feed solution. Thus, bacteria continue to oxidize ferrous to ferric iron only until they run out of acid, and this results in the first stage of chalcocite leaching progressing in a narrow band beyond which the solution contains virtually no acid, and hence leaves the column at high pH (as confirmed by the data in Figure 4a).

In the wake of the first-stage chalcocite leaching front, leaching of the second stage 'covellite' and pyrite continues. Since both of these reactions are relatively slow at ambient temperatures, the rate of ferric consumption is now much lower, and the solution potential climbs to higher levels (600 to 700 mV *vs.* Ag/AgCl) where bacterial oxidation, now limited by low ferrous concentrations, can meet the reduced demand for ferric iron. This explains the relatively rapid increase in potential as observed in the experiments. The reaction proceeds more or less homogeneously over the entire length of the column, and during second-stage chalcocite leaching the overall rate of copper extraction is governed by mineral kinetics rather than reagent limitations.

Figure 2 indicates a slight 'slump' in the Cu leach curve for the tall column experiment around day 30. This phenomenon can be explained with the first-stage chalcocite leach front effect. Experimental evidence suggests that first-stage chalcocite leaching occurs over no more than the length of two columns (60 cm) at any one time (Figure 5). In the 6 m column this would account for only a small portion of the bed. Thus, after acid rinsing, only the top layer is active initially, whereas the rest of the column is idle for lack of acid. Consequently the overall rate of extraction (relative to the whole column charge) becomes very small. As the chalcocite front moves further down, 'covellite' and pyrite are leaching in its wake, and as this zone increases in size, so does the overall portion of the heap actively leaching. The overall rate of extraction thus increases further. Once the chalcocite front breaks through, the entire heap is in second-stage leaching and the overall rate of copper extraction corresponds to that of the 'covellite' leaching.

Beyond 75%, the rate of copper extraction gradually declines (Figure 2). This is seen as a topological effect, where the exposed surface area of the copper minerals is diminished with time. Also, the leaching of mineral grains embedded within ore particles may be limited by the diffusion of reagents and reaction products through narrow pores and cracks, or prevented altogether by occlusion (explaining why no more than 92% of copper has ever been extracted from any column). Fe leaching and generation of acidity, on the other hand, do continue in all experiments (Figure 3). This is an indication that bio-oxidation activity is shifting towards pyrite leaching in the later stages, probably because competition for ferric iron from the copper sulfides is substantially diminished.

4.4. Chalcocite Leaching Under CO_2-Limiting Conditions

As was outlined above, chalcocite heap bioleaching usually employs forced aeration with aeration rates in industrial heaps ranging from 0.08 to 2.0 $Nm^3/m^2/h$. The following calculation demonstrates that this would be sufficient to supply all the oxygen needed for the conversions typically achieved in chalcocite heaps:

> Assume a chalcocite ore containing 0.8% Cu as Cu_2S and 4% FeS_2. The ore is heap bioleached in a 10 m heap at 1,500 kg/m^3 packed density, and is targeted to achieve 80% Cu extraction after 12 months. The pyrite conversion in that time is estimated at 20%. The following overall reactions apply:
>
> (7a) $$Cu_2S + 4Fe^{3+} \rightarrow 2Cu^{2+} + 4Fe^{2+} + S^0$$
>
> (7b) $$FeS_2 + 14Fe^{3+} + 8H_2O \rightarrow 15Fe^{2+} + 2SO_4^{2-} + 16H^+$$
>
> It is further assumed that all Fe^{3+} is generated by bio-oxidation according to
>
> (7c) $$4Fe^{2+} + O_2 + 4H^+ \rightarrow 4Fe^{3+} + 2H_2O$$
>
> Given the feed grades, the initial chalcocite concentration amounts to 0.063 mol/kg and the pyrite concentration to 0.33 mol/kg. Over one year the targeted conversion rate would come to 0.0505 mol/kg/a for chalcocite and 0.0667 mol/kg/a for pyrite. The total demand for Fe^{3+} according to the stoichiometry of equations (7a) and (7b) is then 1.135 mol Fe^{3+}/kg/a or, by equation (7c), 0.284 mol O_2/kg/a.
>
> At the given heap height and packing density, at total of 15,000 kg ore would be placed per m^2 of heap area, and hence a total oxygen supply of 4,257 $mol/m^2/a$ is required. At 21% oxygen content of air this translates into a total air supply rate of about 450 $Nm^3/m^2/a$ or 0.05 $Nm^3/m^2/h$, noting that 'normal' conditions (as indicated by the capital N) for gases refer to 0°C and 1 atm pressure. Allowing for no more than 25% oxygen utilization per pass in the heap as a reasonable margin of safety, this would suggest an aeration rate of 0.2 $Nm^3/m^2/h$.

This calculation would stipulate that aeration at the lower end of the range of industrial operational practice suffices for oxygen supply in a typical chalcocite heap. However, one factor that has been overlooked in these considerations is CO_2 supply. Autotrophic bacteria common to bioleaching rely on CO_2 as their only carbon source for growth. This is either drawn from carbonate minerals in the ore rock material or absorbed from the surrounding air. Kinetic studies of microbial ferrous iron oxidation by *Leptospirillum* and *Acidithiobacillus* species reported in the literature indicate a carbon yield of 0.022 to 0.05 mol C/mol O_2 (Boon, 1996; van Scherpenzeel, 1996; Breed *et al.*, 1999; Breed & Hansford, 1999), or, in other words, the amount of CO_2 needed is about 2.2 to 5% of the amount of O_2 required.

In air, however, the CO_2 content is only about 0.17% of that of O_2 (350 ppm CO_2 vs. 21% O_2). Thus, if all CO_2 needs to be adsorbed from the air, the air supply rate needs to be considerably faster than what is required to satisfy the oxygen demand for the same application. Continuing to above calculation on this basis – assuming an average carbon yield of 0.035 mol C/mol O_2 and allowing for complete CO_2 utilization – would require a minimum air supply rate of 0.933 $Nm^3/m^2/h$ or about 4.5 times the rate estimated for the necessary oxygen supply. Hence it is reasonable to suspect that in heap leach applications with a low aeration rates the supply of CO_2 may in fact be overall rate limiting if there are no carbonates in the ore.

This is supported further by the results of an experiment in a second 6 m tall column on the ore material used in this study. The second column was run under identical conditions to the one described above, except that the aeration rate was 0.25 $Nm^3/m^2/hr$ (without any enrichment with CO_2 – corresponding to what is used at the industrial site, and what, according to above calculation, would be sufficient for oxygen supply). Results from the leaching experiments are shown in Figure 6.

As Figure 6 indicates, copper extraction from both columns proceeds at more or less the identical pace until about day 35 when they begin to diverge. In the column run under standard conditions this coincides with the day when the potential rapidly rises and noticeable quantities of micro-organisms are found in the effluent. As discussed above, this is indicative of the narrow, acid-limited chalcocite front, which is traveling through the column, breaking through. However, neither of these effects is observed in the column run on low aeration. Microbial breakthrough occurs much later (around day 60) and the potential in the effluent never rises above 500 mV throughout the experiment. Although copper extraction follows similar trends to those of the standard experiment, it continues to lag behind by about of 20 days. This column also produced a consistently lower effluent pH, and Fe levels in the effluent were consistently below the 2 g/L in the feed.

The discrepancy between the experiments is explained as follows: The first 30% of copper extraction resulted from flushing of acid-soluble copper and was not affected by aeration rate or CO_2. This is followed by the stage 1 phase during which leaching of chalcocite occurs in a relatively narrow band and is governed principally by the rate of acid supply. However, in the low aeration column CO_2 supply is probably not sufficient to sustain the growth necessary to consume all available acid, as effluent pH in this column stays low. The band would thus be narrower and travel at a slower pace, commensurate with CO_2 availability, as is witnessed by the delayed breakthrough of micro-organisms in the effluent. Microbial growth in the wake of the front will be affected even more by a lack of CO_2 as this zone grows longer. Hence copper production falls behind in the low aeration column while the stage 1 zone is slowly traveling through the column. Once this breaks through (around day 60) the rate of copper leaching becomes similar again in both columns (as indicated by the slopes in Figure 6a), indicating that the rate of stage 2 leaching is overall rate-determining. The oxidation of pyrite and sulfur remains affected

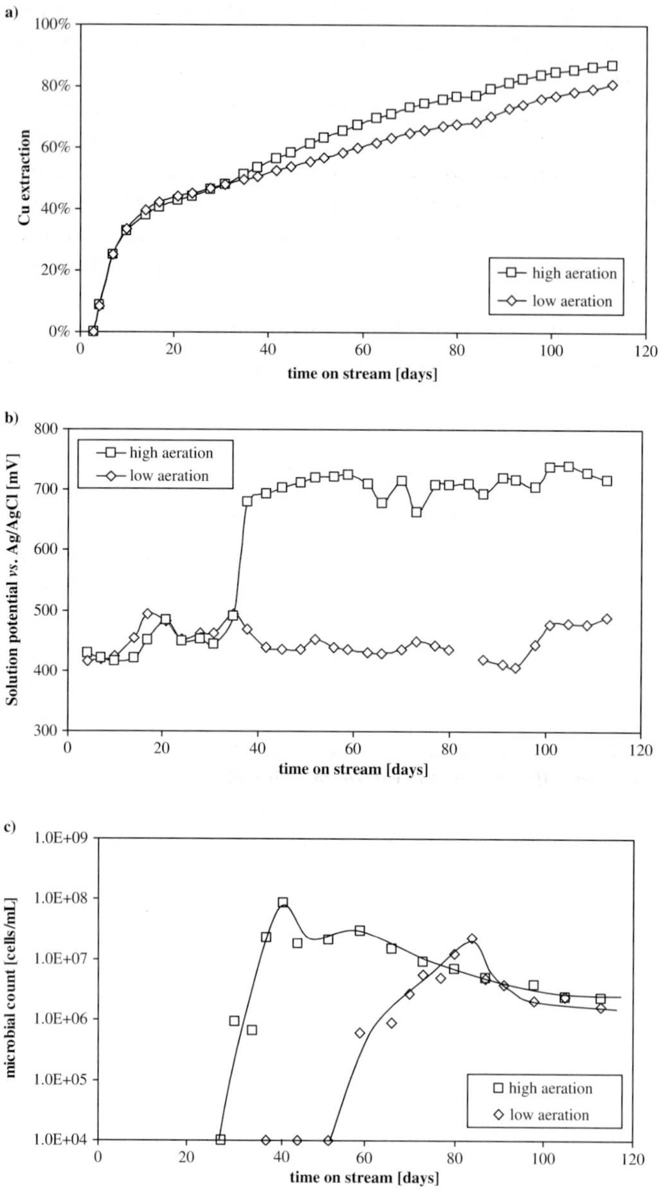

Figure 6. Selected results of the CO_2-limited tall column test: a) copper extraction; b) solution potential; c) microbial count in solution. High aeration refers to 2 m^3/m^2/hr of air enriched to 1% CO_2; low aeration refers to 0.25 Nm^3/m^2/hr of ambient air (350 ppm CO_2)

in the low aeration column as attested to by the higher pH (i.e. less oxidation of elemental or pyritic sulfur) and the lower iron levels in the effluent (jarosite precipitation instead of surplus iron generation from pyrite oxidation).

Therefore it is possible that the lack of CO_2 affects primarily the rapid bio-oxidation of stage 1 chalcocite leaching when bio-oxidation can become rate-limiting. The slower oxidation of second-stage chalcocite remains controlled by its slow mineral oxidation kinetics. Unfortunately the experiment is not fully conclusive in this regard, since the rate of air supply in the low aeration column was also close to that required for oxygen demand at the given rate of leaching (but was not limiting as it would have been for CO_2 demand). Nonetheless, the study can be taken as an indicator that low aeration rates practiced in some heap operations may be limiting the overall rate of leaching more than necessary and should be re-evaluated.

4.5. Chalcocite Leaching In High Salinity Process Solution

All standard column experiments were run with a microbial culture grown in the laboratory under optimal growth conditions before introducing these into the columns, and the feed solution contained only acid and iron. In real heap operations, however, conditions are generally quite far removed from such an ideal situation, and thus microbial survival and efficiency are more seriously challenged (Garcia *et al.*, 1998).

Feed and effluent solution (raffinate and PLS) samples were obtained from the industrial site from which the ore material had been sourced. Microbial activity was low in the PLS and virtually nil in the raffinate. The chemical composition of the raffinate (which is the same as the PLS except that most of the copper has been removed) is given in Table 2. The concentrations of dissolved salts are exceptionally high, with magnesium and aluminium alone accounting for about 95% of the total dissolved salts (TDS) at 148 g/L. Such concentrations far exceed

Table 2. Analyzed composition of the raffinate at the industrial operation (only components >10 mg/L listed)

Element	Concentration [mg/L]	Element	Concentration [mg/L]
Al	12,200	Zn	376
Ca	467	Cl^-	1,300
Co	16.2	F^-	80.1
Cu	216	NO_2^-	28.1
Fe	2,460	NO_3^-	105.9
Mg	10,100	o-PO_4	532
Mn	669	SO_4	116,880
P	221		
K	29.0	pH	1.24
Na	1,670	E (mV vs SHE)	640

tolerance levels reported in the literature for iron-oxidizing micro-organisms: Blight and Ralph (2004) indicate a immediate five-fold decline in growth rate in a 40 g/L Na_2SO_4 solution, and, in an extension of this study, Shiers *et al.* (2005) report a 50% permanent retardation of growth rate at this concentration of Na_2SO_4. Touvinen *et al.* (1971) report an adverse effect of Al concentrations exceeding 10 g/L on the growth of an unspecified *(Acidi-)thiobacillus* strain.

Furthermore, the chloride concentration of 1.3 g/L is also of some concern. Lawson *et al.* (1995) report on adverse effects on microbial growth of chloride concentrations exceeding 1 g/L with complete inhibition at 7 g/L. Shiers *et al.* (2005) report initial retardation of growth rate at 1 g/L and 3 g/L NaCl, but with long-term recovery through adaptation, and permanent retardation at higher concentrations.

The high cation concentrations result from the gradual dissolution of silicate gangue minerals (in the present case primarily chlorite, biotite and plagioclase; Lamberg, 1997) and the fact that the entire solution inventory is continuously recirculated, with only copper removed in solvent extraction and no purge. The chloride concentration relates primarily to the dissolution of atacamite ($CuCl_2{\cdot}3Cu(OH)_2$), a chloride-containing copper oxide mineral which occurs in the ore in trace concentrations.

A microbial culture was gradually enriched and maintained in the original undiluted raffinate in a shaker-incubator at 150 rpm at 30°C. This was then used to inoculate a 4-column experiment which was run with the raffinate as feed solution, but under otherwise identical conditions to the standard experiments reported above. The results from the two 4-column experiments are compared in Figure 7.

Figure 7a shows the copper extraction *vs.* time. In both experiments the first 30% of copper extraction, representing the rinsing of acid-soluble copper, is rapidly flushed from the columns. Over the next 30% of copper extraction, roughly corresponding to stage 1 leaching, the two experiments differ significantly: In the standard experiment this phase is nearly complete within 25 days, whereas the experiment with native culture and raffinate solution requires 40 days. Leaching trends beyond 60% extraction are, however, very similar in both experiments.

Figures 7b and 7c show the progression of pH and solution potentials along the length of the segmented columns. The curves relating to the standard 4-column experiments show (similar to Figure 4a) chalcocite (low potential) leaching in a narrow zone moving down the column, followed by high potential stage 2 leaching.

In the raffinate experiment the stage 1 zone is still observed, but it progresses much more slowly, and the transition from stage 1 to stage 2 leaching is much more gradual. Only some of the available acid is consumed, and the pH remains more or less stable with only a minor peak around day 50, when the low potential front breaks through. Thus it is clear that stage 1 leaching proceeds through the raffinate column within a broad band rather than a narrow zone, and is not controlled by acid supply, as opposed to the standard experiment.

The cause of these discrepancies becomes clear from Figure 8. Plotting the number of bacteria counted in solution as a function of time and depth in the column shows counts lower by an order of magnitude in the column run with native bacteria

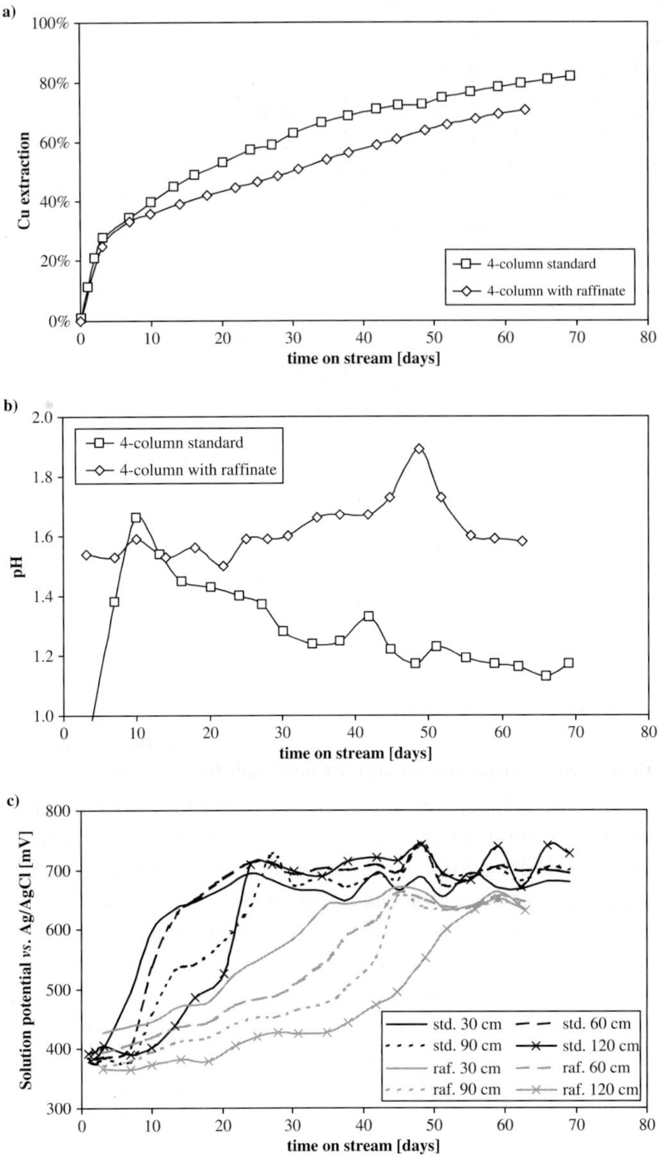

Figure 7. Selected results of the 4-column tests under standard conditions (std.) and with raffinate and native micro-organisms (raf.): a) copper extraction; b) effluent pH; c) progression of solution potential

in industrial raffinate as compared to the standard column (laboratory culture in acid and iron only). Nevertheless, both sets of data show the same progression trends, with numbers moving down the columns in a "growing wave". The propagation rate of this wave corresponds to that of the high potential wave in the wake of

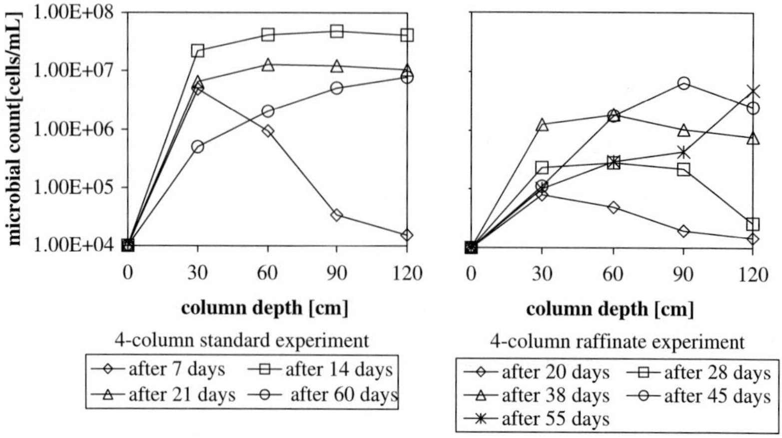

Figure 8. Bacterial counts over the length of the column in experiments Z5 and Z8

Table 3. Growth rate and final yield of growth experiments

Culture	Growth rate (h^{-1})	Doubling time (h)	Final yield (cells/mL)
Laboratory	0.109	6.36	1.10×10^8
Native	0.019	36.3	2.27×10^7

the chalcocite stage 1 leach front. Therefore, it is postulated that the slower copper extraction rate in the raffinate column is linked to the slower propagation of bacteria through the column at much smaller numbers. The rate of copper leaching is hence controlled by bacterial growth kinetics rather than acid supply.

The growth characteristics of the cultures used in the column experiments were investigated in a series of shake-flask experiments. The experimental details are described in Petersen & Dixon (2004). The results clearly suggest that the laboratory culture grows about six times more rapidly than, and to about five times the concentration of, the native culture in raffinate solution under identical conditions (Table 3). It should be noted that the total amount of bio-oxidation (in terms of copper leached) was not established in these tests. Therefore the measured yield in terms of cell numbers does not necessarily relate to the same amount of ferrous iron oxidized.

5. MODELING STUDIES

As is discussed earlier in this chapter, numerous heap bioleach models have been proposed in the literature over the years. A detailed review of these has been given by Dixon (2003). For the purpose of analyzing the results presented in the preceding sections, HeapSim, a comprehensive modeling tool developed by the authors (Petersen & Dixon, 2006a; Ogbonna *et al.*, 2005), was employed. The overall model combines an advection-diffusion model to account for transport of solutes through the heap and to mineral sites in the ore, with a multi-reaction model at the mineral site incorporating gas adsorption, microbial growth and oxidation, and mineral leaching through appropriate kinetic terms. Additional features include an independent heap heat balance and a gas mass balance.

Two aspects of the model are of particular interest in the context of the present discussion. The first relates to the way in which bulk solution flow is modeled. It is assumed that solution flows through discrete channels in the ore bed, which are in contact with a cluster of ore material containing an aqueous phase in the form of stagnant moisture. Thus flow is the predominant mode of mass transport in these channels, whereas within the ore clusters it is by molecular diffusion through the stagnant moisture. The diffusion length R (i.e. the 'radius' of the cluster) is a characteristic model parameter. The larger the cluster is, the further reagents and reaction products have to diffuse through stagnant solution. In narrow columns, such as used in this study, R can be taken to be close to the pipe radius (80% of radius has been found to be a good value, i.e. 4 cm in the present case). In unconfined heaps irrigated by drip emitters it is likely to be much larger, closer to the half distance between two drip points. This has been discussed in some detail and partially confirmed through experiments by Petersen and Dixon (2003).

The other aspect relates to microbial oxidation, which is modeled through Michaelis-Menten kinetics, combining microbial growth with ferrous iron oxidation via the Pirt equation (Petersen & Dixon, 2004; Ojumu *et al.*, 2006). The characteristic parameter in this approach is the optimum microbial growth rate k_g (often also referred to as μ_{max}), which is modified by Monod and resistance terms to account for the effect of substrate concentrations, temperature, *etc*. Typical values for k_g for a healthy laboratory culture and a native culture stressed by high concentrations of Mg and Al sulfates as well as chlorides are reflected in Table 3.

Data from the standard 8-column and tall column tests discussed above were used to calibrate the HeapSim model. This is a somewhat cumbersome process described in more detail elsewhere (Dixon & Petersen, 2004; Petersen & Dixon, 2006b). Once calibrated, the model can then be used to simulate other column experiments simply by changing the relevant operational parameters (such as flow rates, column height, feed composition, *etc.*) but leaving the intrinsic parameters describing mineral and microbial kinetics and mass transfer parameters unchanged. In this fashion the 4-column experiment has been simulated, as is shown in terms of copper extraction in Figure 9, and the fit of the experimental data is very good. Similarly good simulations can be achieved in terms of all measured parameters, such as iron concentration, pH, solution potential, *etc.* (for more details see Dixon & Petersen, 2003).

In similar fashion, the 4-column experiment using native bacteria and raffinate solution can be modeled. Relative to the standard calibration only the microbial growth rate k_g was changed according to the results reflected in Table 3. Also adjusted was the feed concentration of iron, which in the raffinate was at 2.46 g/L

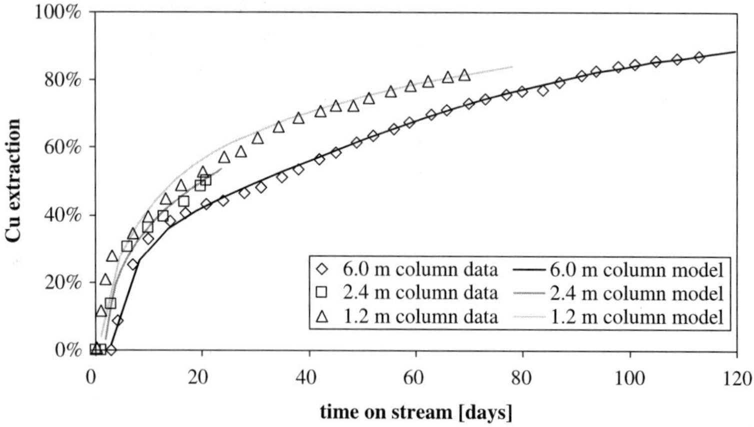

Figure 9. Simulations of 4-column, 8-column and tall column standard experiments

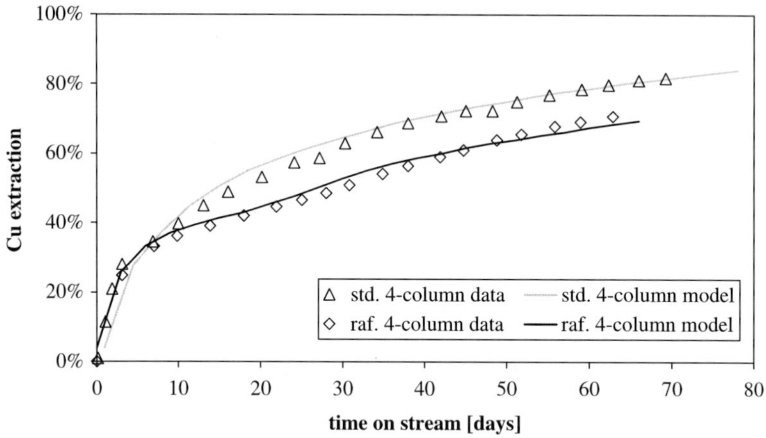

Figure 10. Simulation of 4-column standard and 4-column raffinate experiments

instead of 2.0 g/L in the standard experiments (Table 2). The result, again in terms of the total copper extraction, is shown in Figure 10, together with the experimental data. Again, the model fits the experimental data well over most of the run and clearly captures the retardation of copper extraction in the stressed experiment. Petersen & Dixon (2004) were also able to show that the model accurately reflected the lower microbial counts found, as shown in Figure 8.

As a next step the model was used to predict copper extraction achievable under full-scale operating conditions in a heap. Four scenarios have been modeled and are shown in Figure 11:

- 70 cm dripper spacing (R = 30 cm) and stressed bacteria (k_g = 0.02 h^{-1}) – this corresponds to the actual operating conditions at the mine site at the time of the investigation,
- 70 cm dripper spacing (R = 30 cm) and healthy bacteria (k_g = 0.1 h^{-1}),
- 50 cm dripper spacing (R = 20 cm) and stressed bacteria (k_g = 0.02 h^{-1}), and
- 50 cm dripper spacing (R = 20 cm) and healthy bacteria (k_g = 0.1 h^{-1}).

Here, 'stressed bacteria' refers to the combined effects of high salinity and potentially limited CO_2 supply on microbial growth. These effects are at present not described in the model explicitly, but accounted for through a reduction of the value of the microbial growth constant k_g.

For all simulations the heap was taken at 10 m height, the feed rate at 10 L/m^2/h, and the aeration rate at 0.25 m^3/m^2/h. Feed composition included 2.0 g/L of iron. Also, the model was adjusted to the geo-climatic conditions of the heap location and temperature was left to self-adjust (as opposed to imposed isothermal conditions when modeling column experiments).

Modeling the heap under actual operating conditions (long dripper spacing, stressed bacteria) predicts approximately 75% copper extraction within 400 days (13 months), which corresponds well with reported production figures at the time (Garcia *et al.*, 1999). At the same time this is considerably longer than any of the laboratory experiments reported above. The primary reason is that the large stagnant clusters between solution channels enforce a long diffusion path for the delivery of one key reagent – the acid needed for the bio-oxidation reaction (1). Petersen and Dixon (2003a) have shown that, as a consequence, the zone-wise chalcocite leaching described above proceeds 'sideways' in heaps, away from the solution channels into the stagnant clusters, as opposed to downwards as in columns.

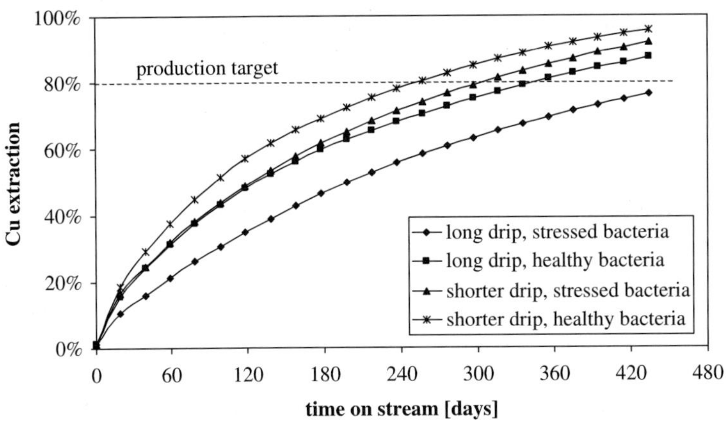

Figure 11. Simulation of full-scale heap scenarios

As the other simulations show, this effect is exacerbated by the retarded microbial growth. Tightening the dripper spacing does improve the overall leach performance, as does the use of a healthy microbial environment. This indicates that the two effects – diffusion through stagnant clusters and retarded microbial growth – are both rate-limiting on overall copper extraction to approximately the same extent. Reduced dripper spacing (around 50 cm) has now become standard practice in heap bioleach operations and reported copper recoveries have shown some improvement.

As Figure 11 indicates, more significant improvement could be made if the microbial health in the heap was also improved by reducing the levels of Mg and Al sulfates and/or chlorides in the circulating solution. Unpublished test work by Petersen has shown that selectively reducing the levels of either $MgSO_4$, $Al_2(SO_4)_3$ or NaCl in a simulated raffinate solution by only 20% improved microbial growth rates of the native culture in each case by a factor of 2-3. It is acknowledged, however, that achieving such a reduction at an industrial scale is not trivial and may in fact be impracticable. Hypothetically thus, the copper extraction rate from the industrial operation could be improved to achieve 80% in as little as 250 days, or in excess of 90% within one year.

6. CONCLUSIONS

Heap bioleaching of chalcocite ores is widely practiced and thus by far the most significant industrial application of bioleaching. Although it appears simple, the process is quite complex, involving sub-processes at the scales of mineral grains, micro-organisms, clusters of particles and the heap as a whole, which can interact in ways that are not always appreciated by intuition alone, and require a comprehensive modeling tool, such as HeapSim, to unravel. The column study on a typical ore material has shown that under optimized conditions high copper extractions should be achievable from heaps within a few months, even at ambient temperature. The fact that much lower extraction rates are the reality in industrial operations is determined by two main factors – poor reagent distribution (especially of acid) within the heap due to localized patterns of flowing and stagnant solution, and low microbial activity due to unfavorable solution chemical conditions. Limited CO_2 supply could also potentially retard microbial growth, but this has not yet been conclusively demonstrated.

A modeling study using the HeapSim tool has shown that these two effects have an overall rate limiting effect of roughly the same magnitude, at least in the context of the operation studied here. Therefore both aspects need to be optimized to achieve significant improvement in heap production. Investigation and optimization of microbial health in the context of heap bioleaching systems have received only limited attention from researchers so far. This is surprising, considering the economic significance of heap bioleaching, and should be addressed in future research efforts.

7. ACKNOWLEDGEMENT

The authors wish to acknowledge the generous financial support by Placer Dome Technical Services Ltd. for the work discussed in this chapter.

8. REFERENCES

Bennet CR, Cross M, Croft TN, Uhrie JL, Green CR, Gebhard JE. 2003. A comprehensive copper stockpile leach model: Background and model formulation. In: Young C, Alfantazi A, Anderson C, James A, Dreisinger DB, Harris B, eds. Hydrometallurgy 2003, Proceedings of the 5th International Symposium Honoring Professor Ian M. Ritchie. TMS Publishers, Warrendale, PA, vol. 1: Leaching and Solution Purification, 315-328.

Blight KR, Ralph DE. 2004. Effect of ionic strength on iron oxidation with batch cultures of chemolithtropic bacteria. Hydrometallurgy 73: 325-334.

Bolorunduro, SA. 1999. Kinetics of leaching of chalcocite in acid ferric sulphate media: Chemical and bacterial leaching. M.A.Sc. Thesis, University of British Columbia.

Boon M. 1996. Theoretical and experimental methods in the modeling of bio-oxidation kinetics of sulfide minerals. PhD Thesis, TU Delft, ISBN 90-9009825-9.

Bouffard SC. 2003. Understanding the Bio-oxidation of Sulfidic Refractory Gold Ores. PhD Thesis, University of British Columbia.

Breed AW, Hansford GS. 1999. Effect of pH on ferrous-iron oxidation kinetics of a predominantly *Leptospirillum ferrooxidans* culture. Biotechnol Bioeng 65:44-53.

Breed AW, Dempers CJN, Searby GE, Gardner MN, Rawlings DE, Hansford GS. 1999. The effect of temperature on the continuous ferrous-iron oxidation kinetics of a predominantly *Leptospirillum ferrooxidans* culture. Biochem Eng J 3:193-201.

Brierley CL. 1997. Mining biotechnology: Research to commercial development and beyond. In: Rawlings DE, ed. Biomining: Theory, Microbes and Industrial Processes. Springer-Verlag, Berlin, 3-17.

Demergasso C, Castillo D, Galleguillos P, Escudero LO, Casamayor E. 2003. Genetic identification and temporal dynamics of microorganisms from a demonstrative dump of a marginal copper sulfide bioleaching project. In: Riveros PA, Dixon DG, Dreisinger DB, Menacho J, eds. Hydrometallurgy of Copper, Proceedings of Copper 2003. CIM-MetSoc, Montreal, Vol. 6.

Dixon DG. 2003. "Heap leach modeling – the current state of the art". In: Young C, Alfantazi A, Anderson C, James A, Dreisinger DB, Harris B, eds. Hydrometallurgy 2003, Proceedings of the 5th International Symposium Honoring Professor Ian M. Ritchie. TMS Publishers, Warrendale, PA, vol. 1: Leaching and Solution Purification, 289–314.

Dixon DG, Petersen J. 2003. Comprehensive modelling study of chalcocite column and heap bioleaching. In: Riveros PA, Dixon DG, Dreisinger DB, Menacho J, eds. Hydrometallurgy of Copper, Proceedings of Copper 2003. CIM-MetSoc, Montreal, 493–516.

Dixon DG, Petersen J. 2004. Modeling the dynamics of heap bioleaching for process improvement and innovation. In: Hydro-Sulfides 2004, Proceedings of the International Colloquium on Hydrometallurgical Processing of Copper Sulfides, April 16 - 19; Santiago, Chile. 13–45.

Garcia C, Binvignat J, Roco J, Campos J. 1998. Bacterial activity in heap leaching using solutions containing very high levels of contaminants. In: Proceedings, Randol Copper Hydromet Roundtable'98, November 17 – 20, British Columbia. Golden CO: Randol International Ltd, Vancouver, 249–254.

Garcia C, Arias H, Campos J, Roco J, San Martin O, Whittaker J. 1999. Bioleaching at Zaldivar. In: Biomine'99 and Water Management in Metallurgical Operations – Conference Proceedings. 1999 August 23 – 24, Perth, Australia, Australian Mineral Foundation Inc., Glenside, Australia, 72-81.

Hawkes RB, Franzmann PD, Plumb JJ. 2006. Moderate thermophiles including *Ferroplasma cupricumulans* sp. nov. dominate an industrial-scale chalcocite heap bioleaching operation. Hydrometallurgy 83: 229-236.

Lamberg P. 1997. Visit to Zaldivar Cu mine. Outokumpu Research – Geoanalytical Laboratory; Report: 97003-GAL-T; Outokumpu, Finland, 30 January (unpublished).

Lawson EN, Nicholas CJ, Pellat H. 1995. The toxic effect of chloride ions on *Thiobacillus ferrooxidans*. In: Jerez JCA, Vargas T, Toledo H, Wiertz JV, eds. Biohydrometallurgical processing, Proceedings of the International Biohydrometallurgy Symposium IBS-95, Vina del Mar, Chile. University of Chile, Santiago de Chile, Vol. 1, 165-174.

Ogbonna N, Petersen J, Dixon DG. 2005. HeapSim – Unravelling the mathematics of heap bioleaching. In: Dry M, Dixon DG, eds. Computational Analysis in Hydrometallurgy, 35th Annual Hydrometallurgy Meeting. CIM-MetSoc, Montreal, 225–240.

Ojumu TV, Petersen J, Searby GE, Hansford GS. 2006. A review of rate equations proposed for microbial ferrous-iron oxidation with a view to application to heap bioleaching. Hydrometallurgy 83: 21-28.

Olson GJ, Brierley JA, Brierley CL. 2003. Bioleaching review part B: Progress in bioleaching: applications of microbial processes by the minerals industries. Appl Microbiol Biotechnol 63:249-257.

Petersen J, Dixon DG. 2003. The dynamics of chalcocite heap bioleaching. In: Young C, Alfantazi A, Anderson C, James A, Dreisinger DB, Harris B, eds. Hydrometallurgy 2003, Proceedings of the 5th International Symposium Honoring Professor Ian M. Ritchie. TMS Publishers, Warrendale, PA, vol. 1: Leaching and Solution Purification, 351-364.

Petersen J, Dixon DG. 2003b. A Modelling Study of the Dynamics of Sulfide Heap Leach Processes with a View to Improved and Novel Application. In: Lorenzen L, Bradshaw DJ, Aldrich C, Eksteen JJ, Wright M, Thom E, eds. Proceedings of the XXII International Mineral Processing Congress. SAIMM, Marshalltown, South Africa, Volume 3, 1231-1240.

Petersen J, Dixon DG. 2004. Bacterial growth and propagation in chalcocite heap bioleach scenarios. In: Tzesos M, Hatzikioseyan A, Remoundaki E, eds. Biohydrometallurgy – A Sustainable Technology in Evolution, Proceedings of the International Biohydrometallurgy Symposium IBS 2003, September 14 – 19, Athens, Greece, National Technical University of Athens, 65–74.

Petersen J, Dixon DG. 2006a. Modelling and Optimization of Heap Bioleach Processes. In: Rawlings DE, Johnson DB, eds. Biomining, Springer-Verlag, Berlin, 153-175.

Petersen J, Dixon DG. 2006b. Modelling and optimisation of zinc heap bioleaching. Hydrometallurgy, in press.

Rawlings DE, Dew D, du Plessis C. 2003. Biomineralization of metal-containing ores and concentrates. TRENDS Biotechnol 21: 38-44.

Ruan R, Wen J, Chen J. 2006. Bacterial heap-leaching: Practice in Zijinshan copper mine. Hydrometallurgy 83: 77-82.

Sand W, Gehrke T, Josza P-G, Schippers A. 2001. (Bio)chemistry of bacterial leaching – direct vs indirect bioleaching. Hydrometallurgy 51: 115-129.

Schnell HA. 1997. Bioleaching of copper. In: Rawlings DE, ed. Biomining: Theory, Microbes and Industrial Processes. Springer-Verlag, Berlin, 21-43.

Shiers DW, Blight KR, Ralph DE. 2005. Sodium sulphate and sodium chloride effects on batch culture of iron oxidising bacteria. Hydrometallurgy 80: 75-82.

Touvinen OH, Niemelä SI, Gyllenberg HG. 1971. Tolerance of *Thiobacillus ferrooxidans* to some metals. Antonie van Leeuwenhoek 37: 489-496.

van Scherpenzeel DA. 1996. The kinetics of ferrous-iron oxidation by Leptospirillum-like bacteria in absence and presence of pyrite and pyrite/arsenopyrite mixtures in continuous and batch cultures. MSc (Eng.) Thesis, University of Cape Town.

SECTION III

GENETICS AND MOLECULAR BIOLOGY

CHAPTER 11

THE USE OF BIOINFORMATICS AND GENOME BIOLOGY TO ADVANCE OUR UNDERSTANDING OF BIOLEACHING MICROORGANISMS

RAQUEL QUATRINI[1], JORGE VALDÉS[1], EUGENIA JEDLICKI[2] AND DAVID S. HOLMES[1]

[1] *Center of Bioinformatics and Genome Biology, Andrés Bello University (UNAB), Santiago, Chile, Life Science Foundation and Millennium Institute of Fundamental and Applied Biology, Santiago, Chile*

[2] *Program of Cellular and Molecular Biology, I.C.B.M., Faculty of Medicine, University of Chile, Santiago, Chile*

1. INTRODUCTION

Commercial biomineral recovery (bioleaching, biooxidation) of copper and gold is a cost effect and environmentally friendly alternative for metal recovery from low grade ores. The complete or nearly complete genome sequences of a number of bioleaching bacteria and archaea are publicly available including genomic information derived from an environmental sample. A study of these genomes and predicted metabolic and regulatory pathways is beginning to provide novel and exciting insights into the metabolism of these microorganisms and how they might work synergistically in tank reactors and in heap bioleaching operations to promote the recovery of metals.

The objective of this chapter is to provide a brief overview of recent progress in the areas of bioinformatics and genome biology as they relate to bioleaching microorganisms. After a short introduction to the principle themes of bioinformatics, metagenomics and genome biology, we will outline different cases of how these approaches have been exploited to reveal novel information. By deconvoluting metabolic potential through bioinformatic analysis new kinds of information can be revealed that not only connect vast amounts of data, but may also capture usable knowledge in the form of biologically valid relations that can subsequently be applied to biotechnological applications such as biomining.

Many questions regarding the biology of microorganisms can normally be addressed by a range of genetic and biochemical experiments. Unfortunately,

E.R. Donati and W. Sand (eds.), Microbial Processing of Metal Sulfides, 221–239.

A. ferrooxidans has proved recalcitrant to standard genetic manipulation and genetic analysis of other bioleaching microorganisms is only recently being developed. There is only one report of transformation in *A. ferrooxidans* and this may be strain specific and not of general use (Kusano *et al.*, 1992) and transduction is unknown. Only recently have techniques for conjugation been established and these remain difficult to control and are of low efficiency (Liu *et al.*, 2000; reviewed in Holmes & Bonnefoy, 2006). Exacerbating the problem is the difficulty of obtaining sufficient cell mass for many biochemical assays. Given these experimental hurdles, metabolic models derived from bioinformatic analyses offer an especially attractive starting point for unraveling the interesting physiology of bioleaching microorganisms.

2. INTRODUCTION TO BIOINFORMATICS, GENOME BIOLOGY, METAGENOMICS AND COMPARATIVE GENOMICS

For the purpose of this chapter bioinformatics is defined as the creation and use of computer algorithms to predict genes, proteins and metabolic pathways, principally in sequenced genomes. Genome biology is the application of these programs to understand the biology of the organism under study or how organisms collaborate in the environment. The distinction is fuzzy at most times but bioinformatics can be thought of as essentially a collection of analytical tools whereas genome biology is more the study of the corpus of information revealed by bioinformatic analyses. Several relevant areas of bioinformatics and genome biology are outlined in Figure1. Bioinformatic algorithms can predict potential genes in a newly sequenced genome and can suggest regulatory sites for these genes, including potential promoters and transcription factor binding sites. Other bioinformatic algorithms can suggest functions for about 60% of the genes in any genome. The power of bioinformatics is that it is capable of making thousands of predictions for any organism and can yield insight into gene function and metabolism in a way that goes far beyond what is possible using only conventional tools of genetics and molecular biology. However, it is what you do with the data that counts and there are bioinformatic applications that can store, compare, and analyze the voluminous quantities of data generated by the use of new technologies.

Another advantage of bioinformatics is that it can reveal answers to questions that are not formulated as hypotheses because of our ignorance of the system under study – we simply do not know enough to ask the right question sometimes – this is the so-called paradigm shift in biology that has taken place in the last ten years. This chapter will illustrate this point. However, bioinformatic predictions should be subjected to experimental validation whenever feasible. However, this is not always possible given the shear number of predictions that can be derived from a bioinformatics analysis and in this chapter we provide an example of how bioinformatics can save time and money and steer the laboratory scientist towards the validation and experimental investigation of certain key predictions while leaving aside, for the moment, other bioinformatic predictions that are more certain or, perhaps, of lesser interest.

Validation of bioinformatic predictions can be accomplished using standard tools of genetics and molecular biology such as RT-PCR (reverse transcription polymerase chain reaction), complementation of mutants in heterologous hosts, the use of spectroscopy coupled with the use of electron transport inhibitors, etc. However, the revolution in genomics has also spawned the need for high throughput analytical techniques such as the analysis of the transcription of all the genes of an organisms (transcriptome) by microarray analysis or all the proteins (proteome) by 2-D gel techniques or by mass spectrometry (Figure 1) and this chapter will also explore a few examples of the use of these techniques.

Bioinformatics is an effective tool for making preliminary inroads into how an otherwise uncharacterized organism functions. It is particularly powerful in cases where it is difficult to implement conventional genetic tools or where no such tools have yet been devised, such as is the case for most bioleaching microorganisms. Bioinformatics can open the black boxes that most bioleaching microorganisms represent to the biologist.

Other potent uses of bioinformatics are in the emerging fields of metagenomics and comparative genomics. Metagenomics is the culture-independent genomic analysis of microbial communities. The term is derived from the statistical

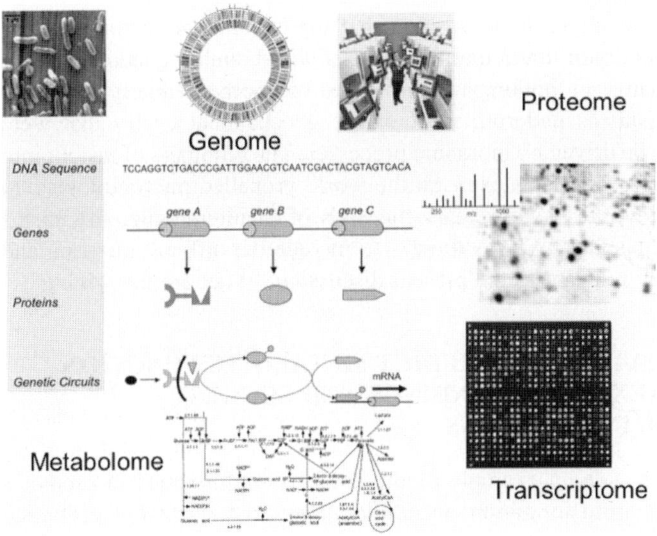

Figure 1. Whole genome sequencing projects (Genome) are enabling the identification of potential genes, proteins and regulatory pathways using a wide range of bioinformatics tools. Further analysis of gene transcription at the mRNA level by microarray analysis (Transcriptome) and proteins by mass spectrophotometry and 2-D gels (Proteome) can suggest metabolic pathways involved in the oxidation of iron and sulfur, microbe-metal interactions, heavy metal resistance and other properties of microorganisms involved in mineral bioleaching

concept of meta-analysis (the process of statistically combining separate analyses) and genomics (the comprehensive analysis of an organism's genetic material) (Schloss *et al.*, 2003). Comparatives genomics is, as its name suggests, the comparison of the genetic information and metabolism between different organisms in order to reveal fundamental insights into biological processes. For example, the comparison between pathogenic and non-pathogenic strains of *Escherichia coli* can suggest what genes and functions make the latter a health hazard. Similarly, a comparison between the iron-oxidizing *A. ferrooxidans* and the non iron-oxidizing *A. thiooxidans* could, in principle, suggest which genes of *A. ferrooxidans* confer on it the capacity to oxidize iron.

An exciting potential of metagenomics is to provide community-wide assessment of metabolic and biogeochemical function. Analysis of specific functions across all members of a community can generate integrated models about how organisms share the workload of maintaining the nutrient and energy budgets of the community. The models can then be tested with genetic and chemical approaches.

Microbiology has experienced a transformation during the last few years that has altered microbiologists' view of microorganisms and how to study them. The realization that most microorganisms cannot be grown readily in pure culture has forced microbiologists to question their belief that the microbial world had been conquered. They have been forced to replace this belief with an acknowledgment of the extent of ignorance about the range of metabolic and organismal diversity. This change fomented a revolution in microbiological thought. At the heart of this revolution was the convincing demonstration that the uncultured microbial world far outsized the cultured world and that this unseen world could be studied (Ward *et al.*, 2005). This change in thinking was prompted by another, equally important realization: microorganisms underpin most of the geochemical cycles that were previously thought to be driven by inorganic processes. The glimmers of insight into the influence that microorganisms exert on the world propelled microbiologists to pursue the uncultured world. In a few years, the study of uncultured microorganisms has expanded beyond asking "Who is there?" to include the difficult question and one that is particularly relevant to the present discussion "What are they doing?"

3. BIOINFORMATICS CAN PREDICT METABOLIC FUNCTION, REVEAL UNEXPECTED FINDINGS AND SUGGEST INDUSTRIAL INNOVATIONS

In this section, we provide an example of the use of bioinformatics to predict a metabolic function (biofilm formation) in *Acidithiobacillus ferrooxidans* in which an unanticipated result (the presence of a galactose uptake system) was revealed by the analysis and subsequently confirmed by experimental approaches. In addition, the chosen example illustrates the occasional failure of bioinformatics to detect a gene encoding an enzymatic activity for which there is experimental evidence of its existence (GalT). This section also shows how bioinformatic analysis can make predictions that might be of relevance to the industrial use of *A. ferrooxidans* in mineral recovery (the addition of galactose or other sugars might stimulate biofilm formation).

A knowledge of the fundamental physical and biological interactions between a bacterium and a mineral surface is central to understanding the intricacies of interfacial phenomena such as bacterial recognition and attachment to specific mineral surfaces and biofilm formation. To obtain its energy and electron requirements from the oxidation of various forms of reduced sulfur and ferrous iron, and so contribute to mineral dissolution, *A. ferrooxidans* must attach and firmly adhere to mineral surface. This is achieved through the production of an extracellular polysaccharide matrix in which the cells divide and eventually develop into complex biofilms.

Whereas the role of biofilm formation in metal solubilization has been actively evaluated in *A. ferrooxidans* (Gehrke *et al.*, 1998; Sand & Gehrke, 2001), few studies have addressed its biochemical and genetic basis. In this respect, bioinformatic analysis of the genome sequence of the type strain of *A. ferrooxidans* ATCC 23270 has proved particularly useful. The occurrence of several clusters of candidate genes encoding enzymes potentially involved in the metabolism of glucose-1-phosphate via the Leloir pathway and in the subsequent conversion to the extracellular polymeric substances (EPS) precursors (Barreto *et al.*, 2005a, 2005b) made it possible to: (1) model *in silico* the first steps in the biofilm formation process in *A ferrooxidans*, (2) suggested potential connections between galactose, EPS formation and general sugar management and (3) generated a number of unanticipated predictions (Figure 2).

Biofilm development is a complex process that follows a regulated sequence of stages involving: transient association with the surface, tight adhesion, aggregation of cells into microcolonies and subsequent growth and maturation (Stoodley *et al.*, 2002). The production and secretion of EPS by microorganisms is accepted as a key mechanism promoting irreversible cell attachment to surfaces and biofilm development. The building blocks in EPS biosynthesis are series of modified sugars (UDP-galactose, UDP glucose, dTDP rhamnose and GDP-mannose) which are synthesized from the simple carbohydrates glucose-1-phosphate (glucose-1-P) and glucose-6-phosphate (glucose-6-P) via three possible pathways: gluconeogenesis, glycogen utilization and the Leloir pathway.

The genes that encode the enzymes that are putatively involved in the biosynthesis of these sugar-nucleotides from glucose-1P, that is GalU (glucose 1-phosphate-pyrophosphorylase), GalE (UDP-glucose 4-epimerase) and the complex RfbaABCD (dTDP-rhamnose synthase), are all present in the genomic sequence of *A. ferrooxidans*. The Leloir pathway is involved in the conversion of galactose to glucose-1-P. Through this pathway galactose is phosphorylated by galactokinase (GalK) to yield galactose-1-P, which is then converted into glucose-1-P by hexose1-P uridyltransferase (GalT), UTP-glucose-1-P uridylyltransferase (GalU) and UDP-glucose epimerase (GalE). The resulting glucose-1-P can then enter the glycolytic pathway or yield the precursor needed for EPS biosynthesis.

Also, a supplementary function catalysed by the enzyme aldose 1-epimerase (GalM) feeds α-galactose into the pathway through the interconversion of the α-and

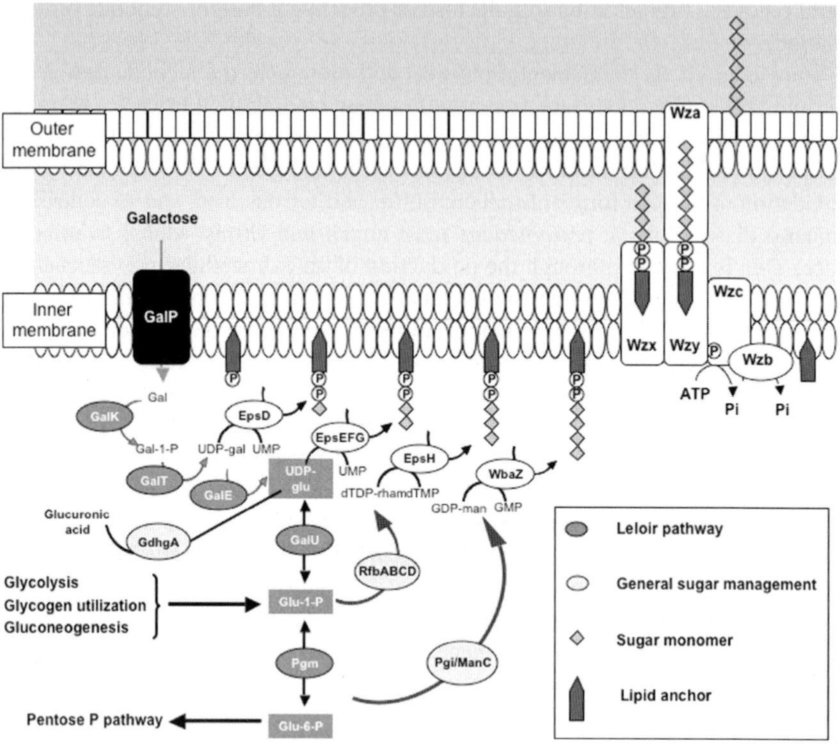

Figure 2. Bioinformatic prediction, with some supporting experimental evidence, for the enzymes and pathways involved in extra-cellular polysaccharides in *A. ferrooxidans*

β-galactose. Candidate genes *galK* and *galM* were also found in the *A. ferrooxidans* genome and were shown to be organized as part of an operon (*gal* operon) together with *galE*, resembling other well-described *gal* operons (Grossiord *et al.*, 2003). As in other genomes the ortholog of *galU* is transcribed from a discrete an independent promoter. All these genes proved to complement the respective *E. coli* mutants supporting the proposed function of the *gal* operon *A. ferrooxidans* (Barreto *et al.*, 2005a). Inspite of some variation in gene composition and gene order (e.g.: *galPgalMTKE* in *L. lactis; galETKMpgm* in *E. coli; galKETRM* in *L. casei* and *galETK* in *K. pneumoniae*), the *gal* operon typically includes a *galT* ortholog. Interestingly, an orthologous *galT* gene could not be detected in the genome sequence of *A. ferrooxidans* by bioinformatic analysis, while an enzymatic activity complementing the growth of a *galT* mutant of *E. coli* (S491) was experimentally detected (Barreto *et al.*, 2005a). The associated gene proved to be distinct from other *galT* orthologs.

In most bacteria where the Leloir pathway is functional, galactose enters the cell via GalP permease. The presence of a putative gene (*galP*) potentially encoding a permease in the *A. ferrooxidans* genome with significant similarity to other known galactose importers was, however, unexpected. The functional proficiency and specific role of this permease seemed questionable given that the organism is considered a strict autotroph and that small sugars are known to inhibit its growth. However, gene expression data demonstrated that *galP* is expressed at the level of RNA (both in iron and in sulfur growing cultures) and that transcription is higher in cells grown in iron in the presence of galactose. Also, ^{14}C-galactose uptake experiments during growth in the presence of sulfur showed that galactose can be taken up by *A. ferrooxidans* and incorporated into EPS (Barreto *et al.*, 2005a, 2005b). Furthermore, the addition of galactose to the growth medium of *A. ferrooxidans*, even though it did not promote growth, stimulated the rate and extent of attachment of cells to pyrite, the amount of EPS formed and the extent of biofilm formation in laboratory conditions (Barreto *et al.*, 2005a,b). This illustrates the paradigm shift in Biology where the powerful new tools of bioinformatics and genome biology allowed the investigators to detect a possible galactose permease where conventional wisdom suggested that none should exist.

The following step in EPS formation is the polymerization of modified sugars on a lipid anchor that is situated in the inner membrane (Broadbent *et al.*, 2003). *A. ferrooxidans* has a candidate operon that potentially encodes for the undecaprenyl pyrophosphate synthetase UppS and the associated anchor formation genes *cdsA*, *dxr* and *pirH*. The modified sugars are attached onto the lipid anchor and polymerized forming chains of sugars of varying length and composition by a variety of glycosyltransferases (GTFs). *A. ferrooxidans* encodes a significant number of different glycosyltransferases, including *epsDEFGH* and *wbaZ*, that could confer a wide range of sugar and linkage specificities. The repeating sugar units of the EPS assembled onto the lipid anchor are predicted to be exported from the cytoplasm to the outer surface of the outer membrane via the Wz protein complex. The activity of these genes has now been validated by protein expression analysis for *A. ferrooxidans* grown on sulfur (Valenzuela *et al.*, 2006).

As in other bacteria, many of the predicted genes involved in EPS biosynthesis of the repeat unit, polymerization and export in *A. ferrooxidans* are clustered in operons. Many of these operons contain genes with redundant functions. This aspect of gene organization is likely to enable the bacteria to exploit different ecological niches and to respond to different environmental stimuli through differential regulation. In fact, the nature of the solid substrate is known to influence the chemical composition of the exopolymers and the mode of adhesion of *A. ferrooxidans* (Gehrke *et al.*, 1998). While sulfur-grown cells exhibit purely hydrophobic surface properties and do not attach to pyrite, positively charged exopolymer-complexed iron(III) ions allow the electrochemical interaction of iron grown *A. ferrooxidans* cells with the negatively charged surface of pyrite particles. Modified glucuronic acid is known to be an important component of EPS in iron grown cells (Schippers & Sand, 1999). Candidate *gdhgA* gene, that potentially encodes an enzyme for the conversion of UDP-glucose to UDP-D glucuronate, is embeded in one of the EPS related operons of *A. ferrooxidans*, the specific activation of which may account for the observation pointed above. Microarray and RT/Q-PCR data support the contention that *A. ferrooxidans* may also be able to differentially regulate its EPS-related operons in response to the nature of the substrate (unpublished results).

Several genes of unknown function were present in the Gal- and EPS-related operons of *A. ferrooxidans*. From a bioinformatics perspective knowing the genetic context of a gene with unknown function is important because it can assist in assigning to it a putative function. An example is the case of an ORF encoding a *luxA-like* gene within the *gal* operon of *A. ferrooxidans*. The predicted gene product exhibits similarity with the LuxA family of bacterial luciferase-like monooxygenases (pfam00296) and with F420-dependent oxidoreductases (COG2141), of which glucose-6-P dehydrogenase from *Rhizobium sp.* is a member (Streit *et al.*, 2004). The similarity to a glucose-6-P dehydrogenase, that catalyzes the conversion of glucose-6-P to 6-phosphogluconolactone, could pinpoint the connection between the glucose-1-P pathway (Leloir pathway) mediated by the *gal* genes and the pentose phosphate pathway. The presence of a *pgm* ortholog (phosphoglucomutase) in the *gal* operon of *A. ferrooxidans* whose role is the conversion of glucose-1-P, the central precursor in nucleotide sugar biosynthesis, to glucose-6-P, an intermediate in sugar breakdown and a major point of entry of carbon into the pentose phosphate pathway, further supports this contention. Glucose-6-P could then be converted to 6-phosphogluconolactone by the hypothetical product of the *luxA*-like gene. Phosphoglucomutase has been shown to play a key role in controlling the flux through the Leloir pathway in yeast, probably due to increased conversion of glucose-1-P to glucose-6-P (Bro *et al.*, 2005). Additionally, if LuxA-like exhibited reverse activity, it could catalyze the conversion of 6-phosphogluconolactone to glucose-6-phosphate, and then it could help channel products of CO_2 fixation towards the formation of EPS precursors.

Bioinformatic reconstruction thus suggests that galactose can be fed to glucose-1-P via the Leloir pathway and diverted towards the formation of EPS precursors, although alternative sources of glucose-1-P such as via gluconeogenesis or glycogen utilization might be called upon to synthesize biofilms when needed. Several aspects of these predictions have been supported by microarray transcript profiling (Appia-Ayme *et al.*, 2006) What actually happens in natural environments, where the source and percentage of the galactose that is available and ends up in biofilms is uncertain, remains to be evaluated. A possible source of environmental galactose, however, might be the assorted heterotrophic microorganisms long known to be associated with *A. ferrooxidans* in bioleaching operations (Bacelar-Nicolau & Johnson, 1999; Marchand *et al.*, 2002).

These considerations raise the question as to whether the addition of galactose, or compounds that contain galactose, could enhance the rate and/or extent of mineral leaching in an industrial operation. The addition of galactose to laboratory cultures promotes the formation of EPS and biofilms (Barreto *et al.*, 2005b), but whether this would occur in an industrial setting and whether this would result in faster rates and/or higher yields of metal recovery are issues still to be explored. Another problem is how the formation of biofilms by *A. ferrooxidans* is regulated. The discovery of a Lux-like quorum sensing system in this microorganism might help explain how it senses cell density in its environment (Rivas *et al.*, 2005; Farah *et al.*, 2005).

4. BIOINFORMATICS CAN HELP FOCUS THE EXPERIMENTAL BIOLOGIST AND CAN PREDICT GENETIC REGULATORY CONNECTIONS

This section provides an example of how bioinformatics can be used to help direct the experimental biologist towards challenging and interesting biological questions and, in this way, can save time and money. It also illustrates the power of bioinformatics to predict genetic regulatory connections.

Prior to the availability of the *A. ferrooxidans* genome sequence, only one gene involved in sulfur assimilation, encoding a possible ATP sulfurylase/kinase activity, had been described in this microorganism (Fry & Garcia, 1989). However, after a bioinformatic analysis of its genome sequence, a much more comprehensive picture of the genes and pathways potentially involved in sulfur (sulfate) uptake and assimilation began to emerge (Valdes *et al.*, 2003). This study also suggested potential interconnections of sulfur metabolism with other core processes such as nitrogen fixation, hydrogen utilization, biofilm formation and amino acid metabolism (Figure 3).

The predicted enzymes that convert sulfate to cysteine in *A. ferrooxidans* exhibit significant similarity with genes in other organisms known to carry out this conversion (Leyh *et al.*, 1998). Further, the potential genes encoding these enzymes are organized in a gene cluster (*cysIJHDN*) that is probably an operon. These bioinformatic predictions are relatively sound and their experimental validation can be placed on a lower priority compared, for example, with the much weaker, but perhaps more interesting, bioinformatic prediction of the link between sulfate and biofilm formation via PAPS or the connection between nitrogen fixation and hydrogen utilization (Figure 3). These predictions could help explain how *A. ferrooxidans* attaches to sulfur minerals and why its genome contains hydrogenases.

Bioinformatics can also be used to infer gene regulation. Regulation of genes and metabolic pathways are important processes to understand because they connect cellular processes with external signals from the environment, which can modulate the function of the microorganism promoting or reducing its capacity to bioleach minerals. In the case of sulfur assimilation in *A. ferrooxidans,* an analysis of the predicted genes and pathways involved suggests that sulfate uptake and assimilation can be controlled in this microorganism by processes that are similar to those experimentally validated in other organisms.

For example, it can be predicted that, if sulfur is channeled by IscS to form Fe-S centers in proteins required for nitrogen fixation, then an increased expression of CysE will promote the formation of O-acetyl-L-serine (Figure 3). This latter compound can spontaneously isomerizes to N-acetyl-serine which is a co-factor

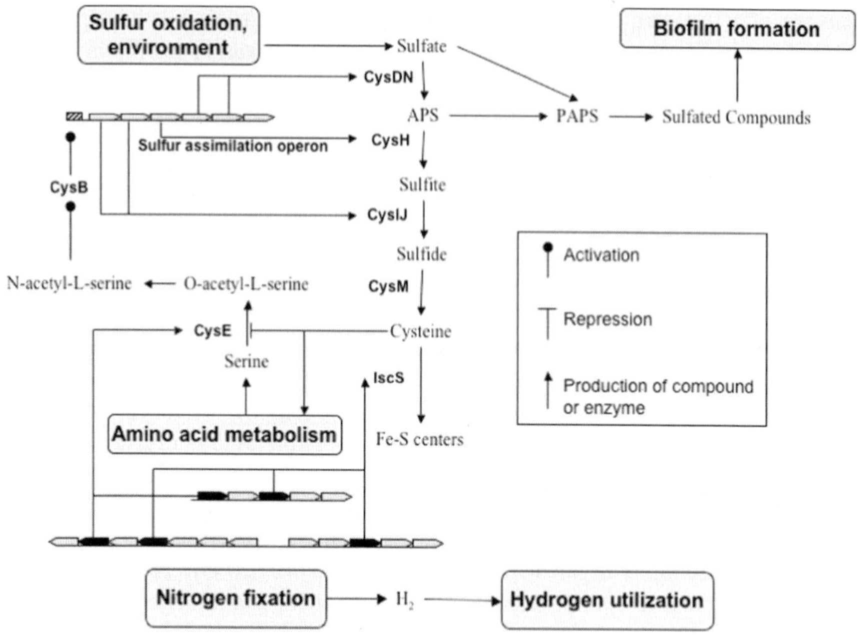

Figure 3. Bioinformatic predictions for sulfate uptake and assimilation in *A. ferrooxidans* with suggested connections to amino acid metabolism, nitrogen fixation and hydrogen utilization (modified from Valdes *et al.*, 2003)

for the positive gene regulator CysB (Lochowska *et al.*, 2001). CysB coupled to N-acetyl-serine is predicted to activate the sulfur assimilation operon which will stimulate the uptake of sulfate from the environment, restoring intracellular levels of sulfur depleted by the need to form Fe-S proteins. Conversely, excess cysteine in the cell limits the amount of O-acetyl-L-serine for CysB activation. A reduction in CysB activity reduces the expression of the sulfate uptake operon lowering, in turn, the assimilation of sulfate until intracellular cysteine levels are restored – a beautiful example of a homeostatic process predicted to be at work in *A. ferrooxidans*.

Recent microarray experiments that compare gene transcription profiles of *A. ferrooxidans* grown in either iron or sulfur medium support this model (Quatrini *et al.*, 2006). Genes proposed to be part of the sulfur assimilation operon *cysIJHDN* are much more expressed in cells grown in iron medium compared to sulfur medium, an observation that is consistent with the idea that sulfate is more limiting in the iron medium.

The challenge now is to integrate the regulation of sulfur assimilation for biosynthetic processes such as amino acid formation and the production of Fe-S centers with its use as an energy and electron source. The regulation of this latter process is not well understood in any organism, making it a difficult, but not necessarily impossible, task for bioinformatics analysis.

5. BIOINFORMATICS CAN MODEL FULL METABOLIC RESPONSES AND PINPOINT UNUSUAL METABOLIC FEATURES THAT ARE LIKELY TO MAKE FUNCTIONAL DIFFERENCES

This section demonstrates the power of bioinformatics to make predictions not only about genes, proteins and pathways but also how it can suggest global regulatory networks that can integrate major cellular responses, in this case, iron uptake and homeostasis. This example also serves to illustrate how an initial question, how does *A. ferrooxidans* regulate its iron metabolism, can lead to new directions that had not been anticipated. It is shown that *A. ferrooxidans* has an unexpected number of iron uptake systems that led to the suggestion, subsequently validated by experiment, that the microorganism can grow at near neutral pH (unpublished results). The discovery of an extensive repertoire of mechanisms for iron uptake also suggests a possible reason for the sensitivity of *A. ferrooxidans* to high ferric iron loads that limit its ability to bioleach minerals.

A. ferrooxidans grows aerobically at extremely acidic pHs (pH 1-2) in environmental situations where it is confronted with high concentrations of iron. Not only is Fe(II) more stable in these conditions (than at neutral pH in aerobic conditions) but Fe(III) is also much more soluble (>0.1 M), exceeding by seven orders of magnitude typical bacterial iron requirements (10-8 M). Furthermore, *A. ferrooxidans* uses FeII as an energy and electron source, thus it is presented with a dilemma: how to benefit from the high iron concentrations that the environment offers fulfilling its energy and micronutrient requirements, whilst simultaneously escaping the potentially harmful effects of eventually high intracellular iron loads? How does the microorganism regulate the use of iron as an energy source versus its need as a micronutrient? These aspects intuitively suggested that *A. ferrooxidans* could exhibit novel mechanisms to cope with iron homeostasis, to ensure a tight homeostatic control and that it might have developed iron uptake mechanisms quite distinct from those of neutrophilic organisms who are faced with a limited iron supply.

Contrary to this presumption *A. ferrooxidans* bioinformatic analysis revealed the presence of a number of potential genes and regulatory pathways involved in iron uptake typical of those found in neutrophilic organisms. Given the presence of such genes, the iron homeostatic response in *A. ferrooxidans* was modeled *in silico* based upon known schemes in other organisms (Figure 4). Through this strategy, several unusual genetic features were revealed that are thought to reflect special requirements for iron uptake and homeostasis in *A. ferrooxidans*.

Bioinformatic analysis revealed the presence of candidate genes potentially encoding two distinct Fe(II) uptake systems, FeoAB and MntH, that may enable *A. ferrooxidans* to take advantage of the readily bioavailable Fe(II) through direct uptake. As in other microorganisms, (where instead FeoB contributes to

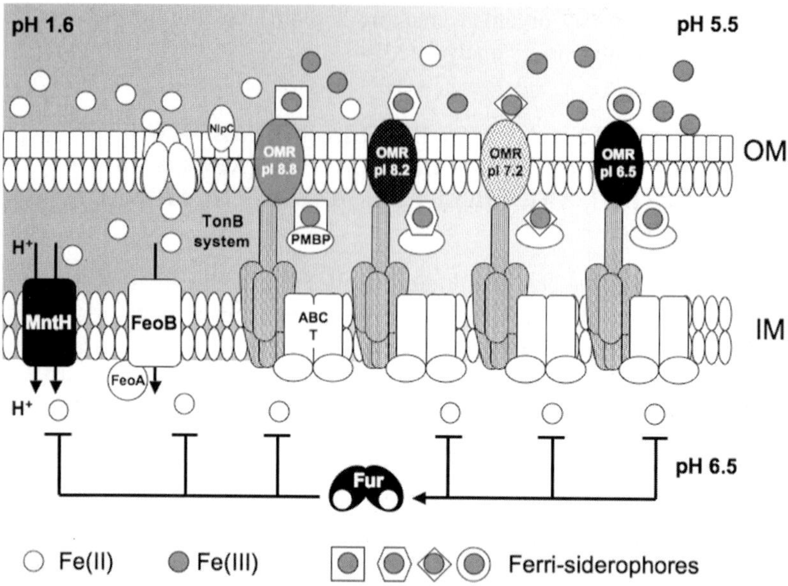

Figure 4. Model of iron uptake and homeostatic mechanisms in *A. ferrooxidans* predicted by bioinformatics analysis. OM, outer membrane; IM, inner membrane; OMR, outer membrane receptor

iron uptake under anaerobic or mildly acidic microaerofilic conditions), the *feoB* gene in *A. ferrooxidans* is organized in operon with a small open reading frame (ORF) termed *feoA* However, it has additionally captured in the same locus a predicted gene encoding a porin-like protein (*porA*). This porin exhibits similarity to OprB (COG3659, pfam04966) which has been implicated in the movement of carbohydrates across the outer membrane but that can also transport other ions (Wylie *et al.*, 1993), suggesting that it may contribute to the uptake of Fe(II). The presence of a Fur box-like sequence upstream of this porin suggests a typical Fur mediated iron-dependent repression of the whole gene cluster (Quatrini *et al.*, 2005a, 2005b). A conserved organization in *Geobacter sulfurreducens* further supports this hypothesis (Rodionov *et al.*, 2004).

The presence of 11 distinct candidate TonB-dependent Fe(III) siderophore outer membrane receptors (OMRs) for high-affinity acquisition of iron in the ferric valence state, came as a surprising finding of the *in silico* metabolic reconstruction. This number rivals or exceeds the complexity found in well-studied neutrophilic bacteria that must scavenge iron from their environment (Quatrini *et al.*, 2005b). Such a finding argues against the simplistic view that *A. ferrooxidans* has an easy time encountering and taking up readily available soluble iron. It is possible that this complexity reflects the capacity of *A. ferrooxidans* to live at higher pHs than those typically associated with its growth in ferrous sulfate (pH 2) and sulfur (pH 3.5), in which it must scavenge less soluble forms of iron. Demonstrations of the capacity of *A. ferrooxidans* to live at pH 4 in sulfur (Pronk *et al.*, 1991; Vian *et al.*, 1986) and, more recently, at pH 5.5 (Barreto *et al.*, 2005a, 2005b) support this point of view. Furthermore, the diverse range of specificities and pIs encountered for *A. ferrooxidans* OMRs supports the contention that different systems are functioning at different pHs and taking up different iron sources. Consistent with this view, it has been found that hydroxamate-type siderophores are capable of forming stable iron complexes at low pH, while iron is easily dissociated from catechol-type siderophores under these conditions (Payne, 1994).

Unlike many bacteria that take up iron via siderophores produced and excreted into the environment by themselves, *A. ferrooxidans* does not carry genes that might be involved in conventional siderophore production. Nevertheless, the fact that it has receptors for dicitrate and heterologous, hydroxamate- and catechol-type siderophores immediately suggests that it can compete efficiently for Fe(III) with siderophore producing organisms within bioleaching consortia or within biofilm forming bacteria sharing common environmental niches (Gonzalez-Toril *et al.*, 2003). The very same feature, might help to explain why *Leptospirillum* strains outgrow *A. ferrooxidans* and dominate the microbial population in the presence of high concentrations of Fe(III) (Rawlings *et al.*, 1999). The fact that *A. ferrooxidans* possesses more predicted Fe(III) uptake transporters than *Leptospirillum* sp. and thus a different inherent ability to take up Fe(III) could render it more susceptible to higher iron concentrations. In this way the high concentrations of Fe(III) evolved from Fe(II) biooxidation would confer a selective advantage to *Leptospirillum* strains over *A. ferrooxidans* (Quatrini *et al.*, 2005b).

Taken together this data point out the great complexity and diversity of iron uptake systems potentially expressed by *A. ferrooxidans* that are probably an accurate reflection of its metabolic plasticity and its ability to survive under significantly different environmental conditions.

6. METAGENOMICS AND COMPARATIVE GENOMICS

This section illustrates how metagenomics is being used to determine the community structure of bioleaching operations. It also provides examples of how a metagenomic project can generate novel insights into microbial community interspecies cooperation and how it can suggest ways to culture previously uncultivated microorganisms. How comparative genomics can reveal information regarding the diversity of microbial iron oxidation mechanisms has been recently reviewed in brief (Holmes & Bonnefoy, 2006).

Metagenomics can be used to address the challenge of studying prokaryotes in the environment that are, as yet, unculturable and which represent more than 99% of the organisms in some environments. This approach builds on recent advances in microbial genomics and in the polymerase chain reaction (PCR) amplification and cloning of genes that share sequence similarity (e.g. 16S rRNA, *nif*, *recA*) directly from environmental samples. Such samples can then be analyzed by techniques such as Denaturing Gradient Gel Electrophoresis (DGGE) and Restriction Fragment

Length Polymorphism (RFLP) gel electrophoresis providing information as to the species diversity in a sample. Alternatively, the sequences can be labeled with a fluorescent marker and used to locate and, in some cases quantitate, the distribution of species in an environmental sample by FISH (fluorescent in situ hybridization) (Handelsman, 2004).

The need to use non-cultivatable techniques to enumerate microorganisms in heap bioleaching operations was first suggested 20 years ago (Yates & Holmes, 1986a, 1986b) and an overview of the process was later published (Holmes, 1991). However, these initial probe approaches were used to enumerate bacteria in cultivation and it was not until recently that the techniques were used to explore the microbial diversity of bioleaching operations and natural communities.

An improved understanding of the microbiology of bioleaching heaps has been identified as key to advancing commercial bioheap operations (Brierley, 2001). However, despite advances in our understanding of the microbiology of stirred tank biooxidation of gold (Rawlings, 2002) and the microbial constitution of some environmental situations (Johnson, 1998; Gonzalez-Toril *et al.*, 2003; Tyson *et al.*, 2004), the microbial ecology of copper bioleaching heaps is still poorly comprehended. Early work, involving counts of live, cultivatable microorganisms and indirect measurements such as oxygen uptake, redox potential, pH, etc., provided an initial description of the bulk biological activity in bioleaching heaps (Brierley, 2001). Additional information has come from the laboratory cultivation of pure and mixed cultures of microorganisms associated with industrial-scale heap leaching processes. This has generated information regarding the identification and properties of *Acidithiobacillus* spp., *Sulfurisphaera*- like, *Leptospirillum spp., Ferroplasma spp.* and *Sulfobacillus*-like microorganisms (Goebel & Stackebrandt, 1994; Rawlings, 2002; Olson *et al.*, 2003; reviewed in Holmes & Bonnefoy, 2006).

More recently, culture-independent approaches based on PCR amplification and denaturing gradient gel electrophoresis (DGGE) and sequencing of 16S rRNA gene fragments from both Bacteria and Archaea have been used to analyze the microbial community inhabiting a low-grade copper sulfide run-of-mine (ROM) bioleaching test heaps (Demergasso *et al.*, 2005a, 2005b; Coram-Uliana *et al.*, 2005). Three important conclusions can be drawn from these studies. First, although it has proved difficult, due to contamination from rock particles, it is possible to isolate DNA of reasonable quality directly from bioleaching heaps without the intervention of bacterial culturing. This prerequisite is essential for any subsequent metagenomic analysis.

Second, it was shown that heap bioleaching, like tank biooxidation (reviewed in Rawlings, 2005) and heap bioleaching (Demergasso *et al.*, 2005a) proceeds in three stages. These stages result from temperature increases due to exothermic oxidation of iron and sulfur: an early stage favoring mesophilic microorganisms (30-40^{0}C) such as *A. ferrooxidans, A. thiooxidans*-like bacteria and Sulfurisphaera-like archaea; a second stage when the temperature begins to rise (40-55^{0}C) when *A. caldus, Leptospirillum* and *Ferroplasma* groups become dominant and a final stage (55-65^{0}C or higher) where *Sulfobacillus*-like bacteria such as

Alicyclobacillus (formerly *Sulfobacillus*, Karavaiko *et al.*, 2005) became dominant and archaea such as *Ferroplasma* thrive. This means that the development and interaction of each of these microbial communities, including possible community biofilm formation in the case of heap bioleaching, must be considered in order to fully understand bioleaching processes and suggest ways by which they can be improved.

Third, the microbial diversity in a bioleaching heap, although not as limited as the Richmond Iron Mountain acid mine drainage biofilm (Tyson *et al.*, 2004), is probably not as complex as naturally occurring acidic environments such as the Rio Tinto where thousands of years of natural selection have promoted a quite diverse ecosystem (Lopez-Archilla *et al.*, 2001). From a practical standpoint, this is important because the restricted diversity of microorganisms suggest that a meaningful interpretation of community structure and function of a bioleaching heap can be derived from a metagenomics analysis.

The best example of a metagenomic project that concerns bioleaching microorganisms, or close relatives of them, is the nearly complete sequencing of the metagenome of a biofilm community in acid drainage of the Richmond Iron Mountain mine (Tyson *et al.*, 2004). This biofilm community is very limited in diversity with *Leptospirillum* group II microorganisms accounting for 75%, *Leptospirillum* group III for 10%, Eukaryotic species for 4% and Archaea 1% (including *Ferroplasma* sp.) of the abundance of organisms respectively. One reason for the reduced diversity could be the very restrictive growth conditions imposed by the low pH (<1). Also, the biofilm was located deep in a mine shaft far from the activity of phototrophs who probably supply most of the fixed carbon in environments such as the Rio Tinto. Carbon limitation will select for chemoautotrophs and reduce the biodiversity possible. Nitrogen limitation, as will be described below, also probably restricts biodiversity.

The metagenomic sequence of Iron Mountain challenged a number of significant hypotheses. First, it appears that *Leptospirillum* group III contains genes with similarity to those known to be involved in nitrogen fixation, suggesting that it provides the community with fixed nitrogen. This was a surprise because the previous supposition was that a numerically dominant member of the community, such as *Leptospirillum* group II, would be responsible for nitrogen fixation. However, no genes for nitrogen fixation were found in the *Leptospirillum* group II genome, leading the authors to suggest that the group III organism, now identified as *L. ferrodiazotrophum* (Tyson *et al.*, 2005) is a keystone species that has a low numerical representation but provides a service that is essential to community function. The identification of *L. ferrodiazotrophum* as the sole nitrogen fixer in the Iron Mountain community also permitted the development of a selective procedure for its isolation and cultivation as a pure strain highlighting how environmental sequence data can provide insights for culturing previously uncultured microorganisms (Tyson *et al.*, 2005).

Furthermore, the prevailing idea that *Ferroplasma* strains, including those found at Iron Mountain, can fix CO_2 has been challenged (Dopson *et al.*, 2005). If it turns out that they are organomixotrophs incapable of fixing CO_2 then some other member of the Iron Mountain community, such as *Leptospirillum*, must be providing the fixed carbon.

Finally, a metaproteome analysis of the Iron Mountain biofilm community, in which proteins were isolated directly from the environment, indicates that a cytochrome predicted to be involved in iron oxidation is a major component of the community proteome (Ram *et al.*, 2005). This protein is encoded by Group II *Leptospirillum* microorganisms that are the dominant community members. Combined metagenomic-metaproteomic approaches are beginning to paint a preliminary picture of autotrophic life at low pH and are likely in the future to yield information regarding the synergistic interactions of microorganisms that will prove useful for improving bioleaching applications.

7. ACKNOWLEDGEMENTS

Work supported in part by Fondecyt 1050063.

8. REFERENCES

Appia-Ayme C, Quatrini R, Denis Y, Denizot F, Silver S, Roberto F, Veloso F, Valdes J, Cárdenas JP, Esparza M, Orellana O, Jedlicki E, Bonnefoy V, Holmes DS. 2006. Microarray and bioinformatic analyses suggest models for carbon metabolism in the autotroph *Acidithiobacillus ferrooxidans*. Hydrometallurgy 83: 273-280.

Bacelar-Nicolau P, Johnson B. 1999. Leaching of pyrite by acidophilic heterotrophic iron-oxidizing bacteria in pure and mixed cultures. Appl Environ Microbiol 65: 585-590.

Barreto M, Quatrini R, Bueno S, Arriagada C, Valdes J, Silver S, Jedlicki E, Holmes D. 2003. Aspects of the predicted physiology of *Acidithiobacillus ferrooxidans* deduced from an analysis of its partial genome sequence. Hydrometallurgy 71: 97-105.

Barreto M, Jedlicki E, Holmes DS. 2005a. Identification of a Gene Cluster for the Formation of Extracellular Polysaccharide Precursors in the Chemolithoautotroph *Acidithiobacillus ferrooxidans*. Appl Environ Microbiol 71: 2902-2909.

Barreto M, Gehrke T, Harneit K, Sand W, Jedlicki E, Holmes DS. 2005b. Unexpected Insights into Biofilm Formation by *Acidithiobacillus ferrooxidans* Revealed by Genome Analysis and Experimental Approaches. In: Harrison STL, Rawlings DE, Petersen J, eds. Proceedings of the 16th International Biohydrometallurgy Symposium, September 25–29, Cape Town, South Africa. Produced by Compress www.compress. co.za, 817-825.

Brierley C. 2001. Bacterial succession in bioheap leaching. Hydrometallurgy 59: 249-255.

Bro C, Knudsen S, Regenberg B, Olsson L, Nielsen J. 2005. Improvement of Galactose Uptake in *Saccharomyces cerevisiae* through Overexpression of Phosphoglucomutase: Example of Transcript Analysis as a Tool in Inverse Metabolic Engineering. Appl Environ Microbiol. 71: 6465-6472.

Broadbent JR, McMahon D, Welker DL, Ober CJ, Moineau S. 2003. Biochemistry, genetics, and applications of exopolysaccharide production in *Streptococcus thermophilus*: a review. J Dairy Sci 86: 407-423.

Coram-Uliana JN, van Hille RP, Kohr WJ, Harrison STL. 2005. Development Of A Method To Assay The Microbial Population In Heap Bioleaching Operations. In: Harrison STL, Rawlings DE, Petersen J, eds. Proceedings of the 16th International Biohydrometallurgy Symposium, September 25–29, Cape Town, South Africa. Produced by Compress www.compress. co.za, 647-656.

Demergasso CS, Galleguillos PA, Escudero LV, Zepeda VJ, Castillo D, Casamayor EO. 2005a. Molecular characterization of microbial populations in a low-grade copper ore bioleaching test heap. Hydrometallurgy, in press.

Demergasso CS, Echeverria A, Escudero L, Galleguillos P, Zepeda V, Castillo D. 2005b. Comparison of fluorescent in situ hybridization (FISH) and catalyzed reporter deposition (Card-Fish) for visualization and enumeration of Archaea and Bacteria Ratio in industrial heap bioleaching operations. In: Harrison STL, Rawlings DE, Petersen J, eds. Proceedings of the 16th International Biohydrometallurgy Symposium, September 25–29, Cape Town, South Africa. Produced by Compress www.compress.co.za, 843-851.

Dopson M, Baker-Austin C, Bond PL. 2005. Analysis of differential protein expression during growth states of *Ferroplasma* strains and insights into electron transport for iron oxidation. Microbiology 151: 4127-4137.

Farah C, Vera M, Morin D, Haras D, Jerez CA, Guiliani N. 2005. Evidence for a functional quorum-sensing type AI-1 system in the extremophilic bacterium *Acidithiobacillus ferrooxidans*. Appl Environ Microbiol. 71: 7033-7040.

Fry IJ, Garcia E. 1989. Cloning and characterization of *Thiobacillus ferrooxidans* genes involved in sulfur assimilation. In: Salley J, McCready RGL, Wichlacz PL, eds. Biohydrometallurgy, CANMET, Ottawa, 171-185.

Gehrke T, Telegdi J, Thierry D, Sand W. 1998. Importance of extracellular polymeric substances from *Thiobacillus ferrooxidans* for bioleaching. Appl Environ Microbiol 64: 2743-2747.

Goebel BM, Stackebrandt E. 1994. Cultural and phylogenetic analysis of mixed microbial populations found in natural and commercial bioleaching environments. Appl Environ Microbiol 60: 1614-1621.

Gonzalez-Toril E, Llobet-Brossa E, Casamayor EO, Amann R, Amils R. 2003. Microbial ecology of an extreme acidic environment, the Tinto River. Appl Environ Microbiol 69: 4853-4865.

Grossiord BP, Luesink EJ, Vaughan EE, Arnaud A, de Vos VM. 2003. Characterization, expression, and mutation of the *Lactococcus lactis galPMKTE* genes, involved in galactose utilization via the Leloir pathway. J Bacteriol 185: 870-878.

Handelsman J. 2004. Metagenomics: Application of Genomics to Uncultured Microorganisms Microbiol Molec Biol Revs 68: 669-685.

Holmes DS. 1991. The Use of DNA Probes for the Identification and Enumeration of Biomining Microorganisms: An Overview for the Non-Specialist. In: Badilla-Ohlbaum R, Vargas T, Herrera L, eds. Bioleaching: From Molecular Biology to Industrial Applications. Editorial Universitaria, Santiago, Chile, 1-12.

Holmes DS, Bonnefoy V. Genetic and Bioinformatic Insights into Iron and Sulfur Oxidation Mechanisms of Bioleaching Organisms. In: Rawlings DE, Johnson B, eds. Biomining: Practice, Theory and Microbiology. Springer (submitted, 2006).

Johnson DB 1998. Biodiversity and ecology of acidophilic microorganisms. FEMS Microbiol Ecol 27: 307-317.

Karavaiko GI, Bogdanova TI, Tourova TP, Kondrat'eva TP, Tsaplina IA, Egorova MA, Krasil'nikova EN, Zakharchuk LM. 2005. Reclassification of '*Sulfobacillus thermosulfidooxidans* subsp. *thermotolerans*' strain K1 as *Alicyclobacillus tolerans* sp. nov. and *Sulfobacillus disulfidooxidans* Dufresne *et al.* 1996 as *Alicyclobacillus disulfidooxidans* comb. nov., and emended description of the genus *Alicyclobacillus*. Int J Syst Evol Microbiol 55: 941-947.

Kusano T, Sugawara K, Inoue C, Takeshima T, Numata M, Shiratori T. 1992. Electrotransformation of *Thiobacillus ferrooxidans* with plasmids containing a *mer* determinant. J Bacteriol 174: 6617-6623.

Leyh TS, Taylor JC, Markham GD. 1988. The sulfate activation locus of *Escherichia coli* K12. cloning, genetic, and enzymatic characterization. J Biol Chem. 263: 2409-2416.

Liu Z, Guiliani N, Appia-Ayme C, Borne F, Ratouchniak J, Bonnefoy V. 2000. Construction and characterization of a *recA* mutant of *Thiobacillus ferrooxidans* by marker exchange mutagenesis. J Bacteriol. 182: 2269-2276.

Lochowska A, Iwanicka-Nowicka R, Plochocka D, Hryniewicz MM. 2001. Functional dissection of the LysR-type CysB transcriptional regulator. Regions important for DNA binding, inducer response, oligomerization, and positive control. J Biol Chem 276: 2098-2107.

Lopez-Archilla AI, Marın I, Amils R. 2001. Microbial community composition and ecology of an acidic aquatic environment: the Tinto River, Spain. Microb Ecol 41: 20-35.

Marchand E, Silverstein J. 2002. Influence of heterotrophic microbial growth on biological oxidation of pyrite. Environ Sci Technol 36: 5483-5490.

Olson GJ, Brierley JA, Brierley CL. 2003. Bioleaching review part B: progress in bioleaching: applications of microbial processes by the minerals industries. Appl Microbiol Biotechnol 63: 249-257.

Payne SM. 1994. Detection, isolation, and characterization of siderophores. Methods Enzymol 235: 329-344.

Pronk JT, Liem K, Bos P, Kuenen JG. 1991. Energy transduction by anaerobic ferric iron respiration in *Thiobacillus ferrooxidans*. Appl Environ Microbiol 57: 2063-2068.

Quatrini R, Appia-Ayme C, Denis Y, Ratouchniak J, Veloso F, Valdes J, Lefimil C, Silver S, Roberto F, Orellana O, Denizot F, Jedlicki E, Holmes DS, Bonnefoy V. 2006. Insights into the iron and sulfur energetic metabolism of *Acidthiobacillus ferrooxidans* by microarray transcriptome Profiling. Hydrometallurgy 83: 263-272.

Quatrini R, Lefimil C, Holmes DS, Jedlicki E. 2005a. The Ferric Iron Uptake Regulator (Fur) from the Extreme Acidophile, *Acidithiobacillus ferrooxidans.* Microbiology 151: 2005-2015.

Quatrini R, Jedlicki E, Holmes DS. 2005b. Genomic Insights into the Iron Uptake Mechanisms of the Biomining Microorganism *Acidithiobacillus ferrooxidans.* J Indust Microbiol Biotechnol 32: 606-614.

Ram RJ, Verberkmoes NC, Thelen MP, Tyson GW, Baker BJ, Blake RC, Shah M, Hettich RL, Banfield JF. 2005. Community proteomics of a natural microbial biofilm. Science 308: 1915-1920.

Rawlings DE. 2002. Heavy metal mining using microbes. Annu Rev Microbiol 56: 65-91.

Rawlings DE. 2005. Characteristics and adaptability of iron- and sulfur-oxidizing microorganisms used for the recovery of metals from minerals and their concentrates. Microbial Cell Factories. 4:13.

Rawlings DE, Tributsch H, Hansford GS. 1999. Reasons why "*Leptospirillum*"-like species rather than *Thiobacillus ferrooxidans* are the dominant iron-oxidizing bacteria in many commercial processes for the biooxidation of pyrite and related ores. Microbiology 145: 5-13.

Rivas M, Seeger M, Holmes DS, Jedlicki E. 2005. A Lux-like quorum sensing system in the extreme acidophile *Acidithiobacillus ferrooxidans*. Biological Res 38: 283-297.

Rodionov DA, Dubchak I, Arkin A, Alm E, Gelfand MS. 2004. Reconstruction of regulatory and metabolic pathways in metal-reducing-delta-proteobacteria. Genome Biol 5: R90.

Sand W, Gehrke T, Jozsa P-G, Schippers A. 2001. (Bio)chemistry of bacterial leaching-direct vs. indirect bioleaching. Hydrometallurgy. 59: 159-175.

Schippers A, Sand W. 1999. Bacterial leaching of metal sulfides proceeds by two indirect mechanisms via thiosulfate or via polysulfides and sulfur. Appl Environ Microbiol 65: 319-321.

Schloss PD, Handelsman J. 2003. Biotechnological prospects from metagenomics. Curr Opin Biotechnol 14: 303-331.

Selkov E, Overbeek R, Kogan Y, Chu L, Vonstein V, Holmes D, Silver S, Haselkorn R, Fonstein M. 2000. Functional analysis of gapped microbial genomes: amino acid metabolism of *Thiobacillus ferrooxidans*. Proc Natl Acad Sci 97: 3509-3514.

Stoodley P, Sauer K, Davies DG, Costerton JW. 2002. Biofilms as complex differentiated communities. Annu Rev Microbiol 56: 187-209.

Streit WR, Schmitz RA, Perret X, Staehelin C, Deakin WJ, Raasch C, Liesegang H, Broughton WJ. 2004. An evolutionary hot spot: the pNGR234b replicon of *Rhizobium* sp. strain NGR234. J Bacteriol 186: 535-542.

Tyson GW, Chapman J, Hugenholtz P, Allen EE, Rachna JR, Richardson PM, Solovyev VV, Rubin EM, Rokhsar DS, Banfield JF. 2004. Community structure and metabolism through reconstruction of microbial genomes from the environment. Nature. 428: 37-43.

Tyson GW, Lo I, Baker BJ, Allen EE, HugenholtzP, Banfield JF. 2005. Genome-Directed Isolation of the Key Nitrogen Fixer *Leptospirillum ferrodiazotrophum* sp. nov. from an Acidophilic Microbial Community. Appl Environ Microbiol 71: 6319-6324.

Valdes J, Veloso F, Jedlicki E, Holmes D. 2003. Metabolic reconstruction of sulfur assimilation in the extremophile *Acidithiobacillus ferrooxidans* based on genome analysis. BMC Genomics 4: 51.

Valenzuela L, Chi A, Beard S, Orell A, Guiliani N, Shabanowitz J, Hunt DF, Jerez CA. 2006. Genomics, metagenomics and proteomics in biomining microorganisms. Biotechnol Adv 24: 197-211.

Vian M, Creo C, Dalmastri C, Gionni A, Palazzolo P, Levi G. 1986. *Thiobacillus ferrooxidans* selection in continuous culture. In: Lawrence RW, Branion RMR, Ebner HG, eds. Fundamental and Applied Biohydrometallurgy. Elsevier, Amsterdam, 395-406.

Ward N, Fraser CM. 2005. How genomics has affected the concept of microbiology. Curr Opin Microbiol 8: 564-571.

Wylie JL, Bernegger-Egli C, O'Neil JD, Worobec EA. 1993. Biophysical characterization of OprB, a glucose-inducible porin of *Pseudomonas aeruginosa*. J Bioenerg Biomembr 25: 547-556.

Yates JR., Holmes DS. 1986a. Molecular Probes for the Identification and Quantification of Microorganisms found in Mines and Mine Tailing. Biotechnol Bioeng Symp 16: 310-319.

Yates JR, Holmes DS. 1986b. Construction and Use of Molecular Probes to Identify and Quantitate Bioleaching Microorganisms. In: Branion RMR, Ebner HG, eds. Fundamental and Applied Biohydrometallurgy, Elsevier Science Publ, Amsterdam, 409-418.

CHAPTER 12

PROTEOMICS AND METAPROTEOMICS APPLIED TO BIOMINING MICROORGANISMS

CARLOS A. JEREZ
Laboratory of Molecular Microbiology and Biotechnology, Department of Biology, Faculty of Sciences, University of Chile, Santiago, Chile
E-mail: cjerez@uchile.cl

1. INTRODUCTION

The use of genomics, metagenomics and high throughput proteomics to study the global regulatory responses that the biomining microorganisms use to adapt to their changing environment is just beginning to emerge. In this chapter, we will concentrate specifically on proteomics and metaproteomics as applied to these microorganisms and how new very interesting advances on the molecular physiology of biomining microorganisms can be obtained. This knowledge will most likely result in later improvements of industrial bioleaching processes.

2. STANDARD PROTEOMICS

Proteomics provides direct information about the dynamic protein expression in tissue or whole cells, giving us a global analysis of the cell proteins and their cellular behaviour. Together with the significant accomplishments of genomics and bioinformatics, systematic analysis of all expressed cellular components has become a reality in the post genomic era, and attempts to grasp a comprehensive picture of biology have become possible.

One important aspect of proteomics is to characterize proteins differentially expressed by dissimilar cell types or cells imposed to different environmental conditions. Two-dimensional polyacrylamide gel electrophoresis (2D-PAGE) (O'Farrel, 1975) in combination with mass spectrometry is currently the most widely used technology for comparative bacterial proteomics analysis (Figure 1) (Gygi *et al.*, 2000; Graves & Haystead, 2002).

2D-PAGE separates a complex mixture of proteins such as those present in a bacterial cell extract based on the isoelectric point of the proteins in the first

E.R. Donati and W. Sand (eds.), Microbial Processing of Metal Sulfides, 241–251.

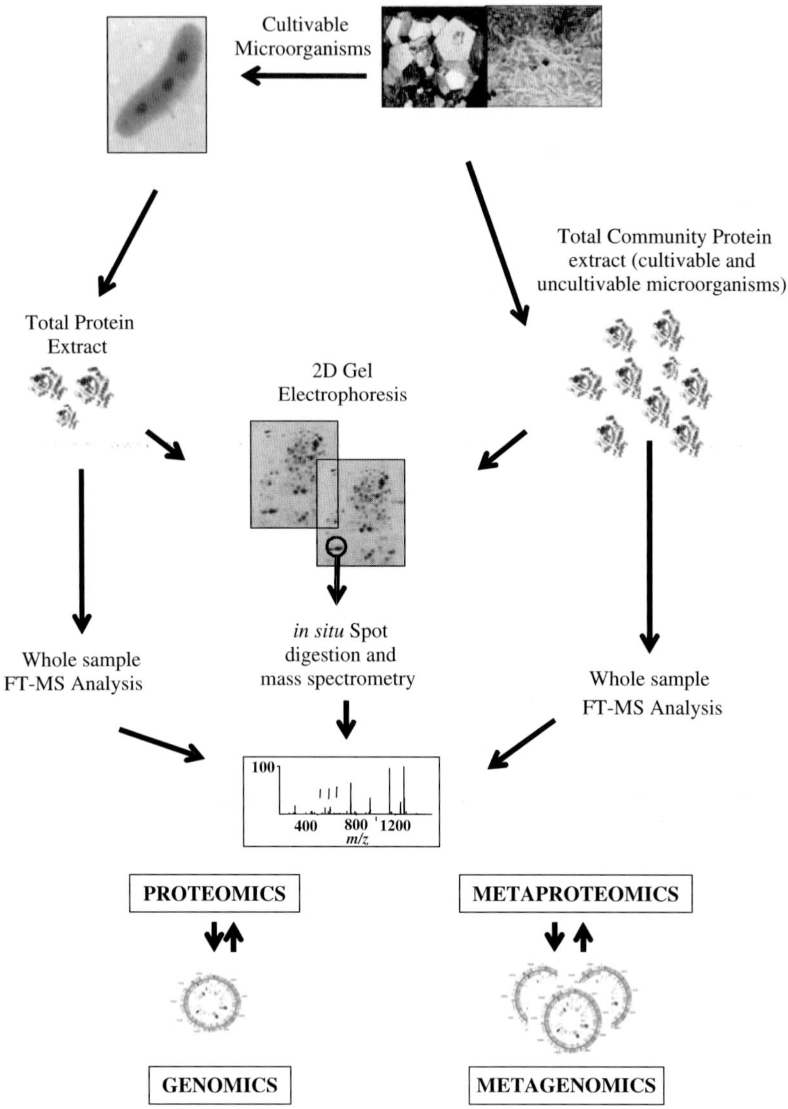

Figure 1. Proteomic and metaproteomic analysis of biomining microorganisms and their relationship with genomics and metagenomics (adapted from Valenzuela *et al.*, 2006)

dimension (isolectrofocusing gel). These proteins or groups of proteins are further resolved in a second dimension (SDS-PAGE), in which the proteins are all negatively charged and are separated based on their molecular masses. Actual 2D-PAGE procedures can resolve around 1000 protein spots in a single run (Graves & Haystead, 2002). Figure 2 shows a typical separation of the proteins present in biomining

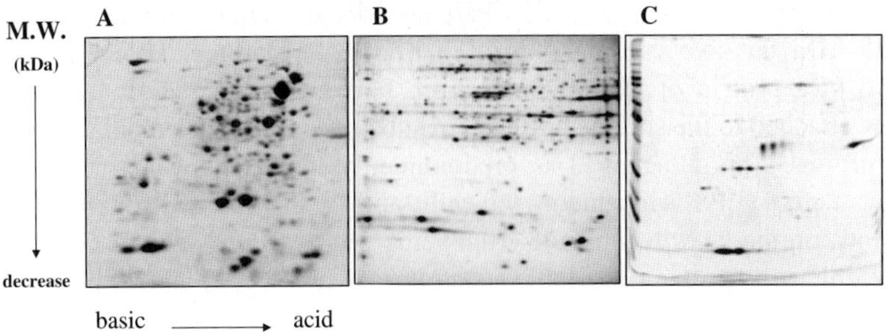

Figure 2. Separation of total proteomes or sub-proteomes from biomining microorganisms by means of 2D-PAGE. **A.** Total cell proteins from *A. ferrooxidans* grown on Fe(II) ions were separated by non-equilibrium 2D-PAGE system and the proteins were stained with Coomassie Blue. **B.** Total cell proteins from *S. metallicus* separated by 2D-PAGE after silver staining. **C.** Periplasmic proteins from *A. ferrooxidans* grown on Fe(II) ions and stained with silver

microorganisms such as *Acidithiobacillus ferrooxidans* (Figure 2A) and *Sulfolobus metallicus* (Figure 2B). One of the limitations of 2D-PAGE is that only the most abundant proteins in the cell can be detected. Therefore, to increase the resolution of the method, it is also possible to analyze a subproteomic cell fraction instead of the total cell proteins as shown for the periplasmic proteins from *A. ferrooxidans* (Figure 2C). With this procedure, only the periplasmic proteins are present instead of the great number of total cell proteins, which improves the separation and identification of the periplasmic proteins.

The high reproducibility of 2D-PAGE is particularly valuable for multiple sample comparisons. In addition, it directly correlates the changes observed at the peptide level to individual protein isoforms.

Several studies have used 2D-PAGE to analyze changes in protein expression of *A. ferrooxidans* under different growth conditions (Valenzuela *et al.*, 2006). Proteins induced under heat shock (Varela & Jerez, 1992), pH stress (Amaro *et al.*, 1991), phosphate limitation (Seeger *et al.*, 1993; Vera *et al.*, 2003) or presence of copper (Novo *et al.*, 2003) have been reported.

A set of proteins that changed their levels of synthesis during growth of *A. ferrooxidans* ATCC 19859 on metal sulfides, thiosulfate, elemental sulfur and Fe(II) ions was characterized by using 2D-PAGE (Ramirez *et al.*, 2004) (Figure 3). N-terminal amino acid sequencing or MS/MS of these proteins allowed their identification and the localization of the corresponding genes in the available genomic sequence of *A. ferrooxidans* ATCC 23270 (http://www.tigr.org). The genomic context around several of these genes suggests their involvement in the energetic metabolism of *A. ferrooxidans*.

Two groups of proteins could be distinguished: proteins highly upregulated by growth on sulfur compounds (and downregulated by growth on Fe(II) ions), and proteins downregulated by growth on sulfur compounds. The highly upregulated proteins were a 44 kDa outer membrane protein, an exported 21 kDa putative thiosulfate sulfur transferase protein, a 33 kDa putative thiosulfate/sulfate binding protein, a 45 kDa putative capsule polysaccharide export protein (WcbC), and a putative 16 kDa protein of unknown function. It is known that most leaching bacteria grow attached to the surface of the solid substrates such as elemental sulfur and metal sulfides. This attachment is predominantly mediated by extracellular polymeric substances (EPS) surrounding the cells, and the amount and composition of EPS depends on the growth substrate (Rohwerder *et al.*, 2003; Sand & Gehrke, 2006). Planktonic cells grown on soluble substrates such as Fe(II) sulfate produce almost no EPS (Gehrke *et al.*, 1998, 2001).

In agreement with these facts, the putative capsule protein that was found induced by cells grown on the solid elemental sulfur in Figure 3 was coded by a putative *wcbC* gene of *A. ferrooxidans* whose context includes the genes *wcbD, wzm* and *wzt*, all of them known in other microorganisms to form a polysaccharide ABC exporter (DeShazer *et al.*, 2001). Other neighbor putative genes appear to be related with the synthesis of EPS (Valenzuela *et al.*, 2006). This is a typical "reverse genetics" approach in which an individual protein differentially expressed in a condition of interest, is isolated from a 2D-PAGE gel and the amino acid sequence of a peptide from the protein is obtained to identify its possible homolog in databases. With this information, its coding gene and genomic context can be searched using the genome DNA sequence (see Figure 1). Depending on these results, a suggested function could be hypothesized. It will be of great importance to demonstrate the expression of putative genes related to EPS synthesis in *A. ferrooxidans* cells grown on different metal sulfides and to find out if these genes are involved in cell attachment and biofilm formation. In this regard, *A. ferrooxidans* is known to form biofilms on solid substrates. Genomic analysis and expression in *E. coli* of several

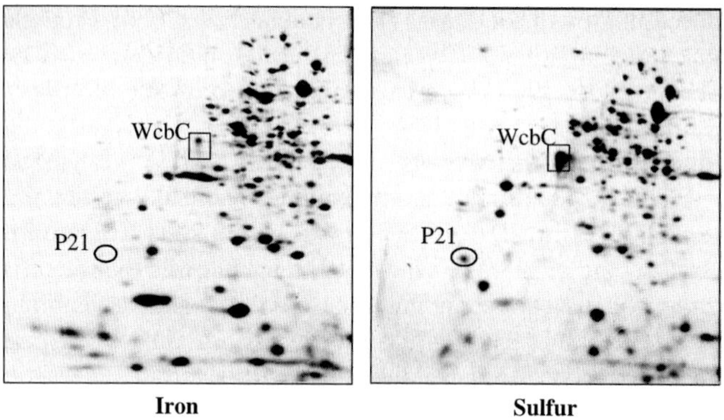

Figure 3. Differential protein expression in *A. ferrooxidans*. Total cell proteins from *A. ferrooxidans* grown in the presence of [^{35}S]-methionine and on Fe(II) ions (A) or on elemental sulfur (B) were separated by 2D-PAGE and the differential expression of several proteins could be detected by autoradiographic analysis. Proteins P21 and WcbC described in the text are indicated

genes from *A. ferrooxidans* related to polysaccharide synthesis, has also recently given support to this gene function (Barreto *et al.*, 2005).

The second group of proteins analyzed by 2D-PAGE by Ramirez *et al.* (2004) were downregulated by growth of *A. ferrooxidans* on sulfur (and upregulated by growth on Fe(II) ions). These proteins were rusticyanin, a cytochrome c552, a putative phosphate binding protein (PstS), the small and large subunits of RuBisCO, a 30 kDa putative CbbQ protein, amongst others. The results suggest in general a separation of the Fe(II) ions and sulfur utilization pathways. Additionally, proteins from both groups (which were upregulated by growth on either Fe(II) ions or sulfur) were together upregulated in case of growth on Fe-containing metal sulfides such as pyrite (FeS_2) and chalcopyrite ($CuFeS_2$). This finding indicates that the two energy-generating Fe(II) ion and sulfur oxidizing pathways are simultaneously induced depending on the kind and concentration of the available oxidizable substrates (Ramirez *et al.*, 2004). In agreement with these results, it has been previously suggested that *A. ferrooxidans* can simultaneously utilize both Fe(II) ions and elemental sulfur as energy sources (Espejo & Romero, 1987).

By using 2D-PAGE, it was possible to identify an exported rhodanese (thiosulfate sulfur transferase)-like protein (P21) (Figure 3) whose levels are increased when *A. ferrooxidans* is grown on metal sulfides and different sulfur compounds but is almost entirely absent during growth on Fe(II) ions (Ramírez *et al.*, 2002, 2004). Unlike cytoplasmic rhodaneses, P21 was located in the periphery of *A. ferrooxidans* cells and was regulated depending on the oxidizable substrate. If P21 and some of the proteins coded by its adjacent genes are involved in thiosulfate metabolism, one should expect an increased expression of these proteins when the cells are grown in pyrite, thiosulfate or sulfur. Protein P21 may not be a periplasmic rhodanese enzyme but rather part of a possible complex in charge of thiosulfate oxidation. This putative complex could be different from the Sox model proposed for sulfur oxidation in many bacteria (Friedrich, 1998) since we did not find any *sox*-like genes in the genome of *A. ferrooxidans* (Ramirez *et al.*, 2004). However, *A. ferrooxidans* harbors duplicated *doxDA* genes that are homologous to the genes encoding thiosulfate:quinone oxidoreductase in *Acidianus ambivalens*, an enzyme oxidizing thiosulfate with tetrathionate as product and ferricyanide or decylubiquinone as electron acceptors (Müller *et al.*, 2004). As pointed out very recently by Friedrich, this gene duplication points to a yet undemonstrated significance in thiosulfate metabolism in *A. ferrooxidans* (Friedrich *et al.*, 2005). The *doxDA-1* is part of the putative transcriptional unit containing *p*21 (Acosta *et al.*, 2005) and its expression determined by DNA macroarrays is enhanced 12-fold in cells grown on elemental sulfur and 3-fold in cells grown on thiosulfate compared to the expression levels of cells grown on Fe(II) ions as shown in Figure 4 and also in Acosta *et al.* (2005) and Valenzuela *et al.* (2006).

These results support the significance of at least one of the two *doxDA* genes in thiosulfate metabolism in *A. ferrooxidans* and are in favor of the idea that acidophilic bacteria oxidize sulfur by a system different from the Sox enzyme system.

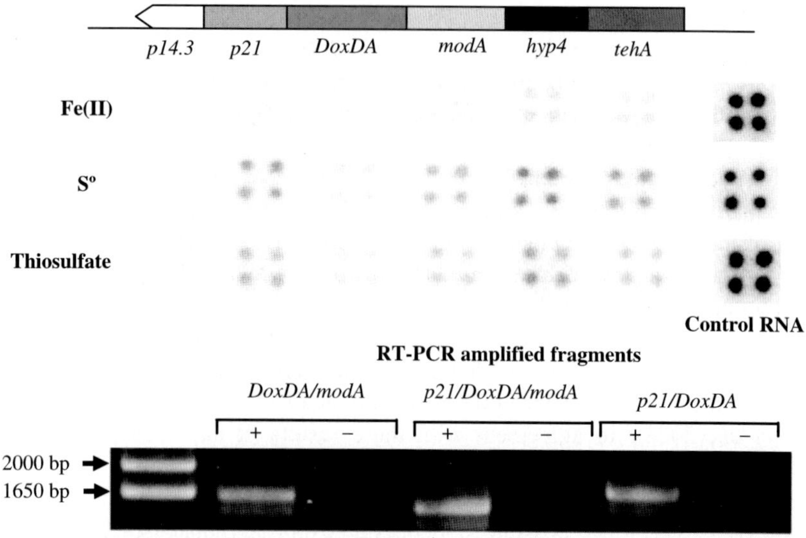

Figure 4. DNA macroarray expression analysis of the gene coding for protein P21 and the genes present in its genomic context. Conditions are similar to those reported by Valenzuela *et al.* (2006)

3. HIGH THROUGHPUT PROTEOMICS AND METAPROTEOMICS

In the past decade, an increasing number of sequenced genomes provides good options for high throughput functional analysis of proteomes (Celestino *et al.*, 2004). Proteomic studies are well advanced for diverse bacteria, such as *E. coli* (Vollmer *et al.*, 2003) and *Bacillus subtilis* (Eymann *et al.*, 2004). Nevertheless, there is still a lack of data for identification of proteins from organisms with unannotated or unsequenced genomes, thus large-scale microbiological proteomics analysis remain challenging. With the development of highly sensitive and accurate computational gene-finding methods such as GeneMark (Borodovsky & Mcininch, 1993) and GLIMMER (Salzberg *et al.*, 1998), new microbial genomes could be explored and scientific knowledge could be acquired.

Solution-based approaches offer unbiased measurements of relative protein expression regardless of their abundance, subcellular localization, or physicochemical parameters. This methodology, however, results in extremely complex samples. For instance, of the 4191 predicted genes in the complete genome of *Escherichia coli*, 2800 of them are believed to be expressed at any time (Corbin *et al.*, 2003). Additional complexity is introduced upon enzymatic digestion, which generates multiple peptide species for each protein (Figure 5). To obtain comprehensive protein expression information in the samples, a chromatographic separation step prior to mass spectrometry (MS) protein analysis is often necessary. HPLC coupled with online electron spray ionization MS (ESI-MS/MS) has been proved to

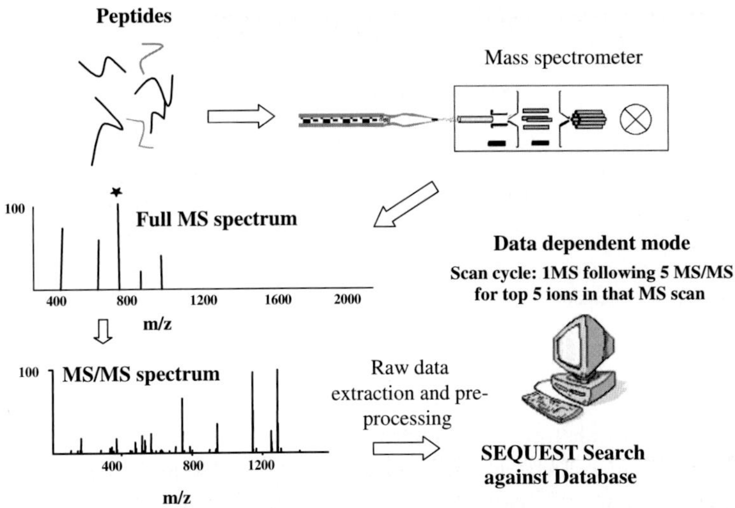

Figure 5. General steps for high throughput protein analysis. A cell extract or a sub-proteome from the microorganism to be studied is subjected to trypsin digestion. The peptides generated are then chromatographically separated by HPLC and subjected to ionization in the mass spectrometer, in which the molecular weights are determined after the separation of the molecular ions according to their mass-to-charge ratio (m/z). The result of ionization, ion separation, and detection is a mass spectrum providing molecular weights or even structural information

be a valid approach for analyzing protein expression in complex samples (Washburn *et al.*, 2001).

Cell lysates were extracted (Gygi *et al.*, 2000) from only 5E10 cells of *A. ferrooxidans* grown up to stationary phase in three different growth media: iron, thiosulfate or sulfur and subjected to trypsin digestion. The resulting peptides were separated on reverse phase column and then analyzed on a ThermoElectron LCQ^{Deca} liquid chromatography ion trap mass spectrometer to confirm the suitability of the samples for further differential analysis (Valenzuela *et al.*, 2006).

The computational method GLIMMER (Salzberg *et al.*, 1998) was utilized to find the genes in the coding region of the *A. ferrooxidans* genome and construct a putative *A. ferrooxidans* protein database. Database search was performed with SEQUEST (Eng *et al.*, 1994), and a peptide sequence was then assigned for each MS/MS spectra. Several peptides identified were manually confirmed and used as markers for the subsequent differential Fourier transform ion cyclotron resonance (FT-ICR) analysis and the targeted LCQ analysis. Differentially expressed proteins were then identified with sequence database search and confirmed manually (Valenzuela *et al.*, 2006).

Table 1 shows part of a LCQ analysis of proteins present in total cell extracts of *A. ferrooxidans*. Some proteins proposed to participate in sulfur oxidation (Rohwerder & Sand, 2003) were present only when *A. ferrooxidans* cells were grown on sulfur compounds.

Table 1. Examples of proteins detected by LCQ/MS analysis of cell-free extracts from *A. ferrooxidans* grown on thiosulfate and elemental sulfur but absent in cells grown on Fe(II) ions[a]

AFO[b]	Predicted protein
1882	sulfide:quinone oxidoreductase (*Thiobacillus denitrificans*)
4310	rhodanese-like protein P21 (*A. ferrooxidans*)
4311	terminal quinol oxidase, subunit, putative doxD-like
4624	subunit of an aa3-type cytochrome oxidase (*A. ferrooxidans*)

[a]unpublished results from Chi, Valenzuela, Hunt and Jerez.
[b]AFO = *A. ferrooxidans* open reading frame.

Alternatively, Fourier transform ion cyclotron resonance mass spectrometer (FT-ICR MS) (Martin *et al.*, 2000) is well suited for the differential analysis of protein expression due to the high mass accuracy and high resolution as well as its inherent wide dynamic range. Peptide charge state can be readily derived with the accurate isotopic peak distribution information provided by FT-MS experiments, and co-eluting species with the same nominal m/z ratio can be resolved. The differentially expressed m/z values identified can be assigned to the peptide sequences and, subsequently, differentially expressed proteins can be identified.

When the total proteins extracted from *A. ferrooxidans* grown on the surface of a sulfur prill were analyzed by FT-MS and compared with those for growth on Fe(II) ions or thiosulfate, the highest scores were obtained for peptides derived from proteins that were identified as TrbB and TrbJ amongst others. These proteomics results were validated by using RT-PCR, real time PCR and DNA macroarrays (Valenzuela *et al.*, 2006). In all cases an increased expression of the *trb* genes analyzed was seen when *A. ferrooxidans* was grown on a solid substrate such as elemental sulfur. These unpublished results strongly suggest the existence of a functional type IV secretion system (Christie *et al.*, 2000) in *A. ferrooxidans*. Although it is not yet clear what the function of that system may be in this bacterium, it is likely related to biofilm formation and exchange of macromolecules such as DNA during conjugation. Obviously, an experimental demonstration of conjugal transfer of DNA between *A. ferrooxidans* strains or between this microorganism and other biomining bacteria will have to be demonstrated.

Recently, the term "metaproteomics" was proposed for the large-scale characterization of the entire protein complement of environmental microbiota at a given point in time (Wilmes & Bond, 2004). These authors applied 2D-PAGE and downstream analysis and mapping to a mixed community of prokaryotic microorganisms after extraction and purification of the entire proteome (Figure 1). The highly expressed proteins were identified by MS. This study is the first report of an application of metaproteomics.

High throughput MS has been used in another for biomining relevant metaproteomic approach to study the community proteomics in a natural acid mine drainage microbial biofilm (Ram *et al.*, 2005). These authors were able to detect 2,033 proteins from the five most abundant species in the biofilm, including 48% of the predicted proteins from the dominant biofilm organism *Leptospirillum* group II, recently described as *Leptospirillum ferriphilum* (Coram & Rawlings, 2002; Tyson *et al.*, 2005). These results are very promising, and therefore should be validated. The authors also determined that one abundant novel protein was a cytochrome central to Fe(II) ion oxidation and acid mine drainage formation in the natural biofilm (Ram *et al.*, 2005). This novel approach together with functional metagenomics will offer integrated studies of microbial communities to find out the role of each community member and how they change under different conditions.

4. CONCLUDING REMARKS

The goal of functional proteomics is to correlate the identification and analysis of distinct proteins with the function of genes or other proteins. With the discovery of a variety of modular protein domains that have specific binding partners, it has become clear that most proteins occur in protein complexes and that the understanding of a function of a protein within the cell requires the identification of its interacting partners (Mann *et al.*, 2001). In the case of a biomining bacterium such as *A. ferrooxidans*, the identification of protein complexes involved in oxidative reactions is of high priority. What complexes are formed by rusticyanin with other proteins in the periplasm? Does the exported rhodanase-like P21 protein form a part of a complex involved in sulfur compound oxidation in the periplasm? What other complexes with oxidative reactions are present in this microorganism? Proteomics may answer these and many other questions what will help to understand better the biomining process. This, together with proper manipulations of the bacterial environment (Rawlings, 2005) will provide further improvements for the bioleaching rates of biomining operations.

Systems microbiology, which is part of systems biology, is a new way to approach research in biological systems. By this approach it may be possible to explore the new properties of microorganisms that arise from the interplay of genes, proteins, other macromolecules, small molecules, and the environment (Buckley, 2004). This is particularly possible today due to the large numbers of genomic sequences which are becoming increasingly available. However, additional genomic sequences of the different biomining microorganisms will be required to define the molecular adaptations to their environment and the interactions between the members of the community. The idea is to consider the microorganism or the community as a whole, integrating fundamental biological knowledge with genomics, proteomics, metabolomics, and other data to obtain an integrated picture of how a microbial cell or a community operates. This knowledge should be usefull to improve the bioleaching process.

5. ACKNOWLEDGEMENTS

The results presented were supported by a grant from FONDECYT (N° 1030767). Sequence data for the *A. ferrooxidans* strain 23270 was obtained from The Institute for Genomic Research website at http://www.tigr.org. I also thank An Chi and Donald Hunt for their expert advice and discussions on high throughput proteomics.

6. REFERENCES

Acosta M, Beard S, Ponce J, Vera M, Mobarec JC, Jerez CA. 2005. Identification of putative sulfurtransferase genes in the extremophilic *Acidithiobacillus ferrooxidans* ATCC 23270 genome: structural and functional characterization of the proteins. OMICS 9: 13-28.

Amaro AM, Chamorro D, Seeger M, Arredondo R, Peirano I, Jerez CA. 1991. Effect of external pH perturbations on in vivo protein synthesis by the acidophilic bacterium *Thiobacillus ferrooxidans*. J. Bacteriol 173: 910-915.

Barreto M, Jedlicki E, Holmes DS. 2005. Identification of a gene cluster for the formation of extracellular polysaccharide precursors in the chemolithoautotroph *Acidithiobacillus ferrooxidans*. Appl Environ Microbiol 71: 2902-2909.

Borodovsky M, Mcininch D. 1993. GeneMark: parallel gene recognition for both DNA strands. Comp Chem 17: 123-133

Buckley MR. 2004. Systems microbiology: beyond microbial genomics. Report of the American Academy of Microbiology. Available from www.asmusa.org.

Celestino PB, de Carvalho LR, de Freitas LM, Dorella FA, Martins NF, Pacheco LG, Miyoshi A, Azevedo V. 2004. Update of microbial genome programs for bacteria and archaea. Genet Mol Res 3: 421-431.

Christie P, Vogel J. 2000. Bacterial type IV secretion: conjugation systems adapted to deliver effector molecules to host cells. Trends Microbiol 8: 354-360.

Coram NJ, Rawlings DE. 2002. Molecular relationship between two groups of the genus *Leptospirillum* and the finding that *Leptospirillum ferriphilum* sp. nov. dominates South African commercial biooxidation tanks that operate at 40 °C. Appl Environ Microbiol 68: 838-845.

Corbin RW, Paliy O, Yang F, Shabanowitz J, Platt M, Lyons CE, Root K, McAuliffe J, Jordan MO, Kustu S, Soupene E, Hunt DF. 2003. Toward a protein profiles of *Escherichia coli*: Comparison to its transcription profile. Proc Natl Acad Sci USA 100: 9232-9237.

DeShazer D, Waag DM, Fritz DL, Woods DE. 2001. Identification of a *Burkholderia mallei* polysaccharide gene cluster by substractive hybridization and demonstration that the encoded capsule is an essential virulence determinant. Microb Pathog 30: 253-269.

Eng JK, McCormack AL, Yates JR. 1994. An approach to correlate tandem mass spectral data of peptides with amino acid sequences in a protein database. J Am Soc Mass Spectrom 5: 976-989.

Espejo RT, Romero P. 1987. Growth of *Thiobacillus ferrooxidans* on elemental sulfur. Appl Environ Microbiol 53: 1907-1912.

Eymann C, Dreisbach A, Albrecht D, Bernhardt J, Becher D, Gentner S, Tamle T, Buttner K, Buurman G, Scharf C, Venz S, Volker U, Hecker M. 2004. A comprehensive proteome map of growing *Bacillus subtilis* cells. Proteomics 4: 2849-2876.

Friedrich CG, Bardischewsky F, Rother D, Quentmeier A, Fischer J. 2005. Prokaryotic sulfur oxidation. Curr Opin Microbiol 8: 253-259.

Friedrich CG. 1998. Physiology and genetics of sulfur-oxidizing bacteria. Adv Microb Physiol 39: 235-289.

Gehrke T, Hallmann R, Kinzler K, Sand W. 2001. The EPS of *Acidithiobacillus ferrooxidans*- a model for structure-function relationships of attached bacteria and their physiology. Water Sci Technol 43: 159-167.

Gehrke T, Telegdi J, Thierry D, Sand W. 1998. Importance of extracellular polymeric substances from *Thiobacillus ferrooxidans* for bioleaching. Appl Environ Microbiol 64: 2743-2747.

Graves PR, Haystead TAJ. 2002. Molecular biologist's guide to proteomics. Microbiol Mol Biol Rev 66: 39-63.

Gygi SP, Corthals GL, Zhang Y, Rochon Y, Aebersold R. 2000. Evaluation of two-dimensional gel electrophoresis-based proteome analysis technology. Proc Natl Acad Sci USA 97:9390-95.

Martin SE, Shabanowitz J, Hunt DF, Marto JA. 2000. Subfemtomole MS and MS/MS peptide sequence analysis using nano-HPLC micro-ESI fourier transform ion cyclotron resonance mass spectrometry. Anal Chem 72: 4266-4274.

Mann M, Hendrickson RC, Pandey A. 2001. Analysis of proteins and proteomes by mass spectrometry. Annu Rev Biochem 70: 437-473.

Müller FH, Bandeiras TM, Urich T, Teixeira M, Gomes CM, Kletzin A. 2004. Coupling of the pathway of sulphur oxidation to dioxygen reduction: characterization of a novel membrane-bound thiosulphate:quinone oxidoreductase. Mol Microbiol 53: 1147-1160.

Novo MT, Garcia Jr O, Ottoboni LM. 2003. Protein profile of *Acidithiobacillus ferrooxidans* exhibiting different levels of tolerance to metal sulfates. Curr Microbiol 47: 492-496.

O'Farrel PH. 1975. High resolution two-dimensional electrophoresis of proteins. J Biol Chem 250: 4007-4021.

Ram RJ, VerBerkmoes NC, Thelen MP, Tyson GW, Baker BJ, Blake RC, Shah M, Hettich RL, Banfield JF. 2005. Community proteomics of a natural microbial biofilm. Science 308: 1915-1920.

Ramirez P, Guiliani N, Valenzuela L, Beard S, Jerez CA. 2004. Differential protein expression during growth of *Acidithiobacillus ferrooxidans* on ferrous iron, sulfur compounds, or metal sulfides. Appl Environ Microbiol 70: 4491-4498.

Ramirez P, Toledo H, Guiliani N, Jerez CA. 2002. An exported rhodanese-like protein is induced during growth of *Acidithiobacillus ferrooxidans* in metal sulfides and different sulfur compounds. Appl Environ Microbiol 68: 1837-1845.

Rawlings DE. 2005. Characteristics and adaptability of iron- and sulfur-oxidizing microorganisms used for the recovery of metals from minerals and their concentrates. Microb Cell Fact 4: 13.

Rohwerder T, Gehrke T, Kinzler K, Sand W. 2003. Bioleaching review part A: progress in bioleaching: fundamentals and mechanisms of bacterial metal sulfide oxidation. Appl Microbiol Biotechnol 63: 239-248.

Rohwerder T, Sand W. 2003. The sulfane sulfur of persulfides is the actual substrate of the sulfur-oxidizing enzymes from *Acidithiobacillus* and *Acidiphilium* spp. Microbiology 149: 1699-1709.

Salzberg SL, Delcer AL, Kasif S, White O. 1998. Microbial gene identification using interpolated Markov models. Nucl Acids Res 26: 2544-2548.

Sand W, Gehrke T. 2006. Extracellular polymeric substances mediate bioleaching/biocorrosion via interfacial processes involving iron(III) ions and acidophilic bacteria. Res Microbiol 157: 49-56.

Seeger M, Jerez CA. 1993. Response of *Thiobacillus ferrooxidans* to phosphate limitation. FEMS Microbiol Rev 11: 37-42.

Tyson GW, Lo I, Baker BJ, Allen EE, Hugenholtz P, Banfield JF. 2005. Genome-directed isolation of the key nitrogen fixer *Leptospirillum ferrodiazotrophum* sp nov from an acidophilic microbial community. Appl Environ Microbiol 71: 6319-6324.

Valenzuela L, Chi A, Beard S, Orell A, Guiliani N, Shabanowitz J, Hunt DF, Jerez CA. 2006. Genomics, metagenomics and proteomics in biomining microorganisms. Biotechnol Adv 24: 197-211.

Varela P, Jerez CA. 1992. Identification and characterization of GroEL and DnaK homologues in *Thiobacillus ferrooxidans*. FEMS Microbiol Lett 77: 149-153.

Vera M, Guiliani N, Jerez CA. 2003. Proteomic and genomic analysis of the phosphate starvation response of *Acidithiobacillus ferrooxidans*. Hydrometallurgy 71: 125-132.

Vollmer M, Nagele E, Horth P. 2003. Differential proteome analysis: two-dimensional nano-LC/MS of *E. coli* proteome grown on different carbon sources. J Biomol Tech 14: 128-135.

Washburn MP, Wolters D, Yates III Jr. 2001. Large-scale analysis of the yeast proteome by multidimensional protein identification technology. Nature Biotech 19: 242-247.

Wilmes P, Bond PL. 2004. The application of two-dimensional polyacrylamide gel electrophoresis and downstream analyses to a mixed community of prokaryotic microorganisms. Env Microbiol 6: 911-920.

CHAPTER 13

CELL-CELL COMMUNICATION IN BACTERIA

A promising new approach to improve bioleaching efficiency?

SUSANA VALENZUELA, ALVARO BANDERAS, CARLOS A. JEREZ AND NICOLAS GUILIANI*

Unit of Bacterial Cell Communication, Laboratory of Molecular Microbiology and Biotechnology, Department of Biology, Faculty of Sciences, University of Chile, Santiago, Chile.

E-mail: nguilian@uchile.cl

1. INTRODUCTION

Changes in bacterial gene expression result from the ability of individual cells to respond to a wide variety of environmental and intracellular signals. Unlike higher organisms, for a long time, bacteria where thought as individual cells incapable of having group behaviour. However, independent studies in Gram-positive and Gram-negative bacteria led to the discovery that microorganisms themselves can produce and respond to extracellular chemical signals (Tomasz, 1965; Nealson *et al.*, 1970). These molecules are now known as autoinducers (AI) and provide the basis for the concept of "bacterial communication".

The general shared mechanism of AI synthesis and response is as follows: 1) Production of the AI molecule by the bacterial cells. 2) Accumulation of the AI in the cell's environment. 3) Coordinated changes in gene expression after reaching an inducing AI concentration threshold.

The term quorum sensing (QS) arises from the fact that accumulation of the AI molecule in the extracellular medium correlates with cell growth (i.e. the AI concentration threshold is related to cell densities) (Fuqua *et al.*, 1994). This has been interpreted as a sensing system that bacterial cells use to measure their total number in the population.

In addition to the unique cases of cell-cell communication systems of *Vibrio harveyi* and *Myxococcus xanthus* (Bassler, 2002), QS systems are classified in three groups (Waters & Bassler, 2005):

1) Gram-negative bacteria use N-acyl homoserine lactones (AHL) as AI molecules. AHLs were the first AI to be characterized. AHL-based QS systems are known as QS type AI-1 and are widespread in Gram-negative bacteria.

E.R. Donati and W. Sand (eds.), Microbial Processing of Metal Sulfides, 253–264.

2) Gram-positive bacteria use small peptides as autoinducers. These are generally processed in the cytoplasm, secreted through an ABC transporter and sensed in the cell's extracellular surface by a two-component system that induces changes in gene expression through a phosphorelay.
3) A third autoinducer system is based on DPD (4,5-dihydroxypentane-2,3-dione) derived molecules, collectively known as autoinducer 2 (AI-2). Due to the high degree of conservation of its components among Gram-negative and Gram-positive bacteria, it is now thought to function as an interspecies communication system.

QS regulates a great variety of bacterial phenotypes and AHLs-mediated gene regulation has been shown to influence exopolysaccharide production and biofilm formation in many proteobacteria (Davies *et al.*, 1998; Huber *et al.*, 2001; Winans & Bassler, 2002; Gonzalez & Marketon, 2003; Labbate *et al.*, 2004).

Therefore, from the perspective of improving biomining efficiency and obtaining a better understanding about the dynamics of the microorganisms population, the characterization of a functional AHL communication system in some biomining and AMD microorganisms acquires a strong relevance.

2. THE QUORUM SENSING TYPE AI-1 PARADIGM

In Gram-negative bacteria, cell-cell communication is mostly mediated by AHLs. It is currently accepted that the type AI-1 QS system is composed of four elements (Figure 1): i) an AHL which is the signalling molecule or autoinducer (AI-1); ii) the AHL synthase protein (I protein family) which synthesizes the AI; iii) a transcriptional regulator (R protein family); and iv) a *cis*-acting DNA palindromic sequence which is the target of the binary complex [R-AHL] (reviewed in Whitehead *et al.*, 2001; Winans & Bassler, 2002; Bassler, 2002; Henke & Bassler, 2004; Waters & Bassler, 2005).

2.1. The AHL Synthases

Three different AHL synthase families have been reported: I protein-type (family 1), LuxM/AinS-type (family 2) and the HdtS-type (family 3) (Gonzalez & Marketon, 2003).

I proteins catalyze the synthesis of AHLs from two substrates, SAM and acyl-ACP. Some I proteins are able to synthesize more than one kind of AHL which can involve AHL with either a ketone or an alcohol at C3, mono unsaturated acyl-chain and variations in acyl-chain length. To date, most of AHLs reported are composed of an acyl-chain with an even number of carbon atoms ranging from 4 to 18. However, few AHLs with an odd chain length have been described in plant, soil and marine bacteria (Gonzalez & Marketon, 2003; Brader *et al.*, 2005; Wagner-Dobler *et al.*, 2005). While it is accepted that AHLs with a small acyl-chain length diffuse freely across the membrane, some reports indicate variations to the paradigm and it has

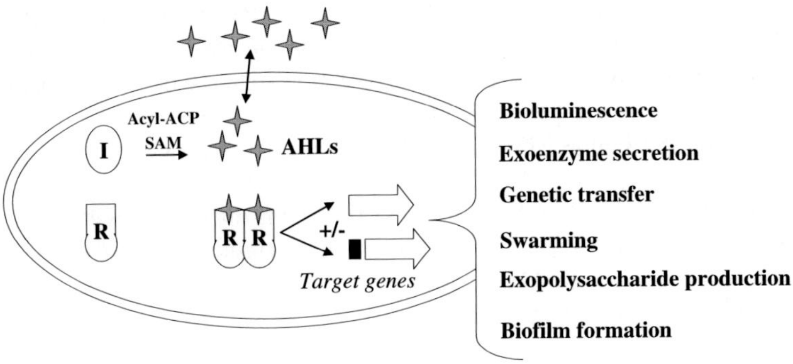

Figure 1. The type AI-1 QS system. AHLs are synthetized inside the bacterial cell by the I protein from acylated acyl carrier protein (acyl-ACP) and S-adenosyl methionine (SAM) substrates and diffuse freely from the cells to external environments. As the cell population increases, external and internal concentrations of AHLs increase too. At the inducing AI concentration, which is reached at high cell density, AHL interacts with the R protein which dimerizes and binds to the promoter region of the target genes (white arrows) to promote or repress their transcription. These genes may or may not possess a consensual palindromic region as the binding site (black box) for AHL/R complex which is also present in the promoter region of the I gene. Some of the physiological functions regulated by the QS type AI-1 are listed. The last four functions are relevant to improving our understanding of the biology of biomining

been proposed that AHLs with large acyl-chains are excreted through an active efflux pump (Pearson *et al.*, 1999).

Studies of the I protein structure suggested that different AHL synthases could have different mechanisms for substrate selection which include hydrophobic pocket restriction and, on the other hand, no theoretical restriction on the length of the acyl-chain, respectively (Watson *et al.*, 2002; Gould *et al.*, 2004). Alternatively, the substrate chain length specificity has been associated with specific residues which appear to be critical for the space accessibility of the acyl-chain (Brader *et al.*, 2005). In addition, bioinformatic studies suggested that a structural spacer located between strand β4 and helix $\alpha 6$ of the I proteins could be related to the selectivity of the acyl-chain length of the AHLs synthesized by the AHL synthases belonging to I family (Mobarec *et al.*, submitted).

2.2. R Protein Family

R proteins are transcriptional regulators composed of two subdomains (for reviews see Whitehead *et al.*, 2001; Miller & Bassler, 2001; Gonzalez & Marketon, 2003). The amino-terminal region of R is involved in the binding of AHLs and crystallography studies allowed the characterization of the AHL-binding cavity (Vannini *et al.*, 2002; Zhang *et al.*, 2002). The carboxy-terminal third includes a helix-turn-helix DNA binding motif. The interaction between R proteins and their cognate AHLs promotes dimerization of the R proteins. The dimeric proteins bind a DNA region formed by an inverted-repeat sequence, or *lux* box-type sequences, centered at about

-40 bp from the transcriptional start site. However, in some cases, palindromic sequences for R-protein binding and autoinduction are absent (Welch *et al.*, 2000; Christensen *et al.*, 2003). The R protein/DNA interaction induces the recruitment of the RNA polymerase to the promoter and therefore the transcription of the target genes, including the I gene which is thereby, in some cases, submitted to an autoinduction.

Most of the described R proteins are activators. However, different reports have characterized R proteins as transcriptional repressors whose binding activity should be reduced by the interaction with AHL (Minogue *et al.*, 2002; Horng *et al.*, 2002; Christensen *et al.*, 2003; Kirke *et al.*, 2004).

3. QUORUM SENSING AND BIOLEACHING

3.1. Quorum Sensing and Bioleaching Microorganisms

Bioleaching and AMD communities are mainly comprised of archaea and bacteria. However, to date all quorum sensing systems have been reported in bacteria and recently, Llamas *et al.* (2005) presented the first evidences for AHL production in an extremophilic bacterium belonging to *Halomonas sp.* Nevertheless, archaea cell-cell communications can not be excluded since Paggi *et al.* (2003) have characterized AHL molecules produced by the haloalkaliphilic archaeon *Natronococcus occultus*.

During the study of the existence of cell-cell communication in acidophilic microorganisms, we analyzed the available genome sequences of the archaea *Sulfolobus sp.* and *Ferroplasma acidarmanus* and the bacterium *Acidithiobacillus ferrooxidans* for the presence of ORFs encoding proteins involved in AI-1 and AI-2 synthesis and sensing. Negative results were obtained for the presence of both QS systems in archaea (unpublished results) and for QS type AI-2 in *A. ferrooxidans*. These results suggest that the type AI-2 communication system may not exist in the biomining population.

On the other hand, positive results were obtained with ortholog genes encoding R proteins and AHL synthases of I-type in *A. ferrooxidans* (Barreto *et al.*, 2003; Farah *et al.*, 2004). In addition, an *hdtS*-like gene has been identified (Farah *et al.*, 2005; Rivas *et al.*, 2005). Farah *et al.* (2005) were the first to report the existence of a functional QS type AI-1 system in the acidophilic bacterium *A. ferrooxidans*. The authors characterized a type AI-1 QS locus and 9 different AHLs which included the two types of functionalities -alcohol or ketone- at C3, and all of them having medium or large acyl-chains with an even number of carbons (Table 1). Interestingly, C_{12}- and C_{14}-AHLs and oxo-C_{12}- and C_{14}-AHLs were detected only in sulfur- and thiosulfate-grown cells. During growth on sulfur, biofilm formation takes place while ferrous iron growth results in only planktonic cells. Whether the difference in AHL production is related to these physiological states is still unknown. In addition, our preliminary results revealed that 3-hydroxy-AHLs are produced by *A. ferrooxidans* cells grown in pyrite (unpublished results).

The type AI-1 QS locus from *A. ferrooxidans* ATCC23270 has the same genetic organization of the QS locus from *Burkholderia sp.* and contains the genes *afeI*,

Table 1. AHLs produced by *Acidithiobacillus ferrooxidans* ATCC 23270

Chemical structure	**Name**	**Energetic Substrate**[a]
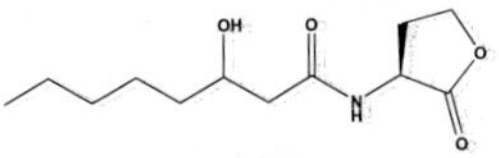	**3-hydroxy-C8-AHL**	**S, T**
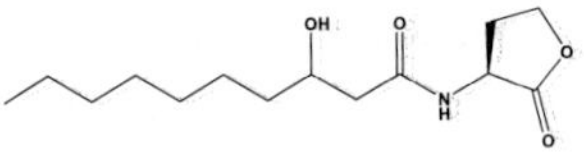	**3-hydroxy-C10-AHL**	**I, S, T**
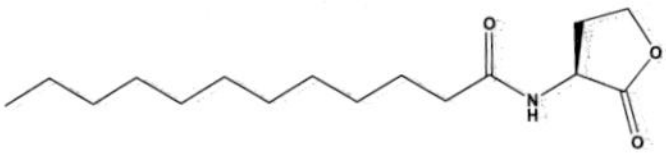	**C12-AHL**	**S, T**
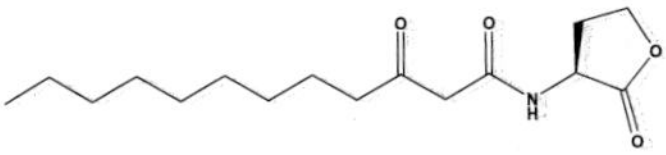	**3-oxo-C12-AHL**	**S, T**
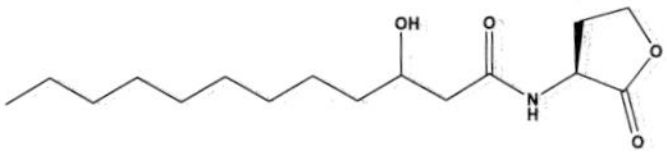	**3-hydroxy-C12-AHL**	**I, S, T**
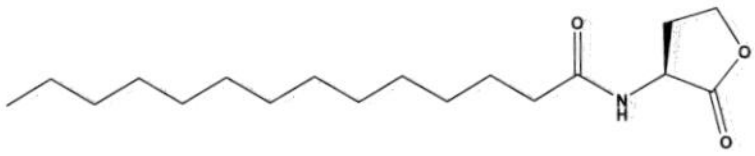	**C14-AHL**	**S, T**
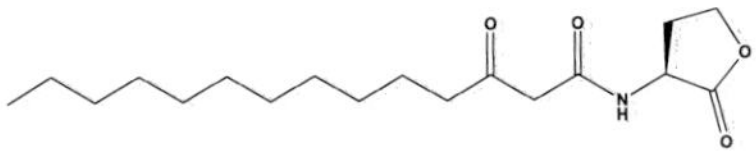	**3-oxo-C14-AHL**	**S, T**
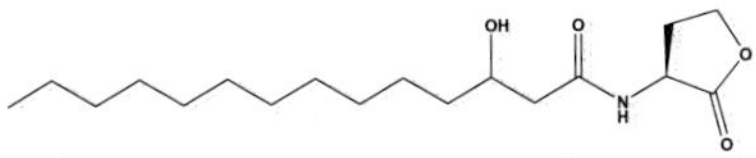	**3-hydroxy-C14-AHL**	**I, S, T**
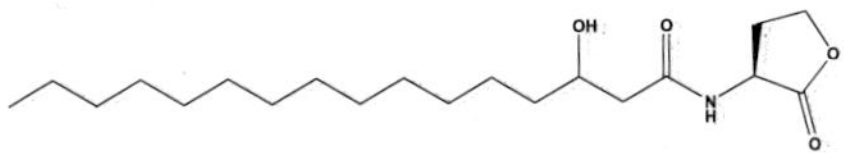	**3-hydroxy-C16-AHL**	**I, S, T**

a: I, Iron; S, Sulfur; T, Thiosulfate.

orf3 and *afeR* which are transcribed in *A. ferrooxidans* and encode for the AHL-synthase AfeI, a protein (Orf3) whose function is unknown and the transcriptional regulator AfeR, respectively. Two *afe* boxes which should correspond to the AfeR binding sites have been characterized upstream of the *afeI* gene (Farah *et al.*, 2005). Recently, the development of an AHLs biosensor system which involves the *afeR* gene, *afe* boxes and *luc* as the reporter gene demonstrated that autoinduction of I protein occurs in *A. ferrooxidans* and that AfeR can regulate positively the transcription of *afeI* gene (Rivas *et al.*, 2005).

On the other hand, various *afe* boxes have been characterized in the genome sequence of *A. ferrooxidans* ATCC 23270 strain (Farah *et al.*, 2005) and genes located downstream have been annotated (unpublished results). This bioinformatic analysis strongly suggests the existence of a type AI-1 QS regulon in *A. ferrooxidans* as is the case in many other bacteria. Depending on the bacterium and the experimental approach used, type AI-1 QS regulon has been found to involve between 3% and 5% of the whole genome (Arevalo-Ferro *et al.*, 2003, Chen *et al.*, 2003; Nouwens *et al.*, 2003; Riedel *et al.*, 2003; Schuster *et al.*, 2003; Wagner *et al.*, 2003). However, further proteomic and transcriptomic studies will be necessary to definitively demonstrate its existence in *A. ferrooxidans* and determine how many genes are directly and/or indirectly, positively and/or negatively regulated by AfeR.

Regarding the transcriptional regulation of *afeI* gene it has been found that the levels of transcription of *afeI* gene were higher in stationary phase than exponential phase (Rivas *et al.*, 2005). Phosphate starvation induces transcription of the *afeI* gene and is concordant with an increase in AHLs production (Farah *et al.*, 2005). In addition, DNA macroarray analysis revealed that *afeI* and *afeR* genes were overexpressed when cells were grown forming a biofilm over sulfur prills and not when grown in iron medium (Farah *et al.*, 2005; Valenzuela *et al.*, 2006), in agreement with Reverse-Transcriptase PCR (Rivas *et al.*, 2005) and Real-Time PCR experiments (unpublished results). All these results, taken together, are in agreement with the finding in other bacteria that the *A. ferrooxidans* QS type AI-1 could be linked to phosphate metabolism, and would influence exopolysaccharide production and biofilm formation.

In addition, our preliminary results from TLC and LC-MS-MS analysis indicate that some other collection strains of *A. ferrooxidans* and some collection strains of *A. thioooxidans* also produce AHLs and hydroxyl-AHLs with large acyl-chain (unpublished results) suggesting that interspecies communication could occur in mixed biofilms with *A. ferrooxidans* and other members of the biomining community.

3.2. QS and Biotechnological Applications

In natural microniches, bacteria compete with other organisms for nutrients. Therefore, since QS regulates key functions for bacterial population survival, bacteria and eukaryotes have developed strategies to interfere with bacterial communication. As it is now well established, QS signaling pathways coordinate numerous physiological functions involved in virulence of many human, animal and plant pathogens. Therefore QS has become a target for the exploration of new prophylactic strategies. The number of studies to identify molecules that could interfere with bacterial communication is growing fast (reviewed in Suga & Smith, 2003; Hentzer & Givskov, 2003; Zhang & Dong, 2004).

The first chemical inhibitor of QS which has been identified is a natural brominated furanone isolated from the macroalga *Delisea pulchra* (Figure 2).

This natural compound interferes with the R protein and it is currently accepted that it displaces the AHL molecules from their binding site. On the other hand, based on mutagenesis experiments, Koch *et al.* (2005) proposed that the receptor for halogenated furanones may be positioned differently in the AHL-binding cavity, but the authors suggest that further work is required to elucidate the precise mechanism for QS inhibition by these compounds. Since the identification of this natural

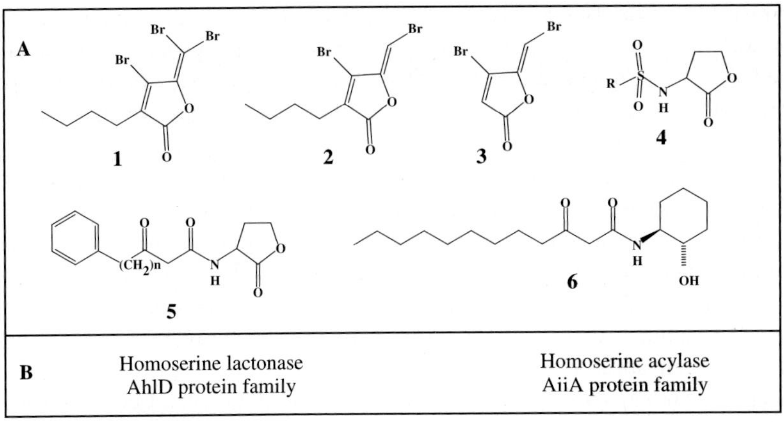

Figure 2. Examples of chemical and enzymatic options for QS interferences. **A**. AHL analogs. Natural antagonists from macroalga *D. pulchra* (1 and 2); Synthetic antagonists (3-5) and agonists (6). **B**. The known AHL-degrading enzyme families

inhibitor, synthesis of AHL analogs allowed the characterization of new antagonist and agonist molecules (Figure 2) (Smith *et al.*, 2003; Hentzer *et al.*, 2002; Reverchon *et al.*, 2002; Castang *et al.*, 2004; Frezza *et al.*, 2006).

A second approach to alter QS signaling has been developed by using an enzymatic way. Some bacteria naturally secrete AHL-degrading enzymes which can be classified into two families (Figure 2): the AHL lactonases and the AHL acylases, which hydrolyze the lactone ring or the acyl-chain of AHL, respectively (reviewed in Dong & Zhang, 2005). In addition, different studies revealed that immortalized human epithelial cell lines possess a proteic QS-quenching activity (Chun *et al.*, 2004) and mammalian sera had a lactonase-like enzyme(s) (Yang *et al.*, 2005). Based on these AHL-degrading activities, tobacco and potato plants have been genetically modified to overexpress AHL acylase and reported to be more resistant to the plant pathogen *Erwinia carotovora* (Dong *et al.*, 2001).

3.3. QS and Bioleaching Applications

To date, the role of QS in bioleaching communities and how this knowledge can affect bioleaching efficiency is still unknown. However, the characterization of a functional QS type AI-1 system in *A. ferrooxidans*, and the identification of various *afe* boxes in its genome strongly suggested the existence of a QS regulon (Farah *et al.*, 2005; Valenzuela *et al.*, 2006). Preliminary positive results about AHL production by another collection strains of *A. ferrooxidans* and some strains of *A. thiooxidans* and, on the other hand, the identification of a QS type AI-1 locus in *Leptospirillum sp.* type III (unpublished results), are opening a very interesting opportunity to explore new ways to eventually improve the bioleaching process.

A. ferrooxidans possesses a flagellum (DiSpirito *et al.*, 1982; Ohmura *et al.*, 1996) as well as genes coding for the type IV secretion apparatus which are transcribed in vivo (Christie & Vogel, 2000; Valenzuela *et al.*, 2006) and type IV pili (Barreto *et al.*, 2003). All these extracelullar appendages are involved in adhesion in many bacteria. It is also known that *A. ferrooxidans* adheres to solid substrates and is able to form biofilm structures during bioleaching (Karamanev, 1992; Blais *et al.*, 1994; Rohwerder *et al.*, 2003). In addition, Barreto *et al.* (2005) have identified a functional operon which is involved in exopolysaccharide synthesis and could participate in biofilm formation. Therefore, as is the case in many Gram-negative bacteria, the QS type AI-1 of *A. ferrooxidans* could regulate some of these physiological functions -and some yet to be defined- which are keys for cellular adhesion and colonization of minerals.

As described above, two approaches have been used to alter QS type AI-1. Nevertheless, for *A. ferrooxidans* and biomining microorganisms in general, the genetic transfer is a limiting step that restricts the design of genetic strategies to build *afeI* (-) mutants or to overexpress genes encoding for AHL-degrading activity. Therefore, the most feasible way available for QS interference in biomining and AMD biofilms or directly in *A. ferrooxidans* is through the chemical approach by using AHL analogs (Figure 3).

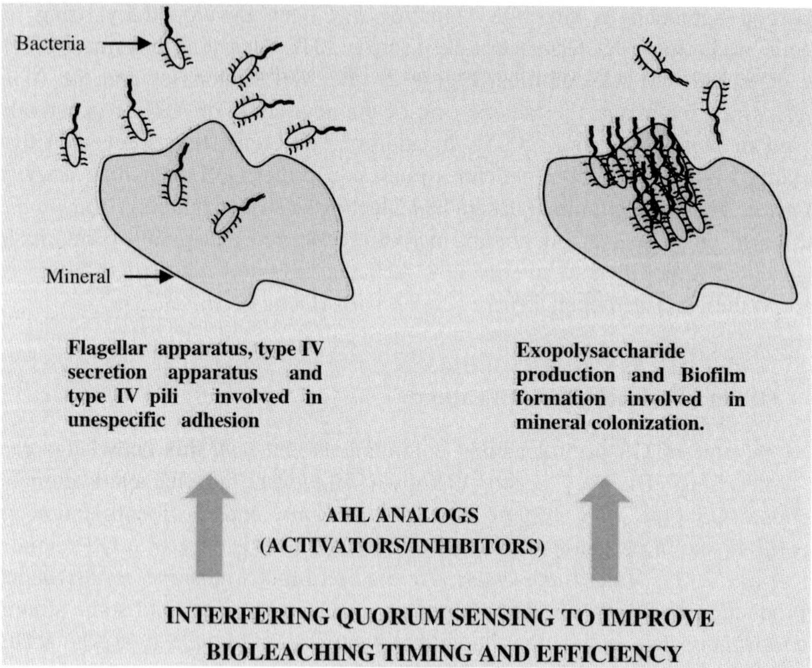

Figure 3. Quorum sensing and biomining improvement: a working model

The first available data revealed that AfeR protein could act as a positive transcriptional regulator (Rivas *et al.*, 2005). However, there is no data to exclude the possibility that AfeR is also able to act as a negative transcriptional regulator, depending on the target genes.

By testing agonist and antagonist molecules and analyzing their effects on unspecific adhesion, exopolysaccharide production and biofilm formation on the minerals surface (Figure 3), it should be possible to study the role of QS in the bioleaching process and eventually improve biomining efficiency by manipulating cell-cell communication. This approach should include studies with single and mixed populations, including bacteria-bacteria and bacteria-archaea communities.

4. ACKNOWLEDGMENTS

This work was supported by grants from FONDECYT (1040676 and 1030767). We acknowledge Dominique Haras and Danièle Morin for LC-MS-MS analysis of AHL extracts. We thank Alain Doutheau for critical reading of the manuscript.

5. REFERENCES

Arevalo-Ferro C, Hentzer M, Reil G, Gorg A, Kjelleberg S, Givskov M, Riedel K, Eberl L. 2003. Identification of quorum-sensing regulated proteins in the opportunistic pathogen *Pseudomonas aeruginosa* by proteomics. Environ Microbiol 5: 1350-1369.

Barreto M, Jedlicki E, Holmes DS. 2005. Identification of a gene cluster for the formation of extracellular polysaccharide precursors in the chemolithoautotroph *Acidithiobacillus ferrooxidans*. Appl Environ Microbiol 71: 2902-2929.

Barreto M, Quatrini R, Bueno S, Arriagada C, Valdés J, Silver S, Jedlicki E, Holmes DS. 2003. Aspects of the predicted physiology of *Acidithiobacillus ferrooxidans* deduced from an analysis of its partial genome sequence. Hydrometallurgy 71: 97-105.

Bassler B. 2002. Small talk: cell-to-cell communication in bacteria. Cell 109: 421-424.

Blais J, Tyagi R, Meunier N, Auclair J. 1994. The production of extracellular appendages during bacterial colonization of elemental sulphur. Process Biochem 29: 475-482.

Brader G, Sjoblom S, Hyytiainen H, Sims-Huopaniemi K, Palva ET. 2005. Altering substrate chain length specificity of an acylhomoserinelactone synthase in bacterial communication. J Biol Chem 280: 10403-10409.

Castang S, Chantegrel B, Deshayes C, Dolmazon R, Gouet P, Haser R, Reverchon S, Nasser W, Hugouvieux-Cotte-Pattat N, Doutheau A. 2004. N-Sulfonyl homoserine lactones as antagonists of bacterial quorum sensing. Bioorg Med Chem Lett 14: 5145-5149.

Chen H, Teplitski M, Robinson JB, Rolfe BG, Bauer WD. 2003. Proteomic analysis of wild-type *Sinorhizobium meliloti* responses to N-acyl homoserine lactone quorum-sensing signals and the transition to stationary phase. J Bacteriol 185: 5029-5036.

Christensen AB, Riedel K, Eberl L, Flodgaard LR, Molin S, Gram L, Givskov M. 2003. Quorum-sensing-directed protein expression in *Serratia proteamaculans* B5a. Microbiology 149: 471-483.

Christie PJ, Vogel JP. 2000. Bacterial type IV secretion: conjugation systems adapted to deliver effector molecules to host cells. Trends Microbiol 8: 354-360.

Chun CK, Ozer EA, Welsh MJ, Zabner J, Greenberg EP. 2004. Inactivation of a *Pseudomonas aeruginosa* quorum-sensing signal by human airway epithelia. Proc Natl Acad Sci USA 101: 3587-3590.

Davies DG, Parsek MR, Pearson JP, Iglewski BH, Costerton JW, Greenberg EP. 1998. The involvement of cell-to-cell signals in the development of a bacterial biofilm. Science 280: 295-298.

DiSpirito A, Silver M, Voss L, Tuovinen O. 1982. Flagella and pili of iron-oxidizing *Thiobacilli* isolated from uranium mine in northern Ontario, Canada. Appl Environ Microbiol 43: 1196-1200.

Dong YH, Wang LH, Xu JL, Zhang HB, Zhang XF, Zhang LH. 2001. Quenching quorum-sensing-dependent bacterial infection by an N-acyl homoserine lactonase. Nature 411: 813-817.

Dong YH, Zhang LH. 2005. Quorum sensing and quorum-quenching enzymes. J Microbiol 43: 101-109.

Farah C, Banderas A, Jerez CA, Guiliani N. 2004. Searching for physiological functions regulated by the Quorum-Sensing autoinducer AI-1 promoted by *afeI/afeR* genes in *Acidithiobacillus ferrooxidans*. In: Tsezos M, Hatzikioseyian A, Remoudaki E, eds. BioHydrometallurgy: a sustainable technology in evolution. Eds National Technical University of Athens, 1361-1368.

Farah C, Vera M, Morin D, Haras D, Jerez CA, Guiliani N. 2005. Evidence for a Functional Quorum-Sensing Type AI-1 System in the Extremophilic Bacterium *Acidithiobacillus ferrooxidans*. Appl Environ Microbiol 71: 7033-7040.

Frezza M, Castang S, Estephane J, Soulère L, Deshayes C, Chantegrel B, Nasser W, Queneau Y, Reverchon S, Doutheaua A. 2006. Synthesis and biological evaluation of homoserine lactone derived ureas as antagonists of bacterial quorum sensing. Bioorg Med Chem 14: 4781–4791.

Fuqua WC, Winans SC, Greenberg EP. 1994. Quorum sensing in bacteria: the LuxR-LuxI family of cell density-responsive transcriptional regulators. J Bacteriol 176: 269-275.

Gonzalez JE, Marketon MM. 2003. Quorum sensing in nitrogen-fixing *rhizobia*. Microbiol Mol Biol Rev 67: 574-592.

Gould TA, Schweizer HP. 2004. Churchill M.E. Structure of the *Pseudomonas aeruginosa* acyl-homoserinelactone synthase LasI. Mol Microbiol 53: 1135-1146.

Henke JM, Bassler BL. 2004. Bacterial social engagements. Trends Cell Biol 14: 648-656.

Hentzer M, Riedel K, Rasmussen TB, Heydorn A, Andersen JB, Parsek MR, Rice SA, Eberl L, Molin S, Hoiby N, Kjelleberg S, Givskov M. 2002. Inhibition of quorum sensing in *Pseudomonas aeruginosa* biofilm bacteria by a halogenated furanone compound. Microbiology 148: 87-102.

Hentzer M, Givskov M. 2003. Pharmacological inhibition of quorum sensing for the treatment of chronic bacterial infections. J Clin Invest 112: 1300-1307.

Horng YT, Deng SC, Daykin M, Soo PC, Wei JR, Luh KT, Ho SW, Swift S, Lai HC, Williams P. 2002. The LuxR family protein SpnR functions as a negative regulator of N-acylhomoserine lactone-dependent quorum sensing in *Serratia marcescens*. Mol Microbiol 45: 1655-1671.

Huber B, Riedel K, Hentzer M, Heydorn A, Gotschlich A, Givskov M, Molin S, Eberl L. 2001. The *cep* quorum-sensing system of *Burkholderia cepacia* H111 controls biofilm formation and swarming motility. Microbiology 147: 2517-2528.

Karamanev D. 1991. Model of the biofilm structure of *Thiobacillus ferrooxidans*. J Biotechnol 20: 51-64.

Kirke DF, Swift S, Lynch MJ, Williams P. 2004. The *Aeromonas hydrophila* LuxR homologue AhyR regulates the N-acyl homoserine lactone synthase, AhyI positively and negatively in a growth phase-dependent manner. FEMS Microbiol Lett 241: 109-117.

Koch B, Liljefors T, Persson T, Nielsen J, Kjelleberg S, Givskov M. 2005. The LuxR receptor: the sites of interaction with quorum-sensing signals and inhibitors. Microbiology 151: 3589-3602.

Labbate M, Queck SY, Koh KS, Rice SA, Givskov M, Kjelleberg S. 2004. Quorum sensing-controlled biofilm development in *Serratia liquefaciens* MG1. J Bacteriol 186: 692-698.

Llamas I, Quesada E, Martinez-Canovas MJ, Gronquist M, Eberhard A, Gonzalez JE. 2005. Quorum sensing in halophilic bacteria: detection of N-acyl-homoserine lactones in the exopolysaccharide-producing species of *Halomonas*. Extremophiles 9: 333-341.

Miller MB, Bassler BL. 2001. Quorum sensing in bacteria. Annu Rev Microbiol 55: 165-199.

Minogue TD, Wehland-von Trebra M, Bernhard F, von Bodman SB. 2002. The autoregulatory role of EsaR, a quorum-sensing regulator in *Pantoea stewartii* ssp. *stewartii*: evidence for a repressor function. Mol Microbiol 44: 1625-1635.

Mobarec JC, Gonzalez A, Perez T, Jerez CA, Guiliani N. Structural prerequisite involved in acyl-chain length selectivity of homoserine lactone synthases belonging to I protein family (submitted).

Nealson KH, Platt T, Hastings JW. 1970. Cellular control of the synthesis and activity of the bacterial luminescent system. J Bacteriol 104: 313-322.

Nouwens AS, Beatson SA, Whitchurch CB, Walsh BJ, Schweizer HP, Mattick JS, Cordwell SJ. 2003. Proteome analysis of extracellular proteins regulated by the *las* and *rhl* quorum sensing systems in *Pseudomonas aeruginosa* PAO1. Microbiology 149: 1311-1322.

Ohmura N, Tsugita K, Koizumi JI, Saika H. 1996. Sulfur-binding protein of flagella of *Thiobacillus ferrooxidans*. J Bacteriol 178: 5776-5780.

Paggi RA, Martone CB, Fuqua C, De Castro RE. 2003. Detection of quorum sensing signals in the haloalkaliphilic archaeon *Natronococcus occultus*. FEMS Microbiol Lett 221: 49-52.

Pearson JP, Van Delden C, Iglewski BH. 1999. Active efflux and diffusion are involved in transport of *Pseudomonas aeruginosa* cell-to-cell signals. J Bacteriol 181: 1203-1210.

Reverchon S, Chantegrel B, Deshayes C, Doutheau A, Cotte-Pattat N. 2002. New synthetic analogues of N-acyl homoserine lactones as agonists or antagonists of transcriptional regulators involved in bacterial quorum sensing. Bioorg Med Chem Lett 12: 1153-1157.

Riedel K, Arevalo-Ferro C, Reil G, Gorg A, Lottspeich F, Eberl L. 2003. Analysis of the quorum-sensing regulon of the opportunistic pathogen *Burkholderia cepacia* H111 by proteomics. Electrophoresis 24: 740-750.

Rivas M, Seeger M, Holmes DS, Jedlicki E. 2005. A Lux-like quorum sensing system in the extreme acidophile *Acidithiobacillus ferrooxidans*. Biol Res 38: 283-297.

Rohwerder T, Gehrke T, Kinzler K, Sand W. 2003. Bioleaching review part A: progress in bioleaching: fundamentals and mechanisms of bacterial metal sulfide oxidation. Appl Microbiol Biotechnol 63: 239-248.

Schuster M, Lostroh CP, Ogi T, Greenberg EP. 2003. Identification, timing, and signal specificity of *Pseudomonas aeruginosa* quorum-controlled genes: a transcriptome analysis. J Bacteriol 185: 2066-2079.

Smith KM, Bu Y, Suga H. 2003. Induction and inhibition of *Pseudomonas aeruginosa* quorum sensing by synthetic autoinducer analogs. Chem Biol 10: 81-89.

Suga H, Smith KM. 2003. Molecular mechanisms of bacterial quorum sensing as a new drug target. Curr Opin Chem Biol 7: 586-591.

Tomasz A. 1965. Control of the competent state in *Pneumococcus* by a hormone-like cell product: an example of a new type of regulatory mechanism in bacteria. Nature 208: 155-159.

Valenzuela L, Chi A, Beard S, Orell A, Guiliani N, Shabanowitz J, Hunt DF, Jerez CA. 2006. Genomics, metagenomics and proteomics in biomining microorganisms. Biotechnol Adv 24: 197-211.

Vannini A, Volpari C, Gargioli C, Muraglia E, Cortese R, De Francesco R, Neddermann P, Marco S. 2002. The crystal structure of the quorum sensing protein TraR bound to its autoinducer and target DNA. EMBO J 21: 4393-4401.

Wagner VE, Bushnell D, Passador L, Brooks AI, Iglewski BH. 2003. Microarray analysis of *Pseudomonas aeruginosa* quorum-sensing regulons: effects of growth phase and environment. J Bacteriol 185: 2080-2095.

Wagner-Dobler I, Thiel V, Eberl L, Allgaier M, Bodor A, Meyer S, Ebner S, Hennig A, Pukall R, Schulz S. 2005. Discovery of complex mixtures of novel long-chain quorum sensing signals in free-living and host-associated marine alphaproteobacteria. Chembiochem 6: 2195-2206.

Waters CM, Bassler BL. 2005. Quorum sensing: Cell-to-cell communication in bacteria. Annu Rev Cell Dev Biol 21: 319-346.

Watson WT, Minogue TD, Val DL, von Bodman SB, Churchill ME. 2002. Structural basis and specificity of acyl-homoserine lactone signal production in bacterial quorum sensing. Mol Cell 9: 685-694.

Welch M, Todd DE, Whitehead NA, McGowan SJ, Bycroft BW, Salmond GP. 2000. N-acyl homoserine lactone binding to the CarR receptor determines quorum-sensing specificity in *Erwinia*. EMBO J 19: 631-641.

Whitehead N, Barnard A, Slater H, Simpson N, Salmond GP. 2001. Quorum sensing in Gram-negative bacteria. FEMS Microbiol Rev 25: 365-404.

Winans S, Bassler B. 2002. Mob Psychology. J Bacteriol 184: 873-883.

Yang F, Wang LH, Wang J, Dong YH, Hu JY, Zhang LH. 2005. Quorum quenching enzyme activity is widely conserved in the sera of mammalian species. FEBS Lett 579: 3713-3717.

Zhang R, Pappas T, Brace J, Miller P, Oulmassov T, Molyneaux J, Anderson J, Bashkin J, Winans S, Joachimiak A. 2002. Structure of a bacterial quorum-sensing transcription factor complexed with pheromone and DNA. Nature 417: 971-974.

Zhang LH, Dong YH. 2004. Quorum sensing and signal interference: diverse implications. Mol. Microbiol 53: 1563-1571.

SECTION IV

OTHER APPLICATIONS

CHAPTER 14

BIOFLOTATION AND BIOFLOCCULATION OF RELEVANCE TO MINERALS BIOPROCESSING

K. HANUMANTHA RAO[1] AND S. SUBRAMANIAN[2]

[1] *Division of Mineral Processing, Luleå University of Technology, SE-97187 Luleå, Sweden*

[2] *Department of Metallurgy, Indian Institute of Science, Bangalore 560 012, India*

1. INTRODUCTION

Minerals bioprocessing, more specifically bio-beneficiation, concerns the enrichment of value minerals from ores and materials using conventional flotation and flocculation methods in the presence of microorganisms. The subject of bio-flotation and bio-flocculation is relatively new and there are already significant works reported in the literature although not overwhelming. Since the adhesion of microbial cells on minerals is a prerequisite to alter the surface chemistry of minerals favorable to either flotation or flocculation, most of the studies dealt on mineral-bacteria interactions in order to understand the mechanisms behind the mineral selectivity achieved in bio-beneficiation processes. This chapter attempts to survey the literature with a special emphasis on *Acidithiobacillus and Bacillus* group bacteria underlying our recent investigations in bioflotation and bioflocculation of metal sulfides. The potential of microorganisms in bio-beneficiation and the research needs to make it a viable technology are briefly outlined.

2. MICROORGANISMS IN MINERAL PROCESSING

Microbial life has been closely intertwined with the geosphere for nearly the entire history of the Earth (Vargas *et al.*, 1998). Microorganisms have a tremendous influence on their environment through the transfer of energy, charge, and materials across a complex biotic mineral-solution interface. Natural phenomena driven by the microbe-mineral interactions are incredibly diverse, including major environment and geochemical processes (Chapelle, 2001). Much of the impetus to study the mineral-bacteria interaction arises from the expected impact on many technological areas, including protection against bacterial infection and biofouling, bioremediations of organic, inorganic and radioactive contaminants, microbial fuel

E.R. Donati and W. Sand (eds.), Microbial Processing of Metal Sulfides, 267–286.

cells, bio-leaching, and bio-flotation. With the gradual depletion of high-grade ores and the finer dissemination of values in the currently processed ore bodies, it has become imperative to develop appropriate technologies to process such lean grade, refractory ores. Recent developments in biotechnology not only offer an attractive alternative to aid the processing of such ores, but also facilitate the remediation of environmental problems resulting from mining wastes. The term "mineral bioprocessing" has been has been aptly coined to describe the use of microorganisms and their direct derivatives in mineral processing, hydrometallurgy and in bioremediation of mineral industry discharges (Smith & Misra, 1991a, 1991b).

Although biohydrometallurgy is a fairly mature field for the leaching of low-grade sulfide ores especially for the recovery of copper, uranium, lead-zinc and precious metals such as gold and silver, the utilization of microorganisms in mineral engineering operations such as flocculation and flotation is of recent origin. Needless to emphasize, there is tremendous potential to use microorganisms as flocculants, flotation collectors, or depressants and in remediation processes, as can be understood by the special conferences dedicated to this subject (Smith & Misra, 1991a; Holmes & Smith, 1995; Hanumantha Rao & Forssberg, 2001). The biomodification of mineral surfaces involves the complex action of microorganism on the mineral surface. There are three different mechanisms by means of which the biomodification can occur:

i) attachment of microbial cells to the solid substrate, ii) oxidation reactions and iii) adsorption and/or chemical reaction with the metabolite products (EPS). Several types of autotropic and heterotrophic bacteria, fungi, yeasts and algae are implicated in minerals biobeneficiation. The major factors that influence the biomodification process include particle size and pulp density of the mineral suspension, inoculum concentration, pre-adaptation of the bacteria to the mineral substrate, contact time of mineral and microbe, pH and nutrient composition of the medium (Yelloji Rao & Somasundaran, 1995).

The bacterial cell surface structure is composed of the cell envelopes, cell wall, cell membrane and surface appendages like flagella, pilli, fimbriae, etc. (Figure 1). The distinction between Gram-positive and Gram-negative bacteria is the presence and absence of the outer most membrane. The peptidoglycan, the main constituent of cell wall, is thick in Gram-positive cell wall compared to a more complicated structure and thinner layer in Gram-negative bacteria. The outer membrane lipopolysaccharides (LPS) are highly hydrophilic and the presence of proteins outside the LPS layer lead to hydrophobic surface. The s-layer protein arrays are found in a wide range of both types of bacteria. The cell surface of all microorganisms carries negative charge, contributed by the phosphate, carboxylate and sulphate groups in the cell wall and capsular macromolecules (James, 1991; Rijnaarts *et al.*, 1995). Thus, the overall physico-chemical properties of the cell surface are a combination of the properties of these individual components, which will impart to mineral particulates up on adhesion.

Microorganisms adhere to mineral surfaces for various reasons. *S. oneidensis* utilize minerals as the terminal electron acceptor in the respiratory cycle. Bacteria

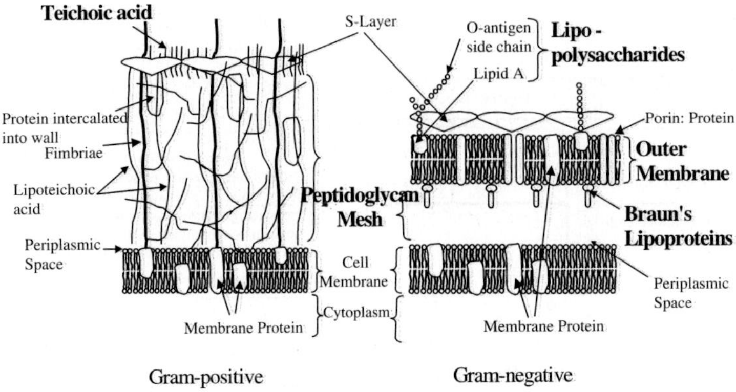

Figure 1. Bacterial cell surface structure (Hammond *et al.*, 1984; Hancock, 1991)

of *Thiobacilli* group recover energy from minerals by the enzymatic oxidation. Common to both these microbiological processes is the bacterial need to access, adhere to, and react with the mineral-water interface. Previous studies (Blake *et al.*, 1994; Deo *et al.*, 2001; Poortinga *et al.*, 2002) have shown that under most physiological conditions the bacterial cell surface carries a net negative charge, while, along with electrostatic forces, hydrophobic, entropic, acid-base, and van der Waals interactions and H-bonding are important in the bacterial adhesion. The microbe attachment to the mineral surface is followed/accompanied by expression of extracellular polymeric substances (EPS) enabling the adhesion, e.g., by trapping near surface or structural ions, changing thereby the charge of the bacterium envelope and/or that of the mineral surface. In addition, EPS can form chemical bonds with the surface and intermediate/promote the nutrition/ respiration chemical reactions. However, the specific mechanisms of adhesion and charge transfer reactions remain a subject of debate (Ehrlich & Brierley, 1990). In particular, the charge transfer mechanism can be direct (also called contact or enzymatic), indirect (mediated by quinine-containing shuttle compounds, or Fe^{3+} ions), and cooperative (includes both direct and indirect mechanisms). Thus, an understanding of the basis of bacterial cell adhesion in relation to their surface chemical characteristics and the resulting surface properties of minerals is of critical importance. The relevance of the physico-chemical surface characteristics of microbial cells to minerals bioprocessing is depicted in Figure 2 (Sharma *et al.*, 2001a).

The feasibility of adherence of a microorganism to a mineral surface will depend on the charge characteristics as well as the hydrophobicity of both the mineral surface and the microorganism. The necessary electrophoretic mobility and surface hydrophobicity values essential for adhesion of various microorganisms to a glass or other surfaces have been well documented by van Loosdrecht and co-workers (1987a, 1987b). Sharma & Hanumantha Rao (2002) have extensively studied and critically evaluated the surface energy of 140 bacterial and seven yeast cell surfaces following Fowkes, Equation of state, Geometric mean and Lifshitz–van der Waals

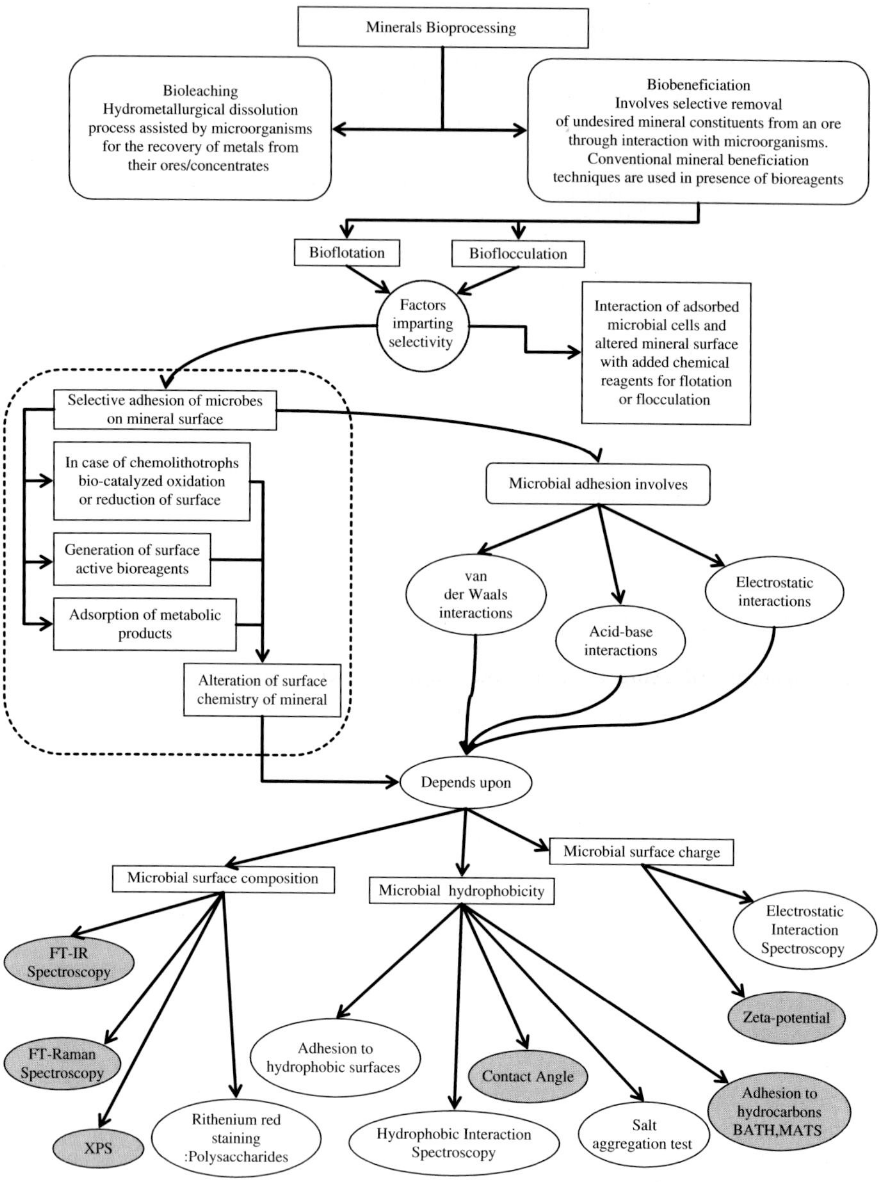

Figure 2. Relevance of physico-chemical surface characterization of microbial cells to minerals bioprocessing (Sharma *et al.*, 2001)

acid–base (LW–AB) approaches. Two independent issues were addressed separately in their analysis. At first, the surface energy and the different components of the surface energy for microbial cells surface were examined. Secondly, the different approaches were evaluated for their internal consistency, similarities and dissimilarities. The Lifshitz–van der Waals component of surface energy for most of the microbial cells was realized to be approximately 40 $mJ/m^2 \pm 10\%$. Equation of state and Geometric mean approaches did not possess any internal consistency and yielded different results. The internal consistency of the LW–AB approach could be checked only by varying the apolar liquid and it evaluated coherent surface energy parameters by doing so. The electron-donor surface energy component remained exactly the same with the change of apolar liquid. This parameter could differentiate between the Gram-positive and Gram-negative bacterial cells. Gram-negative bacterial cells having higher electron-donor parameter had lower nitrogen, oxygen and phosphorous content on their cell surfaces. Among the four approaches, LW–AB was found to give the most consistent results. This approach provided more detailed information about the microbial cell surface and the electron–donor parameter differentiated different type of cell surfaces (Sharma *et al.*, 2002).

3. APPLICATIONS OF MICROORGANISMS AS FLOCCULANTS AND FLOTATION REAGENTS IN MINERAL PROCESSING OPERATIONS

The possibility of using bacteria and fungi as flocculants for Florida phosphatic clays was investigated as early as 1963 (Gary *et al.*, 1963). The flocculation of organic and inorganic solid wastes using the products of lysed bacterial cells has been reported by Bernstein (1972). Pioneering studies by Smith and co-workers (1991) have shown that *Mycobacterium phlei (M. phlei)* was able to flocculate phosphate slimes, hematite and coal. It has been reported that *M. phlei* produce extracellular polymers and surfactants under certain conditions, which can cause flocculation of the microorganisms themselves or of other solids (Harris & Mitchell, 1973; Long & Wagner, 1987; Dugan, 1987; Finnerty & Singer, 1988; Unz, 1987). The unique combination of hydrophobic character and highly negatively charged surface facilitate *M. phlei* to be a good flocculant for dolomitic phosphate slime, fine hematite and fine coal particles (Misra *et al.*, 1991, 1993; Smith *et al.*, 1994; Raichur *et al.*, 1995). The addition of *Bacillus subtilis* and *M. phlei* onto dolomite and apatite was studied by sorption measurements and scanning electron microscopy. It was found that both species adsorb onto dolomite more readily than onto apatite at acidic and near neutral pH values. Thus both bacteria function as depressants in anionic collector flotation of dolomitic phosphate ores, while *Bacillus subtilis* functions as the stronger depressant, especially for dolomite (Zheng & Smith, 1997; Zheng *et al.*, 1998, 2001).

Several studies have shown the potential of microorganisms to be used as flotation collectors or modifiers. The depression of pyrite in coal flotation has been reported (Attia & Elzeky, 1985; Attia *et al.*, 1993; Capes *et al.*, 1973; Elzeky & Attia, 1987; Kawatra *et al.*, 1989; Atkins *et al.*, 1987). It has been suggested that the bacteria and their secretions, mainly polysaccharides and lipids, adsorb on the pyrite surface and render it more hydrophilic, resulting in its depression (Bagdigian & Myerson, 1986). A thermophilic microorganism, *Sulfolobus acidocaldarious*, has been utilized for the removal of pyritic sulfur from coal (Kargi & Robinson, 1985). It is generally postulated that the depression could either result from the microorganisms oxidizing the surfaces of pyrite, a sulfide mineral or from its adsorption onto the mineral, particularly when the microorganism has primarily hydrophilic surface (Smith & Misra, 1991b). Some microorganisms will also function under certain conditions as flotation activators, especially when the adsorbed microorganisms have active surface sites onto which a collector might adsorb.

The influence of simple salts of organic acids namely citrate, oxalate and tartarate and *Aspergillus niger* saccharose and molasses worts on the adsorption of sodium oleate onto salt type minerals such as barite, calcite and magnesite has been studied. It was observed that the adsorption of oleate was decreased on the chosen mineral surfaces in the presence of investigated worts (Sadowski & Golab, 1991). The flocculation of fine hematite, calcite and kaolinite suspensions using the yeast *Candida parapsilosus* and its derivatives has been carried out. It was observed that with the cells and their derivatives, excellent flocculation of calcite could be achieved. However a much greater weight of the whole cells was required for equivalent flocculation as compared to the cell fragments and the soluble fraction of the cells, which presumably consisted of proteins, fatty acids and amphoteric surfactants (Schneider *et al.*, 1994). *M. phlei* bacterial cells have been shown to function as a flotation collector for hematite (Dubel *et al.*, 1992; Smith *et al.*, 1993). The flocculation and flotation of ore by addition of hydrophobic *M. phlei* has been investigated. The flocculation phenomenon has been explained based on the hydrophobic interactions between the coal particles and the bacterium (Raichur *et al.*, 1996). The use of sulphate reducing bacteria (SRB), microbe fat and biomass in the flotation of several sulfide and non-sulfide minerals was reported for the first time by Solojenken and his co-workers (1976; 1979). The SRB was found to depress the flotation of both chalcopyrite and sphalerite but not that of molybdenite and galena. The good selectivity of microbe fat as a collector in the flotation of fluorspar, and biomass as a depressant for associated minerals namely, calcite, barite and quartz in the selective flotation of celestine was demonstrated. The ability of biomass macromolecules to hydrate in aqueous solution and to preferentially adsorb onto the gangue minerals enabled their use as depressants in non-sulfide flotation (Solojenken, 1979). The flocculation of fine fluorite and calcite particles with *Corynebacterium xerosis* has been carried out and high quality flocs obtained (Haas *et al.*, 1999, 2000). The hydrophobic bacterium namely *Rhodococcus opacus* has been investigated as a flotation reagent for the hematite-quartz system, wherein it was possible to float the hematite and depress the quartz (Mesquita *et al.*, 2003).

3.1. Bioflotation and Bioflocculation Studies using Acidithiobacillus' Group of Bacteria

Acidithiobacillus ferrooxidans (*A. ferrooxidans*), a commonly implicated autotrophic, acidophilic, mesophilic microorganism utilizes sulfur, thiosulphate and iron as energy source. They are Gram negative, small rods of 0.5 by 1-3 μm, occurring singly or occasionally in pairs. These bacteria have been extensively utilized in minerals bioprocessing.

The selective depression of pyrite from complex coal-pyrite matrix has been achieved using *A. ferrooxidans* (Dogan *et al.*, 1985, 1986; Harada & Kuniyoshi, 1985; Townsley *et al.*, 1987). The effect of growth medium of *A. ferrooxidans* on pyrite and galena flotation was studied by Misra & Chen (1995). It was observed that pyrite could be significantly depressed by using a combination of biotreatment involving *A. ferrooxidans* and xanthate collector. The mechanism of pyrite depression has been attributed to the oxidation of the elemental sulfur on pyrite surface leading to the formation of hydrophilic jarosite (Misra & Chen, 1995; Misra *et al.*, 1996). The effect of biotreatment with *A. ferrooxidans* on the floatability of galena and sphalerite has been reported (Yelloji Rao *et al.*, 1992a, 1992b, 1997).

The floatabilities of five sulfide minerals namely pyrite, chalcocite, molybdenite, millerite and galena were tested in the presence of *A. ferrooxidans* by Nagaoka and co-workers (1999). It was observed that pyrite was significantly depressed by the bacterium, while the floatabilities of the other sulfide minerals were not affected. It has been postulated that the suppression of pyrite floatability was caused by profuse bacterial addition to pyrite surfaces (Nagaoka *et al.*, 1999). The bioflotation of low grade Sarcheshmeh copper ore has been carried out using *A. ferrooxidans* bacteria, wherein it was found that pyrite was depressed while chalcopyrite and other sulfide minerals were unaffected. About 50% depression of pyrite was observed (Kolahdoozan *et al.*, 2004; Hosseini *et al.*, 2005). The effect of bacterial conditioning and the flotation of copper ore and concentrate from Turkey were investigated using *A. ferrooxidans*. The results indicated an increase in the copper grade to about 22% after bacterial conditioning due to pyrite depression (Yuce *et al.*, 2006). The surface characterization of *A. ferrooxidans* grown under different conditions has been carried out. The changes in the surface charge on the cells have been attributed to the differences in the protein content synthesized by the bacteria exposed to different growth conditions, as revealed by FT-IR and FT- Raman spectroscopic studies. (Sharma *et al.*, 2001a, 2001b, 2003; Hanumantha Rao *et al.*, 1998; Das *et al.*, 1999). The selective bioflotation of chalcopyrite from pyrite has also been achieved using *A. ferrooxidans* (Sharma *et al.*, 1999). The bacterial cells grown in the presence of sulfide minerals such as pyrite, chalcopyrite exhibited higher hydrophobicity than the cells grown in liquid 9K medium. It has been reported that a proteinaceous secretion was synthesized by *A. ferrooxidans* to facilitate adhesion on mineral surfaces (Devasia *et al.*, 1993, 1996).

The selective depression of pyrite from chalcopyrite or arsenopyrite as well as from a combination of chalcopyrite and arsenopyrite has been carried out by biomodulation in the presence of *A. ferrooxidans* (Chandraprabha *et al.*, 2004a, 2004b, 2005). It was observed that after interaction with the bacterial cells, pyrite remained depressed even in the presence of xanthate collectors and the sequence of conditioning with the bacterial cells and collector was found to have a bearing on the selectivity of separation. The initial conditioning of the mineral mixture with the collector followed by treatment with bacterial cells resulted in better selectivity. The extracellular polymeric substances derived from strains of *A. ferrooxidans* have been used as depressants in the bioflotation of sulfides such as pyrite, chalcopyrite, galena and sphalerite. The selective depression of pyrite from chalcopytire and sphalerite from galena has been reported using extracellular polysaccharides (Vogt *et al.*, 2003). The selective separation of sphalerite or arsenopyrite from pyrite has been studied in the presence of *A. ferrooxidans* (Deshpande *et al.*, 2001, 2004). The selective flotation separation of cinnabar from antimonite has been carried out by Lyalikova & Lyubavina (1986) using *A. ferrooxidans*. They suggested that antimonite was oxidized by the bacteria, leading to its depression, while cinnabar was not affected. It has been demonstrated that sulfide minerals such as pyrite and sulfur can be efficiently separated from quartz through selective bioflocculation using *A. ferrooxidans* and *A. thiooxidans* (Natarajan & Das, 2003).

The biomodulation of galena and sphalerite surfaces using *A. thiooxidans* has been studied by Santhiya and co-workers (2001a). It was found that galena was totally depressed in the pH range of 5-11 after bacterial interaction, while the flotation recovery of sphalerite was not affected. The significant differences in the adsorbabilities of the bacterial cells onto galena and sphalerite coupled with the nature of the interaction products, be it the respective sulfates or hydroxides, have been attributed to the contributing factors for the diametrically opposite flotation behavior observed with respect to galena and sphalerite after interaction with the cells. The flotation separation of chalcopyrite from pyrite using *A. thiooxidans* and potassium isopropyl xanthate has been recently studied (Chandraprabha & Natarajan, 2006). The selective bioflocculation of galena from sphalerite has also been successfully achieved (Santhiya *et al.*, 2000).

3.2. Bioflotation and Bioflocculation Studies using *Bacillus Species*

The biobeneficiation of bauxite for the removal of silica has been reported using *Bacillus circulans* (Groudeva & Groudeva, 1983) and *Bacillus mucilaginosis* (Karavaiko *et al.*, 1989; Ogurtsava *et al.*, 1990). The microbial ecology of bauxite ore and water samples from Jamnagar mines, India, has been studied and preliminary investigations showed that calcium and iron impurities could be selectively removed from bauxite using *Bacillus polymyxa* (Anand *et al.*, 1996; Natarajan *et al.*, 1997). It may be noted that recently *Bacillus polymyxa* has been renamed as *Paenibacillus polymyxa (P. polymyxa)* (Ash *et al.*, 1993). *P. polymyxa* is a nitrogen fixing chemo organotroph and a facultative anaerobe. It is known to produce metabolites such as acidic, formic, lactic and succinic acids besides ethanol, 2,3-butanediol and or acetoin (Roberts, 1947; Groudev & Groudeva, 1986; Gottschalk, 1989;

Mankad & Nauman, 1992). It also produces extracellular polysaccharides such as levan that forms the capsule of the organism (Murphy, 1952). The scaling up of bauxite beneficiation from shake flask experiments to demonstration bioreactors has been reported by Modak and co-workers (1999). The important microbiological and engineering challenges in the scaling up process have also been highlighted (Vasan *et al.*, 2001). The interaction of *P. polymyxa* with calcite, hematite, corundum, kaolinite and quartz resulted in the quartz and kaolinite surfaces being rendered more hydrophobic, whereas the other minerals exhibited enhanced hydrophilicity after bacterial interaction. The selective flocculation of quartz from its binary mixtures with corundum, hematite and calcite has been demonstrated (Deo & Natarajan, 1997a, 1997b, 1998). The mechanisms of adhesion of *P. polymyxa* onto the above minerals have been attributed to mainly electrostatic forces while in the case of quartz, chemical forces are additionally involved (Deo *et al.*, 2001). Furthermore, the interaction of metabolic products of *P. polymyxa* resulted in enhanced adsorption of bacterial proteins on quartz and kaolinite and of polysaccharides on hematite, corundum and calcite (Natarajan & Deo, 2001).

The adhesion behavior of *Paenibacillus polymyxa* bacteria on pyrite and chalcopyrite was examined by the surface thermodynamics and the extended DLVO theory approaches. In addition, the bacteria were adapted to pyrite and chalcopyrite minerals, and the adhesion behavior of these bacteria was also investigated. The significance of acid–base interactions in adhesion was assessed. The essential parameters needed for the calculations of interaction energy between bacteria and minerals were experimentally determined. The results illustrated that the bacterial surfaces were more energetic than the mineral surfaces and the bacteria acquired acid–base surface energy component during their adaptation to mineral. The XDLVO approach was found to be more effective in predicting the adhesion behavior than the expectations from the thermodynamic approach. The thermodynamic approach yielded no bacterial adhesion on minerals and this discrepancy was the result of inadequate description of electrostatic interactions. The adhesion predictions by the XDLVO theory were able to partially explain the bioflotation results of pyrite and chalcopyrite. This approach showed that on account of high bacterial surface energy, their aggregation was not feasible. However, due to the hydrophobicity of pyrite and chalcopyrite, their aggregation became possible (Sharma & Rao, 1999, 2003; Sharma *et al.*, 2000).

The selective separation of pyrite from quartz and calcite was achieved through microbially induced flotation and flocculation in the presence of *P. polymyxa* (Patra & Natarajan, 2003). Similarly, chalcopyrite could be selectively separated from quartz and calcite by treatment with *P. polymyxa* and extracellular protein (Patra & Natarajan, 2004). The microbially induced flotation and flocculation of pyrite from sphalerite or galena using bacterial cells and metabolic products of *P. polymyxa* has also been carried out (Patra & Natarajan, 2004a, 2004b, 2006). The bioflocculation of high ash Indian colas using *P. polymyxa* has been studied and the ash analysis of the flocculated portion showed a decrease in ash content by 60% attesting to selective flocculation of coal (Vijayalakshmi & Raichur, 2002).

Adhesion experiments revealed that maximum adhesion of *P. polymyxa* occurred onto the coal samples around the point of zero charge of the coal and the bacterium namely around pH 2 (Raichur & Vijayalakshmi, 2003). *Bacillus subtilis* was also found to be a good bioflocculant for fine coal and both adhesion and settling were found to be independent of pH (Vijayalakshmi & Raichur, 2003).

Extensive studies have been carried out on the bioprocessing of galena and sphalerite minerals in the presence of *P. polymyxa* cells as well as their metabolic products (Santhiya , 2001b). The interaction of *P. polymyxa* with the chosen minerals resulted in significant surface chemical changes on the minerals as well as the bacterial cells leading to shifts in their isoelectric points to less acidic pH values. It is worthy of mention that galena was rendered hydrophilic, while sphalerite was made hydrophobic consequent to the interaction with the bacterial cells and the metabolic products. Further, enhanced adhesion of bacterial cells onto galena surface was observed vis-à-vis sphalerite. The magnitude of adsorption of the various bioreagents was higher onto galena compared to sphalerite. The adsorption density of extracellular polysaccharide for sphalerite exhibited a characteristic maximum in the pH range of 6-7, while in the case of galena the adsorption density was found to increase with increase of pH. The adsorption of extracellular protein onto the chosen minerals was higher below pH 6. The adsorption isotherms of the cells and the various bioreagents for the minerals exhibited Langmuirian behavior. Co-adsorption tests indicated that the amount of extracellular polysaccharide adsorbed was diminished in the presence of the bacterial protein, but the adsorption of the protein was not altered in the presence of extracellular polysaccharide. A major finding in this investigation was that galena could be selectively depressed or flocculated from its synthetic mixture with sphalerite in the presence of bacterial cells, collector and activator, as well as by the metabolite of *P. polymyxa*. Additionally, *P. polymyxa* cells were adapted to galena and sphalerite. The polysaccharide content of the galena-adapted cells was found to be higher than that of the sphalerite-adapted cells. On the other hand, the protein content of the sphalerite-adapted cells was higher than that of the galena-adapted cells. The mechanisms of microbe-mineral interactions have been ascertained through ruthenium red adsorption, protein assay and cell surface hydrophobicity tests. The secreted bacterial protein has been characterized by SDS paging and mass spectroscopy techniques. Consequent to adaptation it was found that galena-adapted cells possessed a slime coating around the cell wall, which was composed of polysaccharides. On the contrary, the sphalerite-adapted cells did not display the slime layer (Figure 3). The adsorption density of ruthenium red onto the bacterial cells revealed that a higher amount of extracellular polysaccharides was present on galena-adapted cells compared to the Bromfield medium grown cells and sphalerite-adapted cells, which had the least amount. The cell surface hydrophobicity tests indicated that sphalerite-adapted cells exhibited the maximum hydrophobicity followed by the Bromfield medium grown cells, while the galena-adapted cells were least hydrophobic in nature. The protein assay tests highlighted that the sphalerite-adapted cells possessed the highest amount of protein whereas the galena-adapted cells contained the least amount. With respect

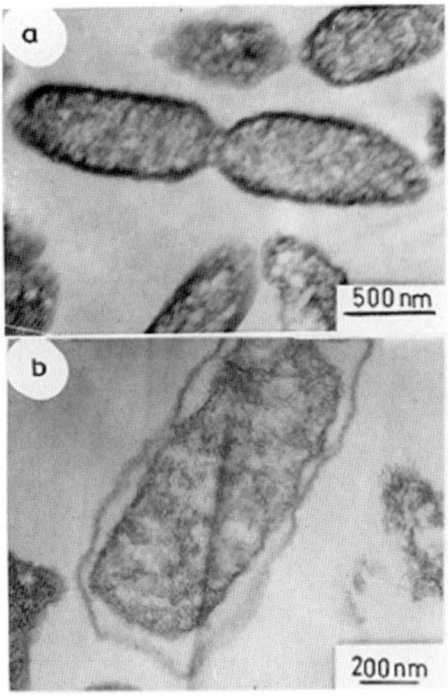

Figure 3. The transmission electron micrographs of (a) sphalerite-adapted and (b) galena-adapted cells

to the differential flotation tests, the effect of the sequence of addition of xanthate collector and bacterial cells onto sphalerite and galena, in the absence and in the presence of copper sulfate activator, showed that the adsorption of xanthate onto galena was diminished in the presence of the bacterial cells, whereas the xanthate adsorption density was unaffected in the pH range of 9-11 on copper activated sphalerite. The changes manifested in the bacterial cell walls consequent to the interaction with galena or sphalerite have resulted in the selective bioflocculation and flotation depression of the former, while the latter is preferentially dispersed or floated (Santhiya *et al.*, 2001b, 2001c, 2001d, 2001e, 2002, 2003, 2005; Subramanian *et al.*, 2003).

The results of the selective bioflotation and selective bioflocculation tests are summarized in Tables 1 and 2 respectively, for the galena-sphalerite system (Santhiya , 2001b). The selectivity index (S.I.) was calculated using the Gaudin's formula (Gaudin, 1987). Based on the results of the extensive differential flotation experiments carried out with different reagent combinations (Table 1), the following salient findings are noteworthy:

i) Good selectivity of separation of galena from sphalerite from a synthetic mixture could be achieved in the presence of *P. polymyxa* cells, activator and collector.
ii) The selectivity of separation was not adequate using sphalerite – adapted cells.
iii) Galena could be selectively separated from sphalerite in the presence of galena-adapted cells, collector and activator.

Table 1. Selective bioflotation test results on sphalerite-galena system using *P. polymyxa* cells, metabolite and extracellular secretions (Collector- 10^{-4} M PIPX; Activator- 10^{-6} M $CuSO_4$)

Experimental conditions	PBS %		ZnS %		S.I
	Float	Tails	Float	Tails	
Flotation with collector and activator only (pH 9–9.5)	98.6	1.2	94.1	2.0	0.76
	98.2	1.4	96.3	2.4	
Flotation with collector and activator after interaction with *P. polymyxa cells* (pH 9–9.5)	3.1	96.3	94.2	4.8	25.0
	4.1	95.2	96.2	3.3	
Flotation with collector and activator after interaction with sphalerite-adapted cells (pH 6.2–6.8)	53.8	45.2	91.3	8.6	2.98
	54.2	45.7	90.7	8.6	
Flotation with collector and activator after interaction with galena-adapted cells (pH 9–9.5)	3.6	94.4	90.3	8.7	17.52
	2.5	96.7	88.2	9.5	
Flotation with collector and activator after interaction with *P. polymyxa metabolite* (pH 9–9.5)	3.7	95.7	94.6	4.9	22.76
	3.9	95.2	95.2	4.3	
Flotation with collector and activator after interaction with ECP (50 ppm; pH 9–9.5)	2.4	96.3	93.8	4.5	31.77
	1.9	97.2	94.7	3.9	
Flotation with collector and activator after interaction with secreted protein (50 ppm; pH 6.2–6.8)	95.2	4.2	96.2	3.8	1.15
	94.3	5.1	94.3	4.3	
Flotation with collector and activator after interaction with ECP (50 ppm) + secreted protein (50ppm) (pH 6.2–6.8)	4.3	93.4	95.2	3.6	26.60
	3.6	94.8	96.3	2.8	

Table 2. Selective bioflocculation test results on sphalerite-galena system using P. polymyxa cells, metabolite and extracellular secretions

Experimental conditions	ZnS %		PbS %	
	Dispersed	Flocculated	Dispersed	Flocculated
Control at pH 9–9.5	70.5	27.8	27.0	71.7
	70.3	29.5	27.0	72.3
Control at pH 6.2–6.8	72.5	27.3	55.4	43.8
	70.8	28.6	56.7	73.2
In the presence of un-adapted *P. polymyxa* cells at pH 9–9.5	93.7	4.9	5.6	94.2
	94.3	4.6	5.8	94.9
In the presence of galena-adapted cells at pH 9–9.5	75.7	23.9	5.6	94.2
	70.3	24.6	5.8	94.1
In the presence of sphalerite-adapted cells at pH 6.2–6.8	89.6	9.8	88.7	9.3
	90.3	8.6	86.4	10.1
In the presence of *P. polymyxa* metabolite at pH 9–9.5	91.8	5.4	6.8	92.6
	91.8	4.7	6.3	92.8
In the presence of 50 ppm ECP at pH 9–9.5	75.3	23.7	5.6	93.7
	72.6	26.8	4.2	95.2
In the presence of 50 ppm secreted protein at pH 6.2–6.8	95.7	6.1	93.4	6.1
	93.2	5.7	94.6	4.8
In the presence of 50 ppm ECP + 50 ppm secreted protein at pH 6.2–6.8	96.7	3.1	7.6	91.3
	95.2	4.2	8.2	90.8

iv) The addition of *P. polymyxa* metabolite enabled selective separation of galena from sphalerite, with and without the addition of collector and activator.
v) Differential flotation tests with the secreted protein did not give an adequate separation.
vi) A high degree of selectivity was achieved using extracellular polysaccharide (ECP), activator and collector.
vii) Selective flotation experiments using a binary combination of ECP and secreted protein yielded a high selectivity index in the presence of collector and activator.

Based on the detailed selective bioflocculation tests carried out on a 1:1 synthetic mixture of galena and sphalerite (Table 2), the degree of selectivity of the reagents could be arranged in the following sequence of decreasing magnitude: *P. polymyxa* unadapted cells ≈ metabolite ≈ ECP+ secreted protein > ECP ≈ galena-adapted cells » sphalerite-adapted cells > secreted protein.

4. SUMMARY

The bioflotation and bioflocculation of sulfide minerals pose unprecedented intellectual and technical challenges. Apart from complex, variable, competitive, and non-equilibrium chemical interactions and intricate composition and structure of biomolecules, electrochemical and electrostatic interactions are triggered at mineral-solution interfaces. Surface species are essentially unstable, evolving in time and under variation of the environment. In the case of mineral-bacteria interfaces, an additional difficulty is associated with sensitivity of the physiological status of microbe, the metabolic reactions, and hence the whole interface phenomenon to any invasive intervention. At the same time, the flotation response can be significantly affected by chemical-structural changes at monolayer fractions. The problems that must be solved before such microorganisms or their metabolic products can be used in commercial operations include culturing the organisms at a cheap cost and reduction in dosages required for various separation processes. The issues to be addressed to transform bioflotation and bioflocculation from a laboratory curiosity to a future technology have been recently highlighted by Smith and Miettinen (2006), with respect to culturing and handling of the microorganisms, bioreagent dosages required for a given industrial application, materials and operating costs for bacterial reagents and capital equipment costs vis-à-vis the existing technologies.

The precise elucidation of the mechanisms of adsorption of both bacterium and reagent (collector) in the presence of the adhered bacteria is necessary. However, this area has not been investigated at the moment. Several hypotheses have been postulated that bacterial adsorption can occur physically (electrostatically or via van der Waals interactions) or chemically, but there is a lack of understanding as to how the reagent adsorption is affected in the presence of the bacteria. In this context, the tremendously complex problem of gaining insight into the mechanisms of adsorption of living organisms is compounded by a not-less-complex surface chemistry in a pulp, in so far as even in the absence of adsorbed bacteria, the sulfide surface species are unstable. Hence exploring bioflotation processes is a real challenge for a researcher.

The interfaces between biological and geological materials, as well as the means to design and manipulate that interface are virtually completely unexplored. Previous studies of interactions of bacteria with metal oxides and sulfides were focused either on the particular biochemical aspects or on the process engineering. The bioflotation/bioflocculation process concerns the mineral response to the bacterium presence, which is essentially interplay between microorganism and the physicochemical properties of the mineral surface, such as the atomic and electronic structure, the net charge/potential, acid-base properties, and wettability of the surface.

Though the mechanisms of adsorption of the microbe's extracellular polymeric substance (EPS) components onto solid surfaces are generally known, they are system specific. These mechanisms need to be investigated thoroughly. The interaction between the biopolymers produced by the bacteria grown in the presence of specific minerals and surfactants also need special attention focusing on the applicability to bioflotation. More specifically, the EPS produced by bacteria during the growth in 1) culture media, 2) in the presence of sulfide minerals, and 3) in the presence of both mineral and flotation reagent, need chemical characterization and structural identification. Once the specificity between biopolymer and sulfide mineral is established, the selective biopolymers can be synthesized for practical application.

5. ACKNOWLEDGEMENTS

The authors express their sincere gratitude to the Swedish Foundation for International Co-operation in Research and Higher Education (STINT), Stockholm, Sweden for financial support for carrying out this investigation. The first author also acknowledges the financial support from the EU BioMinE project.

6. REFERENCES

Anand P, Modak JM, Natarajan KA. 1996. Biobeneficiation of bauxite using *Bacillus polymyxa*: calcium and iron removal. Int J Miner Process 48: 51-60.

Ash C, Priest FG, Collins MD. 1993. Molecular identification of rRNA Group 3 Bacilli using a PCR probe test. Antonie van leeuwenhoek 64: 253-260.

Atkins AS, Bridgewood EW, Davis AJ. 1987. A study of the suppression of pyritic sulfur in coal flotation by *Thiobacillus ferrooxidans*. Coal Prep 5: 1-13.

Attia YA, Elzeky MA. 1985. Biosurface modification in the separation of pyrite from coal by froth flotation in processing and utilization of high sulfur coals. In: Attia YA, ed. Elsevier, Amsterdam, 673-682.

Attia YA, Elzeky MA, Ismail M. 1993. Enhanced separation of pyrite from oxidized coal by froth flotation using surface modification. Int J Miner Process 37: 61-71.

Bagdigian RM, Myerson AS. 1986. The adsorption of *Thiobacillus ferrooxidans* on coal surfaces. Biotechnol Bioeng 28: 467-479.

Bernstein RA. 1972. Waste treatment with microbial nucleo-protien flocculating agent, US Patent No. 3,684, 706.

Blake RC, Elizabeth AS, Howard GT. 1994. Solubilization of minerals by bacteria: Electrophoretic mobility of *Thiobacillus ferrooxidans* in the presence of iron, pyrite and sulfur. App Environ Microbiol 60: 3349-3357.

Capes CE, McIlhinney AE, Sirianni AF, Puddington IE. 1973. Bacterial oxidation in upgrading pyritic coals. Can Min Metall Bull 66: 88-91.

Chandraprabha MN, Natarajan KA, Modak JM. 2004a. Selective separation of pyrite and chalcopyrite by biomodulation. Coll Surf B 37: 93-100.

Chandraprabha MN, Natarajan KA, Somasundaran P. 2004b. Selective separation of arsenopyrite from pyrite by biomodulation in the presence of *Acidithiobacillus ferrooxidans*. J Colloid Interface Sci 276: 323-332.

Chandraprabha MN, Natarajan KA, Somasundaran P. 2005. Selective separation of pyrite from chalcopyrite and arsenopyrite by biomodulation using *Acidithiobacillus ferrooxidans*. Int J Miner Process 75: 113-122.

Chandraprabha MN, Natarajan KA. 2006. Surface chemical and flotation behavior of chalcopyrite and pyrite in the presence of *Acidithiobacillus thiooxidans*. Hydrometallurgy, in press.

Chapelle FH. 2001. Ground-Water Microbiology and Geochemistry, 2nd ed., Wiley, New York.

Das A, Hanumantha Rao K, Sharma PK, Natarajan KA, Forssberg KSE. 1999. Surface chemical and adsorption studies using *Thiobacillus ferrooxidans* with reference to bacterial adhesion to sulfide minerals. Amils R, Ballester A, eds. Biohydrometallurgy and the environment toward the mining of the 21st century, Proceedings of the International Biohydrometallurgy Symposium IBS-99, El Escorial, Spain. Elsevier, Amsterdam, Part A, 697-707.

Deo N, Natarajan KA. 1997a. Surface modification and biobeneficiation of some oxide minerals using *Bacillus polymyxa*, Min Metall Process 14: 32-39.

Deo N, Natarajan KA. 1997b. Interaction of *Bacillus polymyxa* with some oxide minerals with reference to mineral beneficiation and environmental control. Miner Eng 10: 1339-1354.

Deo N, Natarajan KA. 1998. Studies on interaction of *Paenibacillus polymyxa* with iron ore minerals in relation to beneficiation. Int J Miner Process 55: 41-60.

Deo N, Natarajan KA, Somasundaran P. 2001. Mechanisms of adhesion of *Paenibacillus polymyxa* onto hematite, corundum and quartz. Int J Miner Process 62: 27-39.

Deshpande RJ, Subramanian S, Natarajan KA. 2001. Effect of bacterial interaction with *Thiobacillus ferrooxidans* on the floatability of spahlerite and pyrite. In: Padmanabhan NPH, Sreenivas T, Srinivas K, eds. Proceedings of the International Seminar on Mineral Processing Technology. Allied Publishers Limited, New Delhi, 159-161.

Deshpande RJ, Subramanian S, Natarajan KA. 2004. Surface-chemical studies on pyrite and arsenopyrite using *Acidithiobacillus ferrooxidans*. In: Rao GV, Misra VN, eds. Proceedings of the International Seminar on Mineral Processing Technology. Allied Publishers Limited, New Delhi, 668-675.

Devasia P, Natarajan KA, Sathyanarayana DN, Ramananda Rao G. 1993. Surface chemistry of *Thiobacillus ferrooxidans* relevant to adhesion on mineral surfaces. Appl Environ Microbiol 59: 4051-4055.

Devasia P, Natarajan KA, Ramananda Rao G. 1996. Role of bacterial growth conditions and adhesion in the bioleaching of chalopyrite by *Thiobacillus ferrooxidans*. Miner Metall Process 13: 82-88.

Dogan ZM, Ozbayoglu G, Hicyilmaz C, Sarikaya M, Ozcengiz G. 1985. Bacterial leaching versus bacterial conditioning and flotation in desulfurization of coal. In: Proceedings of the XV Int Miner Process Congr. Cannes, France, Vol 2, 304-313.

Dogan ZM, Ozbayoglu G, Hicyilmaz C, Sarikaya M, Ozcengiz G. 1986. Bacterial leaching versus bacterial conditioning and flotation in desulfurization three different coals. In: Lawrence RW, Branion RMR, Ebner HG, eds. Fundamental and Applied Biohydrometallurgy. Elsevier, New York, 165-170.

Dubel J, Smith RW, Misra M, Chen S. 1992. Microorganisms as chemical reagents: the hematite system. Miner Eng 5: 547-556.

Dugan PR. 1987. The function of microbial polysaccharides in bioflocculation and biosorption of metal ions. In: Attia YA, ed. Flocculation in Biotechnology and Separation Systems. Elsevier, Amsterdam, 337-350.

Elzeky MA, Attia YA. 1987. Coal slurries desulfurization by flotation using thiophilic bacteria for pyrite depression. Coal Prep 5: 15-37.

Ehrlich EC, Brierley CL. 1990. Microbial Mineral Recovery. McGraw-Hill, New York.

Finnerty WR, Singer ME. 1988. A microbial surfactant: Genetic engineering and applications. In: Attia A, Moudgil BM, Chander S. eds. Interfacial Phenomena in Biotechnology and Materials Processing, Elsevier, Amsterdam, 75-88.

Gaudin AM. 1987. Principles of Mineral Dressing, 5*th* ed. Tata McGraw Hill, New Delhi, 231-238.

Gary JH, Fled IL, Davis EG. 1963. Chemical and physical beneficiation of Florida phosphate slimes. USBM RI 6163, 33-34.

Groudev SN, Groudeva VI. 1986. Biological leaching of aluminium from clays. In: Biotechnol Bioeng Symp. Wiley, New York, 91-99.

Groudeva VI, Groudev SN. 1983. Bauxite dressing by means of *Bacillus circulans*. Travaux ICSOBA 13: 257-263.

Gottschalk G. 1989. Nutrition of bacteria. In: Starr MP, ed. Bacterial Metabolism, Springer, New York, 1-10.

Haas SR, Nascimento FR, Schneider IAH, Gaylarde C. 1999. Flocculation of fine fluorite particles with *Coryevacterium xerosis*. Revista de Microbiologia 30: 225-230.

Haas SR, Nascimento FR, Schneider IAH. 2000. Flocculation of fine calcite particles with *Coryevacterium xerosis*. In: Proceedings of the XXI Int Miner Process Congr, Rome, Vol A, 57-61.

Hammond SM, Lambert PA, Rycroft AN. 1984. The Bacterial Cell Surface, Kapitan Szabo, Washington DC.

Hancock IC. 1991. Microbial cell surface architecture. In: Mozes N, Handley PS, Busscher HJ, Rouxhet PG, eds. Microbial Cell Surface Analysis, VCH, New York, 22-59.

Hanumantha Rao K, Das A, Sharma PK. 1998. Microbes adsorption on sulfide minerals and bioflotation. In: Proc Konf Mineralteknik, Föreningen Mineralteknisk Forskning, MinFo, Stockhom, 91-110.

Hanumantha Rao K, Forssberg KSE. 2001. Minerals Bioprocessing IV. Int. J Miner Process 62: 1-332.

Harada T, Kuniyoshi N. 1985. Effects of bacterial oxidation on the floatability of pyrite, J Inst Min Metal Inst Japan 101: 719-724.

Harris RH, Mitchell R. 1973. The role of polymers in microbial aggregation. In: Starr MP, Ingraham JL, Raffel S, eds. Annual Reviews of Microbiology, Annual Reviews Inc, Palo Alto, 27, 27-50.

Holmes DS, Smith RW. 1995. Minerals Bioprocessing-II, The Minerals, Metals and Materials Society, Pennsylvania.

Hosseini TR, Kolahdoozan M, Tabatabaei YSM, Oliazadeh M, Noaparast M, Eslami A, Manafi Z, Alfantazi A. 2005. Bioflotation of Sarcheshmeh copper ore using *Thiobacillus ferrooxidans* bacteria. Miner Eng 18: 371-374.

James AM. 1991. Charge properties of microbial cell surfaces. In: Mozes N, Handley PS, Busscher HJ, Rouxhet PG, eds. Microbial Cell Surface Analysis, VCH, New York, 221-261.

Karavaiko GI, Avakyan ZA, Ogurtsava LV, Safanova OF. 1989. Microbiological processing of bauxite. In: Salley J, McGready RGL, Wichlacz L, eds. Biohydrometallurgy, CANMET, Ottawa, 93-102.

Kargi F, Robinson JM. 1985. Biological removal of pyretic sulfur from coal by the thermophilic organism *Sulfolobus acidocaldarious*, Biotechnol Bioeng 27: 41-49.

Kawatra SK, Eisele TC, Bagley ST. 1989. Studies of pyrite dissolution in Pachna tanks and depression pyrite flotation by bacteria. In: Biotechnology in Minerals and Metal Processing, SME/AIME, 55-61.

Kolahdoozan M, Tabatabaei SM, Yen WT, Hosseini Tabatabaei R, Shahverdi AR, Oliazadeh M, Noaparast M, Eslami A, Manafi Z. 2004. Bioflotation of low grade Sarcheshmeh Copper sulfide. Trans Indian Inst Met 57: 485-490.

Long S, Wagner F. 1987. Structure and properties of biosurfactants. In: Kosaric N, Cairns WL, Gray NCC, eds. Biosurfactants and Biotechnology, Marcel Dekker, New York, 21-46.

Lyalikova NN, Lyubavina LL. 1986. On the possibility of using a culture of *Thiobacillus ferrooxidans* to separate antimony and mercury sulfides during flotation. In: Lawrence RW, Branion RMR, Ebner HG, eds. Fundamental and Applied Biohydrometallurgy, Elsevier, New York, 403-406.

Mankad T, Nauman EB. 1992. Effect of oxygen on steady state product distribution in *Bacillus polymyxa* fermentations. Biotechnol Bioeng 40: 413-426.

Mesquita LMS, Lins FF, Torem ML. 2003. Interaction of a hydrophobic bacterium strain in a hematite-quartz flotation system. Int J Miner Process 71: 31-44.

Misra M, Smith RW, Dubel J. 1991. Bioflocculation of finely divided minerals. In: Smith RW, Misra M, eds. Mineral Bioprocessing, The Minerals, Metals and Materials Society, Pennsylvania, 91-103.

Misra M, Chen S, Smith RW, Raichur AM. 1993. *Mycobacterium phlei* as a flotation collector for hematite. Miner Metall Process 10: 170-175.

Misra M, Chen S. 1995. The effect of growth medium of *Thiobacillus ferrooxidans* on pyrite and galena flotation. In: Holmes DS, Smith RW, eds. Minerals Bioprocessing II, The Minerals, Metals and Materials Society, Pennsylvania, 313-322.

Misra M, Bukka K, Chen S. 1996. The effect of growth medium of *Thiobacillus ferrooxidans* on pyrite flotation. Miner Eng 9: 157-168.

Modak JM, Vasan SS, Natarajan KA. 1999. Calcium removal from bauxite using *Bacillus polymyxa.* Miner Metall Process 16: 6-12.

Murphy D. 1952. Structure of levan produced by *Bacillus polymyxa*. Can J Chem 30: 872-878.

Nagaoka T, Ohmura N, Saiki H. 1999. A Novel mineral flotation process using *Thiobacillus ferrooxidans*, Appl Env Microbiol 65: 3588-3593.

Natarajan KA, Modak JM, Anand P. 1997. Some microbiological aspects of bauxite mineralisation and beneficiation. Miner Metall Process 14: 47-53.

Natarajan KA, Deo N. 2001. Role of bacterial interation and bioreagents in iron ore flotation. Int J Miner Process 62: 143-157.

Natarajan KA, Das A. 2003. Surface chemical studies on '*Acidithiobacillus*' group of bacteria with reference to mineral flotation. Int J Miner Process 72: 189-198.

Ogurtsava LV, Karavaiko GI, Avakyan ZA, Korenevskii AA. 1990. Activity of various microorganisms in extracting elements from bauxite. Microbiology 58: 774-780.

Patra P, Natarajan KA. 2003. Microbially induced flocculation and flotation for pyrite separation from oxide gangue minerals. Miner Eng 16: 965-973.

Patra P, Natarajan KA. 2004a. Microbially induced flocculation and flotation for separation of chalcopyrite from quartz and calcite. Int J Miner Process 74: 143-155.

Patra P, Natarajan KA. 2004b. Microbially induced flocculation and flotation of pyrite and sphalerite. Coll Surf B 36: 91-99.

Patra P, Natarajan KA. 2006. Surface chemical studies on selective separation of pyrite and galena in the presence of bacterial cell and metabolic products of *Bacillus polymyxa*, J Colloid Interface Sci 298: 720-729.

Poortinga AT, Bos R, Norde W, Busscher HJ. 2002. Electrical double layer interactions in bacterial adhesion to surfaces. Surf Sci Reports 47: 1-32.

Raichur AM, Misra M, Smith RW. 1995. Differential adhesion of hydrophobic bacteria onto coal and associated minerals. Coal Prep 16: 51-63.

Raichur AM, Misra M, Bukka K, Smith RW. 1996. Flocculation and flotation of coal by adhesion of hydrophobic *Mycobacterium phlei*. Coll Surf B 8: 13-24.

Raichur AM, Vijayalakshmi SP. 2003. The effect of nature of raw coal on the adhesion of bacteria to coal surface. Fuel 82: 225-231.

Rijnaarts HM, Norde W, Lyklema J, Zehnder A. 1995. The isoelectric point of bacteria as an indicator fort he presence of cell surface polymers that inhibit adhesion. Coll Surf B 4: 191-197.

Roberts JL. 1947. Reduction of ferric hydroxide by strains of *Bacillus polymyxa*. Soil Sci 63: 135-140.

Sadowski Z, Golab Z. 1991. Biomodification of mineral surface properties by *Aspergillus niger*. In: Smith RW, Misra M, eds. Mineral Bioprocessing, The Minerals, Metals and Materials Society, Pennsylvania, 81-90.

Santhiya D, Subramanian S, Natarajan KA. 2000. Surface chemical studies on galena and sphalerite in the presence of *Thiobacillus thiooxidans* with reference to mineral beneficiation. Miner Eng 13: 747-763.

Santhiya D, Subramanian S, Natarajan KA, Hanumantha Rao K, Forssberg KSE. 2001a. Biomodulation of galena and sphalerite surfaces using *Thiobacillus thiooxidans*. Int J Miner Process 62: 121-141.

Santhiya D. 2001b. Investigations in to the surface chemistry of galena, sphalerite and sulfur minerals using *Thiobacillus thiooxidans* and *Bacillus polymyxa*. Doctoral thesis, Indian Institute of Science, Bangalore, India.

Santhiya D, Subramanian S, Natarajan KA. 2001c. Surface chemical studies on sphalerite and galena using *Bacillus polymyxa*. Part I: Microbially induced mineral separation. J Colloid Interface Sci 235: 289-297.

Santhiya D, Subramanian S, Natarajan KA. 2001d. Surface chemical studies on sphalerite and galena using *Bacillus polymyxa*. Part II: Mechanisms of microbe-mineral interaction. J Colloid Interface Sci 235: 298-309.

Santhiya D, Subramanian S, Natarajan KA. 2001e. Role of microorganisms in surface chemical changes on sphalerite and galena. In: Ciminelli VST, Garcia Jr. O. Biohydrometallurgy: Fundamentals, Technology and Sustainable Development. Elsevier Science BV, Amsterdam, Part B, 13-22.

Santhiya D, Subramanian S, Natarajan KA. 2002. Surface chemical studies on sphalerite and galena using extracellular polysaccharides isolated from *Bacillus polymyxa*. J Colloid Interface Sci 256: 237-248.

Santhiya D, Subramanian S, Natarajan KA, Indi SS. 2003. Surface chemical studies on galena and sphalerite using mineral-adapted *Bacillus polymyxa*. In: Lorenzen L, Bradshaw DJ, eds. Proceedings of the XXII Int Miner Process Congr, The South African Institute of Mining and Metallurgy, South Africa, Vol. 2, 902-911.

Santhiya D, Subramanian S, Natarajan KA. 2005. Effect of bacterial proteins and polysaccharides on the selective separation of sphalerite from galena. In: Harrison STL, Rawlings DE, Petersen J, eds.

Proceedings of the 16th International Biohydrometallurgy Symposium, September 25–29, Cape Town, South Africa. Produced by Compress www.compress.co.za, 325-333.

Schneider IAH, Misra M, Smith RW. 1994. Bioflocculation of fine suspensions by *Candida parapsilosis* and its sonication products. In: Mulukutla PS, ed. Reagents for Better Metallurgy, TMS, Warrendale, 197-208.

Sharma PK, Das A, Hanumantha Rao K, Forssberg KSE. 1999. *Thiobacillus ferrooxidans* interaction with sulfide minerals and selective chalcopyrite flotation from pyrite. In: Parekh BK, Miller JD, eds. Advances in Flotation Technology, SME/AIME, 147-165.

Sharma PK, Hanumantha Rao K. 1999. Role of a heterotrophic *Paenibacillus polymyxa* bacteria in the bioflotation of some sulfide minerals. Miner Metal Process 16: 35-41.

Sharma PK, Hanumantha Rao K, Natarajan KA, Forssberg KSE. 2000. Bioflotation of sulfide minerals in the presence of heterotrophic and chemolithotrophic bacteria. In: Massacci P, ed. Proceedings of the XXI Int Miner Process Cong, Developments in Mineral Processing, Elsevier, Vol. B, B8a 93-103.

Sharma PK, Hanumantha Rao K, Forssberg KSE, Natarajan KA. 2001a. Surface chemical characterization of *Paenibacillus polymyxa* before and after adaptation to sulfide minerals. Int J Miner Process 62: 3-25.

Sharma PK, Das A, Hanumantha Rao K, Natarajan KA, Forssberg KSE. 2001b. Surface characterisation of *Thiobacillus ferrooxidans* cells grown under different conditions. In: Ciminelli VST, Garcia Jr. O. Biohydrometallurgy: Fundamentals, Technology and Sustainable Development. Elsevier Science BV, Amsterdam, Part B, 589-98.

Sharma PK, Hanumantha Rao K, Forssberg KSE. 2002. Analysis of bacterial cells surface free energy by contact angle measurements. In: Subramanian S, Natarajan KA, Rao BS, Rao TRR, eds. Proceedings of the Int Seminar Miner Process Tech, Navbharath Enterprises Publishers, Bangalore, 426-434.

Sharma PK, Hanumantha Rao K. 2002. Analysis of different approaches for evaluation of surface energy of microbial cells by contact angle goniometry. Adv Colloid Interface Sci 98: 341-463.

Sharma PK, Hanumantha Rao K. 2003. Adhesion of *Paenibacillus polymyxa* on chalcopyrite and pyrite: surface thermodynamics and extended DLVO theory. Coll Surf B, 29: 21-38.

Sharma PK, Das A, Hanumantha Rao K., Forssberg KSE. 2003. Surface characterization of *Acidithiobacillus ferrooxidans* cells grown under different conditions. Hydrometallurgy 71: 285-292.

Smith RW, Misra M. 1991a. Mineral Bioprocessing, The Minerals, Metals and Materials Society, Pennsylvania.

Smith RW, Misra M. 1991b. Mineral Bioprocessing-An overview. In: Smith RW, Misra M, eds. Mineral Bioprocessing, The Minerals, Metals and Materials Society, Pennsylvania, 3-26.

Smith RW, Misra M, Dubel J. 1991. Mineral bioprocessing and the future. Miner Eng 4: 1127-1141.

Smith RW, Misra M, Chen S. 1993. Adsorption of a hydrophobic bacterium onto hematite: Implications in the froth flotation of hematite. J Soc Ind Microbiol 11: 63-67.

Smith RW, Chen S, Misra M. 1994. Hydrophobic bacteria as flocculating agents for mineral suspensions. In: Moudgil BM, Somasundaran P, eds. Dispersion and aggregation: Fundamentals and applications, Engineering Foundation, New York, 499-506.

Smith RW, Miettinen M. 2006. Microorganisms in flotation and flocculation: Future technology or laboratory curiosity? Miner Eng 19: 548-553.

Solojenken PM, Lyubavina LL, Larin VK, Bergelson LD, Dyatlovitskaya EV,. 1976. A new collector in non-sulfide ore flotation. Bulletin Nonferrous Metall 16: 21-31.

Solojenken PM. 1979. Floatability and leaching of low grade ores with reagents of biological origin. In: Proceedings of the Int Conf Advances in Chemical Metallurgy, BARC, Bombay, India, 37/1-37/13.

Subramanian S, Santhiya D, Natarajan KA. 2003. Surface modification studies on sulfide minerals using bioreagents. Int J Miner Process 72: 175-188.

Townsley CC, Atkins AS, Davis AJ. 1987. Suppression of pyrite sulfur during flotation by *Thiobacillus ferrooxidans*. Biotech Bioeng 30: 1-8.

Unz RF. 1987. Aspects of bioflocculation. In: Attia YA, ed. Flocculation in Biotechnology and Separation Systems, Elsevier, Amsterdam, 351-368.

van Loosdrecht MCM, Lyklema J, Norde W, Schraa G, Zehnder AJB. 1987a. The role of bacterial cell wall hydrophobicity in adhesion. Appl Environ Microbiol 53: 1893-1897.

van Loosdrecht MCM, Lyklema J, Norde W, Schraa G, Zehnder AJB. 1987b. Electrophoretic mobility and hydrophobicity as a measure to predict the initial steps of bacterial adhesion. Appl Environ Microbiol 53: 1898-1901.

Vargas M, Kashefi K, Blunt-Harris EL, Lovely DR. 1998. Microbiological evidence of Fe (III) reduction on early Earth. Nature 395: 65-68.

Vasan SS, Modak JM, Natarajan KA. 2001. Some recent advances in the bioprocessing of bauxite. Int J Miner Process 62: 173-186.

Vijayalakshmi SP, Raichur AM. 2002. Bioflocculation of high-ash Indian coals using *Paenibacillus polymyxa*. Int J Miner Process 67: 199-210.

Vijayalakshmi SP, Raichur AM. 2003. The utility of *Bacillus subtilis* as a bioflocculant for fine coal. Coll Surf B 29: 265-275.

Vogt V, Gock E, Sand W. 2003. Sulfide ore flotation with extracellular polymeric substances (EPS) as biological depressant reagents. In: Lorenzen L, Bradshaw DJ, eds. Proceedings of the XXII Int Miner Process Congr, The South African Institute of Mining and Metallurgy, South Africa, Vol 2, 997-1006.

Yelloji Rao MK, Natarajan KA, Somasundaran P. 1992a. Effect of bacterial conditioning of sphalerite and galena with *Thiobacillus ferrooxidans* on their floatability. In: Smith RW, Misra M, eds. Minerals Bioprocessing, TMS, Warrendale, 105-120.

Yelloji Rao MK, Natarajan KA, Somasundaran P. 1992b. Effect of biotreatment with *Thiobacillus ferrooxidans* on the floatability of sphalerite and galena. Miner Metall Process 9: 95-100.

Yelloji Rao MK, Somasundaran P. 1995. Biomodification of mineral surfaces and flotation. In: Matis KA, ed. Flotation Science and Engineering, Marcel Dekker Inc, New York, 455-472.

Yelloji Rao MK, Natarajan KA, Somasundaran P. 1997. Growth and attachment of *Thiobacillus ferrooxidans* during sulfide mineral leaching. Int J Miner Process 50: 103-210.

Yuce AE, Mustafa Tarkan H, Zeki Dogan M. 2006. Effect of bacterial conditioning and the flotation of copper ore and concentrate. African J Biotechnol 5: 448-452.

Zheng X, Smith RW. 1997. Dolomite depressants in the flotation of apatite and collophane from dolomite. Miner Eng 10: 537-545.

Zheng X, Smith RW, Mehta RK, Misra M, Raichur AM. 1998. Anionic flotation of apatite from dolomite modified by the presence of bacteria. Miner Metall Process 15: 52-56.

Zheng X, Arps PJ, Smith RW. 2001. Adhesion of two bacteria onto dolomite and apatite: their effect on dolomite depression in anionic flotation. Int J Miner Process 62: 159-172.

CHAPTER 15

HYDROGEN SULFIDE REMOVAL FROM GASEOUS EFFLUENTS

JOSÉ MANUEL GÓMEZ AND DOMINGO CANTERO

Biological and Enzymatic Reactors Research Group, Departamento de Ingeniería Química, Tecnología de Alimentos y Tecnologías del Medio Ambiente, Facultad de Ciencias, Universidad de Cádiz Campus Universitario de Puerto Real. 11510 Puerto Real (Cádiz) Spain

E-mail: josemanuel.montesdeoca@uca.es

1. INTRODUCTION

The development of clean processes that enable us to maintain the standards of living achieved without damaging irreversibly our natural environment, constitutes one of the greatest challenges facing humanity for this 21st century.

A good example of this are the efforts being made on the reduction of carbon dioxide emissions, which are responsible for the global warming of the planet, and of the oxides of sulfur and nitrogen that are responsible of the phenomenon called acid rain that destroys forests and poisons lakes (Del Olmo *et al.*, 2005).

Legislation that is increasingly strict in respect of the permitted levels of sulfur compounds in gaseous emissions is becoming an important driving force for the development and implementation of processes for reducing these emissions. Specifically the European Union has issued directives, which over time have been incorporated in the national legislation of the member countries, that require drastic reductions of the content of sulfur oxide in emissions from newly-installed combustion equipment and plants, and at the same time demand very significant reductions for existing installations: specifically, for the year 2003, these latter were required to bring the sulfur content in such emissions down to 60% of the levels prevailing in 1980 (E.U. 98/70/EC, 1998).

Crude petroleum contains significant quantities, of the order of 0.5 to 50 g/L of sulfur, and its use as fuel causes the formation and release of sulfurated compounds. Currently the pressure of international legislation is obliging refineries to reduce this content, from 3 to 0.5 g/L of sulfur in the past decade and to levels of 0.05 g/L currently (Anabtawi *et al.*, 1992). Among the processes involved in the refining of crude is a process of desulfurization that gives rise to the formation of H_2S; this

E.R. Donati and W. Sand (eds.), Microbial Processing of Metal Sulfides, 287–309.

compound must be prevented from release by transforming it into elemental sulfur and eliminating it from the gaseous emissions.

The emission of sulfur compounds into the environment is undesirable because of their toxicity, corrosive properties, the unpleasant odour and the high oxygen demand dictate strigent control of its release into the environment. H_2S is one of the more toxic pollutants which can be pose severe risks to human health and to industrial facilities, particularly in the pulp and paper and petrochemical industries and in the anaerobic treatment of wastewater (Burgess *et al.*, 2001).

Methods for sulfide removal in common use today are physicochemical which involve direct oxidation, liquid redox and liquid adsorption processes, ozone oxidation and even incineration are commonly employed to desulfurize gases containing hydrogen sulfide (Burgess *et al.*, 2001). The main problem of these conventional methods is their high cost, since they use catalysis and consume large amounts of energy (Buisman *et al.*, 1989). These processes often result in the elimination of by-products such as sulfur dioxide, which is a strong atmospheric pollutant, transforming the problem of water pollution into one of air pollution (Jensen & Webb, 1995).

So, the use of microorganisms capable of oxidizing H_2S and producing elementary sulfur or sulfate from a complete and/or incomplete metabolism has been considered a potential alternative for the large-scale treatment of this gas (Sublette *et al.*, 1987; Pagella *et al.*, 1996; Janssen *et al.*, 2001). In addition to high H_2S removal efficiencies, technologies that use microorganisms have several advantages: no catalyst and no oxidants necessary; no chemical sludge to be disposed; little biological sludge is produced; low energy consumption; possible reuse of sulfur; little if any sulfate or thiosulfate discharge and fast process.

Several microorganisms, e.g., *Chromatiaceae, Chlorobiaceae, Xanthomonas* sp., *Pseudomonas* sp., and especially *Thiobacillus* (some of these species were later reclassified as *Acidithiobacillus* and *Halothiobacillus*) have proved be able to oxidize H_2S in biological reactors (Jensen & Webb, 1995). These species are chemoautotrophic and can, therefore, oxidize reduced forms of sulfur (including H_2S) as a source of energy for their growth. One of them, the acidophilic *A. ferrooxidans* can also oxidize Fe^{2+} to Fe^{3+} and metabolize metallic sulfides into their corresponding metal (Leduc *et al.*, 1994). The ability of these bacteria to oxidize Fe^{2+} to Fe^{3+} is a potential mechanism for the removal of H_2S.

Several years ago, these biological processes have not been commercially viable, due to various factors that limit its application; but now, it is possible to find high rate biotechnology based desulfurization installations working for a wide range of applications. This progress has been possible due to the specific conditions for the biocatalyst are well-known; the bacteria are highly active, stable and safe; the bioreactor design has reached the point of a standard product that can almost be bought of the shelf and the diffusion of H_2S is no problem as it is highly soluble to the chemically enhanced absorption (Janssen *et al.*, 2001).

In this chapter, we are going to present key processes for hydrogen sulfide removal and discuss recent developments about their application in the industry with a special emphasis to those biological processes in which *Thiobacillus* or *Acidithiobacillus* can be used.

2. METHODS FOR THE ELIMINATION OF H_2S

Numerous different methods exist for the treatment of gaseous effluents that contain both acid compounds H_2S and CO_2.The selection of an optimum method for the elimination is not easy and, in many cases, to obtain the degree of purification of gas desired, it is necessary to combine different processes.

For the evaluation of the methods of elimination, the following variables, among others, must be taken into account:

- Concentration of H_2S in the gas to be treated.
- Pressure, temperature and composition of the gas.
- Ratio of H_2S to CO_2.
- Volume of gas to be processed.
- Total H_2S load
- Gas specifications for the process output.
- Economic considerations.
- Environmental implications.

The following section gives brief details of the processes for the elimination of H_2S that present the most interest in respect of their industrial application.

In particular, references are made to the processes of direct oxidation, amine plants, redox processes, non-regenerative processes, and biological processes.

2.1. Amine Plants

Treatment of gas with solutions of alkanolamines is the most commonly used method to eliminate H_2S and CO_2 from gaseous effluents, since this method is very appropriate for the purification of large volumes of gases.

Processes with alkanolamines have come to be the predominant choice for the treatment of gases in oil refineries and for the purification of natural gas. Nevertheless, they present some problems that are difficult to solve: corrosion, losses of amines, formation of foams, and obstruction of the equipment.

The amines that are of most interest for treatments of this type are: monoethanolamine, diethanolamine and methyldiethanolamine (Frazier & Kohl, 1950; Miller & Kohl, 1953; Kohl, 1951).

The basic plant required for this process consists of an absorber where the acid gas is brought into contact with the solution of amines in countercurrent flow, and an exhaustion column where the acid gas previously retained in the effluent is separated out. This residual acid gas leaving the exhaustion column can be sent directly to a plant that converts the H_2S to elemental sulfur like a Claus plant.

2.2. Processes of Direct Oxidation of H_2S

These processes are based on the catalytic oxidation of hydrogen sulfide to elemental sulfur (Kohl & Nielsen, 1997). The first method developed in this line is due to the scientist Carl Friedrick Claus who, in 1883, developed a patent for a catalytic process of oxidation of H_2S to S, employing bauxite as catalyst. Currently, alumina and bauxite are the most commonly used catalysts.

More than a gas purification process in the strict sense, the principal objective of this process is the recovery of sulfur from acid gases with high concentrations of H_2S.

The process of Claus can be summarized to the following reactions:

(1) $H_2S + 1/2O_2 \rightarrow H_2O + S$ (overallreaction)

(2) $H_2S + 3/2O_2 \rightarrow H_2O + SO_2$

(3) $2H_2S + SO_2 \leftrightarrow 2H_2O + 3S^0$

Basically the plant has two recovery cycles. In the first place the acid gas, together with a stoichiometric quantity of air, is fed into an oven where one third of the H_2S is burned to SO_2. This is the thermal stage where reactions 1 and 2 take place. It is performed at around 900-1200°C. The gases that leave the oven are cooled to 500°C in the steam recovery boiler (Kohl & Nielsen, 1997). About 60 to 70% of the H_2S can be converted this way.

Next the gases enter the catalytic stage in which reactions 2 and 3 take place, at a working temperature of 450°C. In this phase the lower the temperature the greater the degree of conversion achieved; however, it is not convenient to work at below the dew point, because it can cause sulfur liquid to be deposited on the bed of the catalyst. It is therefore an advantage to employ several catalytic stages in series, with condensation and elimination of the resulting sulfur. Recoveries of 92-95% can be obtained in a plant with two catalytic reactors.

One of the limitations on the degree of conversion achieved in this process is due to side reactions taking place as a result of the presence of CO_2 and light hydrocarbons in the feed gas, with carbonyl sulfide and carbon disulfide being formed as by-products. These compounds are formed in appreciable quantities (17%) and therefore this should be considered in the design and efficiency of the operation in the plant.

Lastly it is important to state that one of the most serious problems that this type of process presents is the de-activation of the catalyst caused by the deposition of sulfates.

Based on the Claus method, new processes have been under development with the aim of improving the elimination yields and reducing the emission of acid gases to the atmosphere.

The optimization of this process has been directed towards obtaining more specific catalysts, on the basis of the following considerations:

Table 1. Processes of direct oxidation for the elimination of H_2S

	Selectox (Hass, 1981)	MODOP (Ketner, 1987)	Superclaus 99 (Lagas, 1988)	Superclaus 99.5 (Goar, 1991)
H_2 S (mg/L)	30-75	7-120	10-45	10-45
Catalyst	"Selectox-33"	TiO_2	Alumina/silica	molybdenum/cobalt
Temperature (^{o}C)	370	160	400	400
Conversion (%)	80-90	90 (one stage)	99	99.5

- Greater selectivity for H_2S.
- Suppression of the formation of undesirable compounds like SO_2 and SO_3.
- Minimization of side reactions.
- Reduction of sensitivity to water.

Among the more recent methods that apply direct oxidation for the recovery of sulfur, that have been developed on an industrial scale to purify gaseous effluents from a Claus plant, the following processes deserve mention: "Selectox", "Super Claus 99", "Superclaus 99.5", and "MODOP". The most representative characteristics of these processes are shown in the following (Table 1).

The Superclaus 99 process is capable of converting more than 85% of the H_2S to elemental sulfur per stage.

A hydrogenation stage is introduced in the Superclaus 99.5 process between the Claus reactor and the Superclaus 99 reactor. The capital investment is between 5% and 20% more than that for a conventional Claus plant.

2.3. Redox Processes

Processes based on redox reactions present clear advantages over those described so far particularly in the purification of gases with low concentrations of H_2S and CO_2 and for relatively small or variable flow rates. These advantages include:

- High rates of H_2S elimination.
- They operate at ambient temperature; therefore energy requirements are minimized.
- The working pressure extends from atmospheric up to more than 70 bar.
- The treatments work well at different flow rates and different gas compositions.
- Costs are lower than for systems based on the utilization of amines and for non-regenerative systems.

In general terms, a catalyst (M) that is alternatively reduced and oxidized is utilized in these processes. The process can be represented by the following reactions (Neumann and Lynn, 1984; Freund *et al.*, 1994).

$$(4) \qquad H_2S + 2M^n \rightarrow S^0 + 2H^+ + 2M^{n-1}$$

$$(5) \qquad ½\, O_2 + 2M^{n-1} + 2H^+ \rightarrow 2M^n + H_2O$$

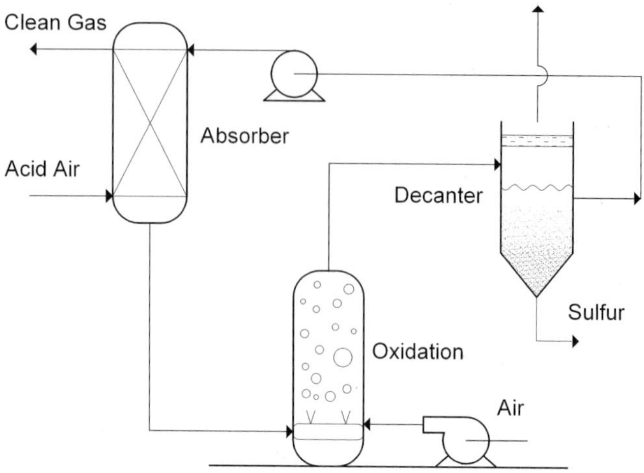

Figure 1. Flow diagram of a redox process

Figure 1 shows a simple diagram of a redox system for the treatment of gases in liquid phase.

The acid gas is introduced in an absorber where the oxidation reaction takes place in the presence of the catalyst. The solution containing the sulfur and the reduced catalyst passes to a reactor where the catalyst is regenerated through oxidation with air. Then the effluent passes to a settler in which the sulfur is collected, and the cycle of treatment is starts again.

The principal disadvantages of this type of process are: the low capacity of absorption of the solutions for H_2S and oxygen, which means that high liquid flow rates must be utilized; the difficulty of separating the precipitated sulfur from the liquid mixture; and the need to dissipate the heat generated by the oxidation of H_2S.

In (Table 2) the principal groups of processes based on redox technology are listed. It includes the approximate date when the process came on the market or was described for the first time in the literature, and its current state of acceptance.

The processes that utilize iron chelates have, in the last decade, been gaining acceptance against other redox technologies, principally because the solution generated is non-toxic. Two processes have competed for control of the market: the "LO-CAT" process (Hardison, 1992) and the "SulFerox" process (Fong *et al.*, 1987).

The economics of these processes depends fundamentally on the following factors:

- The cost involved in the circulation of the catalytic solution at the rates necessary.
- The energy requirements of the circulation pump and the air supply.
- The cost of the chemical reagents.
- The costs for separating and handling the sulfur.
- Costs for maintenance and cleaning of the installation.

Table 2. Processes of oxidation in liquid phase for the elimination of H_2S

Catalyzer	Commercial name	Date	Current situation
Polythionate	Koppers C.A.S.	1945	I
Iron oxide	Ferrox	1926	C
	Gluud	1927	I
	Burkheiser	1953	I
	Manchester	1953	I
Iron cyanide	Fischer	1931	I
	Staatsmijnen-Otto	1945	I
	Autopurification	1945	I
Thioarsenate	Thylox	1929	I
	Giammarco-Vetrocoke	1955	C
Naphtoquinones and/or vanadate	Perox	1950	C
	Takahax	1964	A
	Stretford	1963	A
	Hiperion	1986	C
	Sulfolin	1985	A
	Unisulf	1981	C
Iron chelates	Cataban	1972	I
	LO-CAT	1978	A
	Sulfint	1980	A
	SulFerox	1 986	A

Current situation:
I= Inactive
C= On the market but currently not promoted
A= Active in the market.

One of the principal disadvantages of utilizing iron as the catalyst is the loss of the chelating agent in the purged solution; also it is very corrosive, which can considerably increase the capital cost of the plant.

2.4. Non-Regenerative Processes

Included in this group are those processes that do not provide the recovery of the reagent employed in the elimination of the contaminant. The majority are absorption and adsorption processes, specific for the treatment of gaseous effluents whose content in H_2S ranges between 0.15 and 1.5 mg/l. These processes are generally used for small loads of H_2S.

These systems normally operate in discontinuous regime, and may be divided into two categories, in function of the type of sorbent utilized, solid or liquid.

Among the solid sorbents most often employed are the following:

- *Iron oxide*. Iron oxide reacts with the hydrogen sulfide to form iron sulfide. The iron oxide can be recovered if air is utilized to oxidize the sulfide to elemental sulfur, but this is not common practice since it interferes with the porosity of the bed and therefore with the system's capacity for elimination (Anerousis & Whtitman, 1984).

- *Zinc oxide.* Zinc oxide reacts with the hydrogen sulfide to form zinc sulfide, which is a fairly stable compound. This method is used to eliminate traces of H_2S in synthesis gases (Carnell, 1986).
- *Alkaline solids.* Strongly alkaline solids like sodium hydroxide, for example, are very effective agents for the elimination of H_2S, by means of acid-base reactions (Kohl & Nielsen, 1997).
- *Adsorbents.* One of the most frequently used is activated carbon (Kohl & Nielsen, 1997).

With respect to the sorbent liquids, the following can be distinguished:

- *Iron oxide sludges.* The chemistry of this type of process is identical to that of the processes with solid iron oxides. The principal advantage they present over the latter is that the exhausted absorbent can be eliminated and replaced with greater facility (Sivalls, 1982).
- *Alkaline solutions.* The processes of this type normally employ solutions of sodium hydroxide in water. During the operation all the acid gases are absorbed in the solution, although some selectivity may exist for hydrogen sulfide and carbon dioxide (Astarita *et al.*, 1983).
- *Zinc oxide sludges.* This method is more expensive than the preceding one. Generally a weak acid anion is added, which forms soluble salts of zinc, increasing the activity of the zinc in solution (Manning *et al.*, 1981).
- *Oxidizing solutions.* A wide variety of oxidizing agents have been used, including dichromate, permanganate, peroxide, nitrite and oxygen (Dobbs, 1986).
- *Aldehydes.* These processes have been developed on the basis of the rapid reaction between an aldehyde, for example formaldehyde, and H_2S, giving rise to non-volatile organic compounds of sulfur. One of the principal problems that this process presents is that the reaction products obtained have a very unpleasant smell; also the solution tends to flocculate and become polymerized, which can cause obstructions in the equipment. This problem, together with the difficulty of eliminating the residues generated, is probably the principal reason why this process has not found a wide area of application (Schaack & Chan, 1989).
- *Triazines.* Certain triazines, such as 1,3,5 tri-(2-hydroxyethyl)- hexahydro-S-triazine, react rapidly with H_2S to give by-products that are soluble in water. The advantage of this type of process is that the by-products formed are not corrosive (Dillon, 1991).

2.5. Biological Methods

The continuous research being undertaken to find increasingly more economic methods has led to the utilization of microbiological processes for the purification of H_2S content in gases.

Most of the microbiological processes developed operate at ambient temperature and atmospheric pressure, thus reducing energy costs to the minimum.

In continuation some of these processes are described, in function of the micro-organism utilized.

2.5.1. *Chlorobiaciae and Chromatiaciae*

Microorganisms belonging to the families *Chlorobiaciae* and*Chromatiaciae* have been employed for the desulfurization of gaseous effluents that contain H_2S, H_2 and CO_2 (Maka & Cork, 1990). These processes are based on the biological catalysis of the Van Niel photosynthesis reaction:

$$(6) \qquad 2n\ H_2S + n\ CO_2 \xrightarrow{h\upsilon} 2n\,S + n(CH_2O) + n$$

Of particular importance in this respect is the work carried out by Kim and Chang (1991) with cells of *Chlorobium thiosulfatophilum* immobilized in a matrix of alginate; these authors achieved rates of oxidation of 0.13 g/Lh of H_2S, with a percentage of elimination of 99.9%. The principal disadvantage of the use of these systems is the requirement of radiant energy and, therefore, the need for large transparent surfaces.

2.5.2. *Xanthomonas*

The first reference work on the use of heterotrophic bacteria for the elimination of hydrogen sulfide was published by Cho *et al.* (1992). The bacteria *Xanthomonas sp.* was utilized in discontinuous regime to eliminate H_2S from gaseous effluents. The maximum specific rate of elimination achieved was 0.13 g/Lh. The product that this bacteria gives rise to was identified as a polysulfide, from which a neutral pH is obtained.

Among the advantages presented by this process are the following:

- The biomass can be cultivated easily, since it grows rapidly in the medium.
- The polysulfides obtained as the final product are preferentially sulfates; therefore no deterioration of the bacterial activity occurs since the pH does not vary. However, the specific rate of elimination of hydrogen sulfide by *Xanthomonas sp.* is lower compared with that obtained by utilising purified strains of the genus *Thiobacillus sp.* (Cho *et al.*, 1991, 1992).

2.5.3. *Thiobacilli, Halothiobacilli and Acidithiobacilli*

Within this family, bacteria formerly belonged to the genus *Thiobacillus* have been utilized for the direct elimination of H_2S from gaseous effluents. Many authors have proposed different designs, although very satisfactory results have not been obtained.

- Gadre (1989) utilized these microorganisms to eliminate the hydrogen sulfide content in biogas. For concentrations of H_2S of 2%, this author obtained rates of oxidation of 0.10 g/Lh of H_2S, although the elimination yields were very low (69.5%).

- Berzaczy *et al.* (1990) carried out the elimination of CS_2 and H_2S from the exit gases of a cellulose production plant (on a pilot plant scale), using a fixed bed reactor in which microorganisms of the genus *Thiobacillus* were immobilized on a medium of polypropylene. The specific rates of elimination achieved were: 0.0070 g/Lh of H_2S and 0.0070 g/Lh of CS_2.
- Sublette & Sylvester (1987a, 1987b, 1987c), working with the bacteria *T. denitrificans* in a continuous stirred tank reactor and under anaerobic conditions, obtained eliminations of H_2S in excess of 97%. However, the maximum rate obtained was only 0.08 g/Lh of H_2S.
- Candehead and Sublette (1990) made a study of the oxidation of H_2S by different species of *Thiobacillus*, *Halothiobacillus* and *Acidithiobacillus*: *T. thioparus*, *T. versutus* (recently reclassified as *Paracoccus versutus*), *H. neopolitanus* and *A. thiooxidans*. The purpose of this research was to determine if any of these strains offered clear advantages in the process of aerobic oxidation of H_2S over the species *T. denitrificans*. The rate of oxidation was determined in a discontinuous reactor supplied with a gas feed containing 1.5 mg/L of H_2S, 95 mg/L of CO_2 and 1.2 10^3 mg/L of N_2. None of the microorganisms tested presented significant advantages over the species *T. denitrificans* under the same conditions of operation.

Neumann *et al.* (1990) patented a method for the elimination of H_2S from biogas utilizing cells of *A. ferrooxidans*, immobilized in a fixed bed reactor. They employed peat, wood chips and compost as medium. The gas was fed continuously to the reactor, together with air to oxidize the hydrogen sulfide to S or SO_4^{2-}. To avoid the risks of explosion, it was necessary to keep the concentration of dissolved oxygen in the purified gas below 42 mg/L. These authors obtained very high yields, for a concentration of H_2S in the feed of 15 mg/L; however they do not mention either the flow rates utilized, or the rates of oxidation obtained. This process has been applied and built numerous times for vent air treatment and other very low load H_2S removal.

Currently, some alternatives based on Redox technology have been developed for the elimination of hydrogen sulfide from biogas by means of *A. ferrooxidans*. One of the most advanced processes under development is that known as BIO-SR (Imaizumi, 1986; Magota & Shiratori, 1988; Satoh *et al.*, 1988). This process was originally developed by Dowa Mining Company Ltd. for the elimination of H_2S from natural gas. The system does not use iron chelates, operates at low pH, and employs bacteria to improve the kinetic regeneration of the solution of reduced iron.

Recently, in 1993, Paques BV (Netherlands) has been developed a new technology based on the sulfur cycled that is marketed under the name Thiopaq®. This process is marketed by Shell/Paques for the oil and gas industry, Univeral Oil Products (UOP, Chicago) for refineries and by numerous licensees for the biogas and landfill gas purification application. Over 50 installations are in operation now, of which 6 installation in the oil and gas industry since sales for that industry started in 2002.

Basically, Thiopaq® unit is useful for biological removal of H_2S from sour gases by integrating a wet scrubber with a HS^- oxidizing bioreactor (Janssen *et al.*, 1997). In the scrubber, the H_2S-containing gas is contacted with a slightly alkaline scrubbing solution in a counter current mode. Hydroxide is consumed to absorb H_2S gas under the formation of hydrogen bisulfide by the reaction:

$$H_2S + OH^- \rightarrow HS^- + H_2O \quad (7)$$

The scrubbing liquid, containing the dissolved sulfide, then goes to the bioreactor. In this biological step elemental sulfur is formed, accompanied by the regeneration of hydroxide ions by this reaction:

$$HS^- + \tfrac{1}{2}\, O_2 \rightarrow S^0 + OH^- \quad (8)$$

The formation of elemental sulfur is preferred since the effluent can be re-used as a washing liquid for H_2S absorption. Compared to conventional caustic soda scrubbing processes, this leads to a reduction of soda consumption by over 95%. The air discharged from the bioreactor does not contain any H_2S and can be discharged directly into the atmosphere. No chelating compounds are needed, and only limited use of caustic soda is required, which means the operating costs are very low (amounting to 60$/tonne S^o).

One recent application of this technology is in H_2S removal from high pressure natural gas, a joint development by Shell and Paques (Janssen *et al.*, 2000). In a demonstration plant, 1500 Nm^3d^{-1} of natural gas with an H_2S content of up to 8% by volume was desulfurized at high gas pressures. The H_2S content in the treated gas was below 1 ppmv, while all absorbed sulfide was oxidized.

Actually this biological process is marketed and sold as a serious competition for the iron based processes shown in Table 2.

3. REMOVAL OF H_2S BY ABSORPTION IN ACID SOLUTIONS OF $Fe_2(SO_4)_3$

In this section we are going to describe other alternative of biological process for hydrogen sulfide removal based on *Acidithiobacillus*. The main difference with Thiopaq process is that absorption takes places in an acid solution and the re-used of the reagent is due to the action of this type of bacteria. So, this process is a direct application of the bioleaching properties of this acidophilic bacterium.

The basic process consists of putting a solution of ferric sulfate into contact with a gas containing H_2S in an absorber (Gómez *et al.*, 1996). The solution absorbs the hydrogen sulfide and oxidizes it to elemental sulfur, in accordance with the following chemical equation:

$$H_2S_{(g)} + Fe_2(SO_4)_{3(aq)} \rightarrow S_{(s)} + 2FeSO_{4(aq)} + H_2SO_{4(aq)} \quad (9)$$

The elemental sulfur is removed from the medium by means of a separation operation and the resulting solution is passed through a bioreactor, where the microorganism performs the oxidative process according to the following equation:

$$(10) \qquad 2FeSO_{4(aq)} + H_2SO_{4(aq)} + ½O_2 - A.ferrooxidans \rightarrow Fe_2(SO_4)_{3(aq)} + H_2O$$

The oxidized solution that results from this biological stage can be utilized again in the absorption stage, thus minimizing the consumption of oxidizing reagent.

One of the advantages of this type of process is that the reaction between the hydrogen sulfide and the ferric sulfate is very rapid and complete, with the result that toxic residual gases are not produced. In addition, apart from sulfur, no dangerous wastes are generated, so this considerably reduces the costs of complementary treatments.

Figure 2 contains a diagram of a plant employed for reducing the hydrogen sulfide content in a flow of biogas produced in anaerobic digesters; the cleaned gas is subsequently used as fuel in a co-generation installation.

The system, has been tested in a conventional wastewater treatment plant (Mesa *et al.*, 2004), comprises a scrubbing tower in which a solution of ferric sulfate is put into contact with the biogas. The solution absorbs the hydrogen sulfide in the gas and oxidizes it to elemental sulfur, while the ferric sulfate is reduced to ferrous sulfate. Once the sulfur is removed from the medium, the solution passes to a bioreactor where the ferrous sulfate is oxidized by the injection of air and the

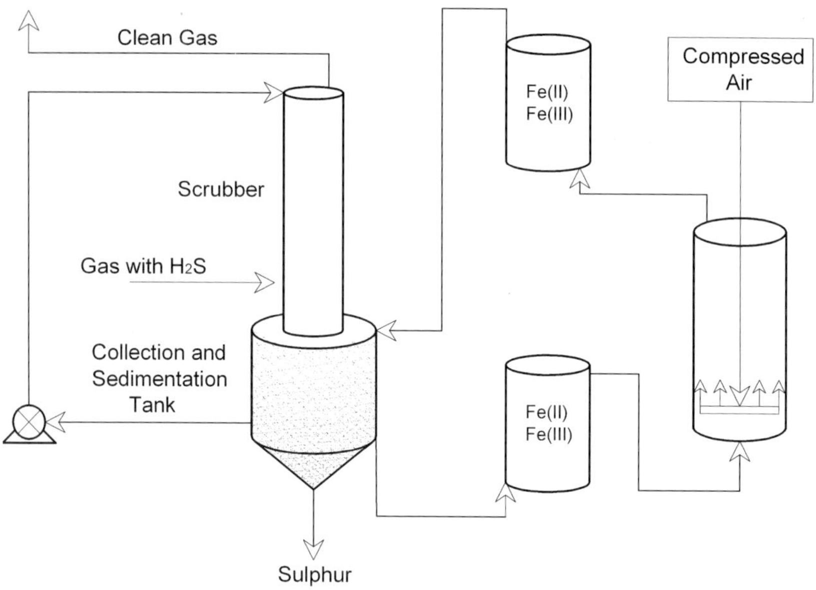

Figure 2. Diagram of the process for H_2S removal

action of the microorganism *A. ferrooxidans*, which is immobilized on polyurethane foam; in this way the reagent is recovered.

This system has been inserted in the gas and sludge line of the wastewater plant, with a co-generation plant that produces 11,520 kWh of electricity per day; this plant provides more than 60% of the total electrical energy consumption of the wastewater plant, and also produces the heat required to keep the anaerobic digesters at 35 ^{o}C.

The mean flow rate of biogas treated is 300 Nm^3/h with a concentration of hydrogen sulfide of up to 7,000 mg/Nm^3.

Actually, this process is going to be marketed by a spin-off enterprise and different units are in construction for hydrogen sulfide removal in wastewater plants.

3.1. Key Design Criteria for the Absorber

The kinetics of absorption of hydrogen sulfide in aqueous solutions of ferric sulfate was investigated by Asai *et al.* (1990). These authors evaluated the influence of the concentration of ferric sulfate and the pH on the rate of absorption.

The highest rates of absorption were obtained for a range of concentrations between 40 and 80 g/L of $Fe_2(SO_4)_3$; the maximum value reached corresponds to 60 g/L, equivalent to a concentration of Fe(II) of 16.5 g/L.

Further, it was observed that the pH significantly affected the rate of absorption, with the highest rates being reached for a pH of 2.

The absorption of hydrogen sulfide is controlled, according to Asai *et al.* (1990), by the irreversible reaction with the monohydrated ferric species $[FeOH^{2+}]$:

$$H_2S + 2FeOH^{2+} \rightarrow S + 2Fe^{2+} + 2H_2O \tag{11}$$

Experimentally it is found that the rate of reaction follows a first order kinetic with respect to the H_2S and with respect to the $[FeOH^{2+}]$. Therefore, the global order of the reaction is equal to 2.

The species $[FeOH^{2+}]$is in solution together with other solvated compounds of ferric iron, and there is a complex system of simultaneous equilibria, for which the pH appears to be the controlling parameter. Relating the possible chemical equilibria of the system with the equilibrium constants already known (Smith & Martell, 1989), and ignoring the species that appear in small concentrations, the most important compounds in solution are ferric hydroxides, complexes with the anion hydrogenosulfate $[HSO_4^-]$ and those related to the anions dihydrogenophosphate $[H_2PO_4^-]$ and hydrogenophosphate $[HPO_4^{2-}]$. The equilibria involving these species are cited in Table 3.

3.2. Reactors and Support Materials Utilized for the Immobilization of Biomass

Many bioreactors require the biocatalysts (microorganisms, enzymes,…) to be immobilized on a solid medium in order to reduce the cellular washing, and to

Table 3. Equilibrium equations for the species of Fe(III) in solution (Pagella *et al.*, 1996)

$Fe^{3+} + OH^- \leftrightarrows FeOH^{2+}$	$Fe^{3+} + HPO_4^2 \leftrightarrows FeHPO_4^+$
$Fe^{3+} + 2OH^- \leftrightarrows Fe(OH)_2^+$	$Fe^{3+} + H_2PO_4 \leftrightarrows FeH_2PO_4^{2+}$
$Fe^{3+} + 3OH^- \leftrightarrows Fe(OH)_3$	$H^+ + SO_4^2 \leftrightarrows HSO_4^-$
$2Fe^{3+} + 2OH^- \leftrightarrows Fe_2(OH)_2^{4+}$	$H^+ + PO_4^3 \leftrightarrows HPO_4^{2-}$
$3Fe^{3+} + 4OH^- \leftrightarrows Fe_3(OH)_4^{5+}$	$H^+ + HPO_4^2 \leftrightarrows H_2PO_4^-$
$Fe^{3+} + HSO_4^- \leftrightarrows FeHSO_4^{2+}$	$H^+ + H_2PO_4^- \leftrightarrows H_3PO_4$
$Fe^{3+} + 2H\ SO_4^- \leftrightarrows Fe(HSO_4)_2^+$	$H^+ + OH^- \leftrightarrows H_2O$

increase the concentration of biocatalyst in the process. Usually, there are two general types of immobilization: entrapment and adsorption.

In the particular case of *A. ferrooxidans*, the technique for immobilization involves biofilms constituted by jarosite and bacteria; this type of structure exerts a very powerful influence on the process, which includes generating an increase of the volumetric rate of oxidation of Fe (II) of up to 10 times (Nikolov and Karamanev, 1987). However, certain anomalous behaviors have been found in respect of the biofilms of *A. ferrooxidans* that cannot be explained with the classic structure of cells that form the structure of the microbial film with bridges of extracellular polymeric substances (Karamanev, 1991).

The specific nature of these anomalies is indicated by the following observations:

- It is observed that the rate of the biological reaction (r_S) against the rate of dilution (D) reaches a maximum, then later falls to a value of the rate of dilution at which the rate of substrate consumption remains constant.
- A large quantity of free cells leaves the biofilms, even at rates of dilution much higher than the rates of dilution that are critical for cellular washing. Karamanev, 1991, found that, during the oxidation of Fe (II) in continuous regime by *A. ferrooxidans* in an inverse fluidized bed reactor, there were large numbers of free cells that were abandoning the reactor at a concentration of $5x10^7$-10^8 cells/mL, even for rates of dilution as high as 1.08 h^{-1}(a rate of dilution ten times greater that the critical rate of dilution for the cellular washing of free cells, 0.097 h^{-1} obtained by Nikolov and Karamanev, 1987). These cells abandon the reactor individually, without forming aggregates.
- The rate of growth of the microorganism in the biofilm is much less than that of the cells in suspension, since a value for the specific rate of growth in the biofilm of only 0.0014 $h^{-1}$is obtained. This datum indicates that the rate of bacterial growth falls by a factor of more than 70 after the fixing.

The only explanation for all these effects is that large quantities of bacteria that arise from the division of the cells in the biofilm leave the film and become resuspended in the liquid. In this case, the rate of growth of the biofilm should depend principally on the rate of formation of the jarosite, which is formed by a chemical process and must be independent of the rate of growth of the microorganisms. Effectively, the production of Fe (III) by the bacteria is much faster than its conversion into jarosite (Olem & Unz, 1977; Pogliani & Donati, 2000) and, therefore, is not the controlling rate.

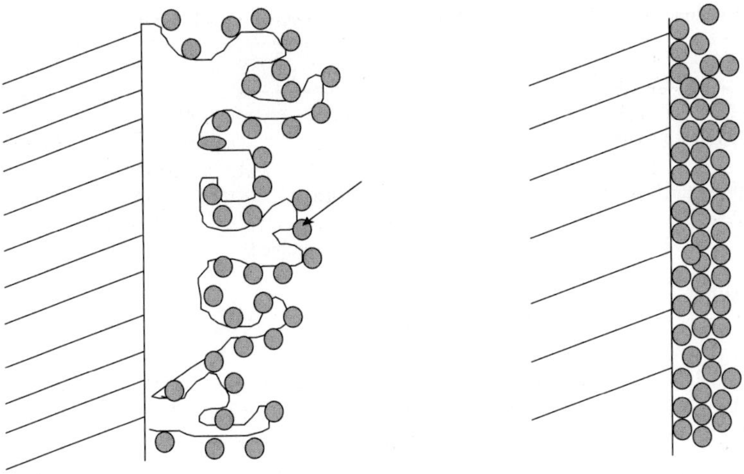

Figure 3. Structures of a biofilm of *A. ferrooxidans* and of a classic biofilm
Left: Structure of jarosite. Right: Cells. (Nemati *et al,* 1998)

In addition, it is assumed that *A. ferrooxidans* adheres to the jarosite by adsorption forces, since it has not been found that this bacteria produces any extracellular polymeric substances of organic type (Karamanev, 1991). For this reason, *A. ferrooxidans* covers the surface of the pores of the jarosite as a single layer of cells (Figure 3). When a cell divides, only one of the new cells remains fixed to the surface, and the other floats free. This type of biofilm can be termed a "biofilm with adsorbed cells".

Due to the pores of the jarosite being open and distributed uniformly throughout the volume of this material, the quantity of cells is proportional to the volume of jarosite. This finding explains why the rate of oxidation is proportional to the volume of biofilm (Karamanev, 1991). The presence of large quantities of bacteria in the effluent liquid for high values of the rate of oxidation can be easily explained by this type of biofilm structure.

There are many types of support utilized for the immobilization of *A. ferrooxidans*; in Table 4 some of these are listed together with the devices utilized. In the light of the results presented in Table 4, it can be stated that the most appropriate supports are activated carbon, polyurethane foam and resin spheres.

With respect to the reactors, very diverse types have been utilized but those that offer the best results are the fixed bed and trickle bed reactors.

3.2.1. Rotating Biological Contactors

Rotating biological contactors (rotating discs) form part of one of the most widely studied and employed devices in the biological treatment of liquid residues. Figure 4 presents a diagram of a biological contactor of rotating discs.

Table 4. Different supports utilized for the immobilization of *Acidithiobacillus ferrooxidans*

Reference	Support	Reactor	r_s(g/Lh)
Nikolov et al. (1987)	PVC	Rotating discs	1.46
Grishin et al. (1989)	Glass spheres	Fixed bed	6.2 (V_r) 8.1 (V_m)
	Spheres of resin	Fixed bed	22.5 (V_r) 29.3 (V_m)
	Activated carbon	Fluidized bed	0.54 (V_r) 0.9 (V_m)
		Fixed bed	60 (V_r) 78 (V_m)
Karamavev et al. (1991)	Expanded polystyrene	Inverse fluidized bed	0.45 (V_r)
García et al.(1989)	Sulfurated mineral	Fixed bed	0.44 (V_r)
Carranza et al. (1990)	Sand	Fixed bed	0.37(V_r) 10-17 (V_m)
	Copper mineral	Fixed bed	0.44(V_r) 11-16 (V_m)
	Activated carbon	Fixed bed	1.05(V_r) 31-62 (V_m)
Sonta et al. (1990)	Glass wool	Fixed bed	2.3
Armentia et al. (1992)	Polyurethane foam	Mobile bed	1.56
Nemati and Webb (1997)	Polyurethane foam	Trickle bed reactor	31.7 (V_m)
Gomez et al. (2000)	Nickel fibre	Fixed bed	1.7 (V_r)
Sonta etal. *(1990)*	Various	Fixed bed	3.0 (V_r)

Note: V_r: volume of the reactor V_m: volume of culture medium

These units consist of a set of discs that rotate concentrically on a semicylindrical trough, in such a way that, at all times, approximately the half of the surface of each disc is immersed in the liquid contained in the trough. Thus, the bacterial film is, for part of the time, immersed in contact with its source of energy, and for the other part, is exposed to the environment in which CO_2 and $O_2$are most easily diffused.

The O_2 and CO_2 are dissolved during the rotation of the discs in the liquid film and penetrate into the bacterial film together with the dissolved Fe(II) and other nutrients, whereas the products of metabolism and the ferric iron are diffused by the liquid film into the solution.

The advantage of these reactors is that the effect of shearing on the biofilm is reduced. They can be used in both discontinuous and continuous regime. In addition they provide excellent aeration. However they present the disadvantage of requiring considerable energy in operation, and considerable maintenance costs.

Of the numerous studies made on the application of rotating biological contactors (rotating discs) to the oxidation of ferrous iron (Olem & Unz, 1977; Wichlacz, 1981; Olem and Unz, 1980; Nakamura *et al.*, 1986), that conducted by Nikolov *et al.*, (1986) is particularly notable since these authors obtained rates of oxidation of 1.46 g/Lh.

3.2.2. *Fluidized Bed*

The fluidized bed reactors, in which the bacterial population is fixed to the particles of the bed, constitutes one of the most promising and effective techniques in the

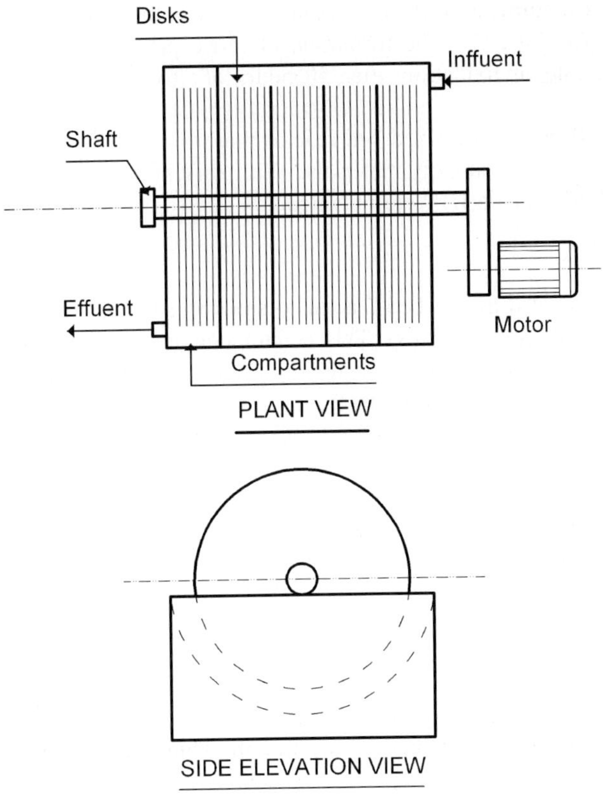

Figure 4. Biological contactor of rotating discs

utilization of bioreactors. Their principal advantages include: a greater area of biofilm-liquid interface; high interface rates; and good matter transfer.

However, the extensive use of these systems is limited by the uncontrolled changes of biomass that occur on the particles of the support; among the consequences of these effects are changes in the hydrodynamic conditions of each particle, and in general, of all the expanded bed (Atkinson, 1981).

Nikolov and Karamanev (1987, 1988) developed a bioreactor of inverse fluidized bed, very stable and with easy control of the thickness of the particles of the bed, for the oxidation of ferrous iron by *A. ferrooxidans*, and were able to maintain a constant bacterial film of 80 microns film thickness.

3.2.3. *Fixed Bed*

With respect to the utilization of fixed bed reactors with different supports, of particular note are the high rates of oxidation achieved by Grishin and Tuovinen (1988), utilizing activated carbon as support material. However, these authors have not taken into account the catalytic effect of carbon in the oxidation of iron. One of the problems presented by the utilization of this type of support is the obstruction of the pores of the materials caused by the formation of ferric precipitates; this leads to a decrease of the rate of oxidation after 50 hours of process operation (Kai *et al.*, 1990).

Sonta and Shiratori (1990) patented a fixed bed reactor that utilizes glass wool, zeolites, bentonite and alumina as support materials. With this system rates of oxidation of the order of 3 g/Lh of Fe(II) were obtained.

3.2.4. Trickle Bed Reactors

Nemati & Webb 1996 employed a trickle bed reactor to study the kinetics of the oxidation of ferrous iron by *A. ferrooxidans*). This reactor contained as support 2200 units of polyurethane foam of 6 mm size occupying a volume of 981 cm^3.

A diagram of this type of reactor is shown in figure 5. The microorganism is immobilized on the polyurethane foam units that are placed in a glass column of 0.5 m height and 0.05 m diameter.

In the lower part of the column there is a chamber for collecting the culture medium, and located in the upper part is a perforated dish for distributing the liquid.

Part of the medium collected in the lower chamber is recirculated to the bioreactor by means of a peristaltic pump; this is a way of increasing the time of residence of the reagent in the reaction zone (the bed of the biocatalyst) and of providing a uniform temperature inside the reactor.

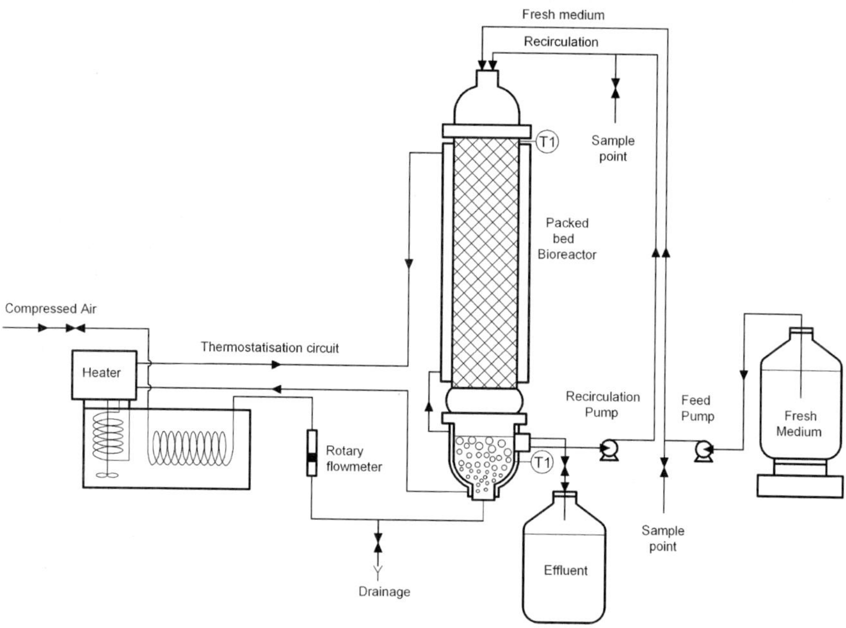

Figure 5. Flow diagram of a trickle bed reactor: Nemati & Webb (1996)

Table 5. Oxidation rates (Nemati and Webb, 1996)

Fe(II) (g/L)	D (h^{-1})	Conversion %	r_s g/Lh
5	<1	>70%	0.69
	>3	40-45	4.13
10	<1	>70	1.38
	>3	20-30	4.6
20	>1.5	<10	1.15-1.61

The fresh medium is pumped continuously to the top of the column and the effluent is collected by gravity in a tank provided for this purpose.

In this type of reactor the adhesion of the microorganism is very irregular and varies locally as a result of the physical detachment of the microbial film from the surface of the support.

Normally the area wetted by the liquid film that flows through a filter under the action of gravity is much less than the biologically active area. This leads to the growth taking place in the regions where the nutrients are provided, and therefore the increase of the film thickness causes the flow to be displaced laterally.

The result is a complex interaction between the kinetic of the microorganism, the thickness of the microbial film, and the area wetted.

The rates of oxidation obtained with this reactor for different concentrations of ferrous iron in the feed, are given in table 5. These have been calculated in function of the volume occupied by the supporting units.

4. FUTURE ASPECTS

The elimination of contaminants from gaseous effluents using biological methods has great application potential, this is doubtless because these processes do not require very demanding conditions of pressure and temperature; no reagents are needed apart from the nutritive requirements of the microorganisms; they are processes for which the policy of zero discharge can be applied, and therefore they can be considered very respectful of the natural environment. The difficulties faced in various commercial applications are due to their relatively low rates of reaction and to the inhibitory effect that high concentrations of these contaminants have on cellular growth.

The use of immobilized biomass is considered a very promising method for increasing the efficiency of these processes, and its application has in many cases represented a very notable increase of the rates of reaction. In this context it is very important to reduce the problems presented due to the deposition on the inert support material of precipitates that make the diffusional transport of nutrients, principally oxygen, more difficult, and that can even produce blockage of the bioreactor. Currently research efforts are directed towards acquiring better knowledge of the parameters that control the efficiency of the process, with the objective of improving their efficiency when operating for long periods of time.

Further, it is necessary to improve the extent of our knowledge in enzymology and microbiology, and it is evident that deeper knowledge of the cellular metabolism and other genetic aspects would enable us to obtain increased rates of reaction and therefore improve its application for commercial processes.

Biofiltration is an example of these advances, and this technology start to be utilized for the reduction of compounds that produce bad odors entrained in the gases released to the atmosphere from waste water treatment plants. The microbiological studies that have generated more knowledge of the means by which microorganisms operate, improvements in the supports utilized so that they offer enhanced retention of humidity, high porosity, low cost and more mechanical strength, have made it possible to begin utilizing these techniques for the elimination of higher concentrations of impurities contained in gases, and in other applications such as the reduction of organic contaminants that are difficult to biodegrade like benzene, toluene and chlorated organic molecules.

5. ACKNOWLEDGEMENTS

This chapter was funded by Education and Science Ministry through the Project PPQ2002-0217.

6. REFERENCES

Anabtawi JA, Al-Jarallah AM, Aitani AM. 1992. Potential for diesel fuel production by hydroprocessing of middle distillates. Energy Sources 14: 155-167.

Anerousis JP, Whtitman SK. 1984. An updated examination of gas sweetening by the iron sponge process. SPE 13280, Houston, Texas.

Armentia H, Webb C. 1992. Ferrous sulfate oxidation using *Thiobacillus ferrooxidans* cells immobilized in polyurethane foam support particles. Appl Microbiol Biotechnol 36: 697-700.

Asai S, Konishi Y, Yabu T. 1990. Kinetics of absorption of hydrogen sulfide into aqueous ferric sulfate solutions. AIChE J 36: 1331-1338.

Astarita G, Savage DW, Longo JM. 1983. Gas treating with chemical solvents. John Wiley & Sons, New York.

Atkinson B. 1981. Immobilized biomass – a basis for process development in wastewater treatment. In: Cooper PF, Atkinson B, eds. Biological fluidized bed treatment of water and wastewater. Ellis Horwood, Chichester, 22-34.

Berzaczy L, Niedermayer E, Kloimstein L, Windperger A. 1990. Biological exhaust gas purification in the rayon fiber manufacture. Chem Biochem Eng Q 2: 201-203.

Berzaczy L, Etzenberger W, Kloimstein L, Niedermayer E, Schmidt A, Windsperger A. 1990. Waagner Biro Aktiengesellschaft, Glanzstoff Austria GesmbH. Procedure for microbiological transformation of sulfur containing harmful components in exhaust gases. US Patent 4,968,622.

Burgess JE, Parsons SA, Stuetz RM. 2001. Developments in odor control and waste gas treatment biotechnology: a review. Biotechnol Adv 19: 35-63.

Candehead P, Sublette KL. 1990. Oxidation of hydrogen sulfide by *Thiobacilli*. Biotechnol Bioeng 35: 1150-1154.

Carnell P. 1986. Gas sweetening with a new fixed bed absorbent. In: Proceedings of Laurance Reid Gas Conditioning Conference. University of Oklahoma, Norma OK, 3-5.

Carranza F, García MJ. 1990. Kinetic comparison of support materials in the bacteria ferrous iron oxidation in a packed-bed columns. Biorecovery 2: 15-27.

Cho K, Hirai M, Shoda M. 1992. Degradation of hydrogen sulfide by *Xanthomonas* sp. strain DY44 isolated from peat. Appl Environ Microbiol 58: 1183-1189.

Cho K, Zhang L, Hirai M, Shoda M. 1991. Removal characteristics of hydrogen sulfide and methanethiol by *Thibacillus sp.* J Ferm Bioeng 71: 44-49.

del Olmo C, Alcon A, Santos VE, García-Ochoa F. 2005. Modelling the production of a *Rhodococcus erythropolis* IGTS8 biocatalyst for DBT biodesulfuration: influence of media composition Enzyme Microb Technol 37: 157-166.

Dillon ET. 1991. Gas sweetening with a novel and selective alkanolamine. In: Proceeding of the 17th GPA Annual Convention.

Dobbs JB. 1986. Gas sweetening with an effective one-step process. In: Proceedings of 1986 Gas Conditioning Conference, University of Oklahoma, Norma OK.

European Union 1998. EU, Directive 98/70/EC.

Fong HL, Kushner DS, Scott RT. 1987. Gas desulfurization using Sulferox. In: Proceedings of Laurence Reid Gas Conditioning Conference, University of Oklahoma, Norman OK, 2-4.

Frazier HD, Kohl AL. 1950. Selective absorption of H_2S from gas streams. Ind Eng Chem 42: 2282-2292.

Freund MS, Shreve GA, Wilisch WCA, Lewis NS. 1994. H_2S Chemistry involved in liquid redox processes: mechanisms of reactions, new ligands and new sensors. In: American Institute of Chemical Engineers, Atlanta, Georgia, 17-21.

Gadre RV. 1989. Removal of hydrogen sulfide from biogas by chemoautotrophic fixed-film bioreactor. Biotechnol Bioeng 34: 410-414.

García MJ, Palencia I, Carranza F. 1989. Biological ferrous iron oxidation in packed-bed columns with low-grade sulfide mineral as support. Process Biochem 24: 84-87.

Goar BG, Lagas JA, Borsboom J, Heijkoop G. 1991. Superclaus Updates: How the process is performing worldwide. In: British Sulfur's 19th International Conference, New Orleans.

Gómez JM, Caro I, Cantero D. 1996. Kinetic equation for growth of *Thiobacillus ferrooxidans* in submerged culture over aqueous ferrous sulfate solutions. J Biotechnol 48: 147-152.

Gómez JM, Cantero D, Webb C. 2000. Immobilization of *Thiobacillus ferrooxidans* cells on nickel alloy fibre for ferrous sulfate oxidation. Appl Microbiol Biotechnol 53: 722-725.

Grishin SI, Tuovinen OH. 1988. Fast kinetics of Fe^{+2} oxidation in packed-bed reactors. Appl Environ Microbiol 54: 3092-3100.

Hardison LC. 1992. Update on LO-CAT process developments and commercial experience. In: GRI Liquid Redox Sulfur Recovery Conference, Austin

Hass RW, Ingalls MN, Trinker TA, Goar BG, Purgason RS. 1981. Process meets sulfur recovery needs. Hydro Process 60: 104-107.

Imaizumi T. 1986. Some industrial applications of inorganic microbial oxidation in Japan. Biotechnol Bioeng Sym 16: 363-371.

Janssen AJH, de Hoop K, Buisman CJN. 1997. The removal of H_2S from air at a petrochemical plant. In: Proceedings of International Symposium Biological Waste Cleaning, Maastricht..

Janssen AJH, Arena B, Kijlstra WS. 2000. New developments of THIOPAQ technology for the removal of H_2S from gaseous steams. In: Proceedings of Sulfur'2000, San Francisco.

Janssen AJH, Ruitenberg R, Buisman CJN. 2001. Industrial applications of new sulfur biotechnology. Water Sci Technol 44: 85-90.

Jensen AB, Webb C. 1995. Treatment of H_2S containing gases: a review of microbiological alternatives. Enzyme Microb Technol 17: 2-10.

Kai T, Takahashi T, Shirakawa Y, Kawabata Y. 1990. Decrease in iron oxidizing activity of *Thiobacillus ferrooxidans* adsorbed on activated carbon. Biotechnol Bioeng 36: 1105-1109.

Karamanev DG. 1991. Model of the biofilm structure of *Thiobacillus ferrooxidans.* J Biotechnol 20: 51-64.

Karamanev DG, Nikolov LN. 1988. Influence of some physicochemical parameters on bacterial activity of biofilm: ferrous iron oxidation by *Thiobacillus ferrooxidans*. Biotechnol Bioeng 31: 295-299.

Ketner R, Liermann N. 1987. MODOP [Mobil Oil Direct Oxidation Process] – a new process to reduce emissions of Claus units. Erdgas Kohle 103: 520-524.

Kohl AL. 1951. Selective H_2S Absorption-A review of available processes. Petrol Processing 6: 26-31.

Kohl A, Nielsen R. 1997. Gas purification. Gulf Publishing Company.

Lagas JA, Borsboom J, Berben PH. 1988. Superclaus-The Answer to Claus plant limitations. In: Canadian Chemical Engineering Conference, Edmonton, Canada.

Leduc LG, Ferroni FG. 1994. The chemolithotrophic bacterium *Thiobacillus ferrooxidans*. FEMS Microbiol Rev 14: 103-120.

Magota H, Shiratori Y. 1988. Dowa Mining. Treatment of sour natural gas containing hydrogen sufide. Japanese Patent 63,205,124.

Maka A, Cork DJ. 1990. Introduction to the sulfur microorganisms and their applications in the environment and industry. Dev Ind Microbiol 31: 99-102.

Manning WP, Rehm SJ, Schmuhl JL. 1981. Method of removing hydrogen sulfide form gas mixtures. US Patent 4,276,271.

Mesa MM, Andrades JA, Macías M, Cantero D. 2004. Biological oxidation of ferrous iron: study of bioreactor efficiency. J Chem Technol Biotechnol 79: 163-170.

Miller FE, Kohl AL. 1953. Selective absorption of hydrogen sulfide. Oil Gas J 51: 175-183.

Nakamura K, Noike T, Matsumoto J. 1986. Effect of operation conditions on biological Fe^{+2} oxidation with rotating biological contactors. Water Res 20: 73-77.

Nemati M, Webb C. 1996. Effect of ferrous iron concentration on the catalytic activity of immobilized cells of *Thiobacillus ferrooxidans*. Appl Microbiol Biotechnol 46: 250-255.

Nemati M, Harrion STL, Hansford GS, Webb C. 1998. Biological oxidation of ferrous sulfate by *Thiobacillus ferrooxidans*: a review on the kinetics aspects. Biochem Eng J 1: 171-190.

Neumann DW, Lynn S. 1984. Oxidative absorption of H_2S and O_2 by iron chelate solutions. AIChE J 30: 62-67.

Neumann W, Röckauf H, Volk N, Michael A, Forkmann R. 1990. Method for removal of hydrogen sulfide from a combustible waste gas. European patent EP 402,704.

Nikolov LN, Karamanev DG. 1987. Experimental study of the inverse fluidized bed biofilm reactor. Can J Chem Eng 65: 214-217.

Nikolov LN, Mehochev D, Dimitrov D. 1986. Continous bacterial ferrous iron oxidation by *Thiobacillus ferrooxidans* in rotating biological contactors. Biotechnol Lett 8: 707-710.

Olem H, Unz RF. 1977. Acid mine drainage treatment with rotating biological contactors. Biotechnol Bioeng 19: 1475-1491.

Olem, H. and Unz, R.F. 1980. Rotating-disc biological treatment of acid mine drainage. J Wat Pollut Control Fed 52: 257-269.

Pagella C, Silvestri P, De Faveri DM. 1996. H_2S gas treatment with *Thiobacillus ferrooxidans*: overall process performance and the chemical step. Chem Eng Res Des 74: 123-132.

Pagella C, Zambelli L, Silvestri P, De Faveri DM. 1996. Hydrogen sulfide gas treatment with *Thiobacillus ferrooxidans*. Process performance and stability. Chem Eng Technol 19: 378-385.

Pagella C, Perego P, Zilli M. 1996. Biotechnological H_2S gas treatment with *Thiobacillus ferrooxidans*. Chem Eng Technol 19: 78-88.

Pogliani C, Donati E. 2000. Immobilization of *Thiobacillus ferrooxidans*: importance of jarosite precipitation. Process Biochem 35: 997-1004.

Satoh H, Yoshizawa J, Kametani S. 1988. Bacteria help desulfurize gas. Hydrocarbon Proc Int Ed 67: 76D–76F.

Schaack JP, Chan F. 1989. Hydrogen sulfide scavenging-1. Formaldehyde-methanol, metallic oxide agents head scavengers list processes. Oil Gas J 87: 51-55.

Selvaraj PT, Litle MH, Kaufman EN. 1997. Biodesulfurization of flue gases and other sulfate/sulfite waste streams using immobilized mixed sulfate-reducing bacteria. Biotechnol Progress 13: 583-589.

Shennan JL. 1996. Microbial attack on sulfur-containing hydrocarbons: implications for the biodesulfurization of oils and coals. J Chem Technol Biotechnol 67: 109-123.

Sivalls CR. 1982. Slurrisweet acid gas treating process- In: Proceedings of Gas Conditioning Conference, University of Oklahoma, Norman OK.

Smith RM, Martell AE. 1989. Critical stabilility constants. In: Inorganic Complexes, Vol 4, Plenum Press, New York.

Sonta H, Shiratori T. 1990. Dowa Mining Co. Ltd. Method of treating H_2S containing gases. US Patent 4,931,262.

Sublette KL. 1987. Aerobic oxidation of hydrogen sulfide by *Thiobacillus denitrificans*. Biotechnol Bioeng 29: 690-695.

Sublette KL, Sylvester ND. 1987a. Oxidation of hydrogen sulfide by *Thiobacillus denitrificans*: desulfurization of natural gas. Biotechnol Bioeng 29: 249-257.

Sublette KL, Sylvester ND. 1987b. Oxidation of hydrogen sulfide by continuous cultures of *Thiobacillus denitrificans*. Biotechnol Bioeng 29: 753-758.

Sublette KL, Sylvester ND. 1987c. Oxidation of hydrogen sulfide by mixed cultures of *Thiobacillus denitrificans* and heterotrophs. Biotechnol Bioeng 29: 759-761.

Wichlacz PL. 1981. Fixed film biokinetics of ferrous iron oxidation. Biotechnol Bioeng Symp 11: 493-504.

INDEX